SAGE was founded in 1965 by Sara Miller McCune to support the dissemination of usable knowledge by publishing innovative and high-quality research and teaching content. Today, we publish over 900 journals, including those of more than 400 learned societies, more than 800 new books per year, and a growing range of library products including archives, data, case studies, reports, and video. SAGE remains majority-owned by our founder, and after Sara's lifetime will become owned by a charitable trust that secures our continued independence.

Los Angeles | London | New Delhi | Singapore | Washington DC | Melbourne

Advance Praise

After an edifying and well-received book, *Indian Lobbying and its Influence in US Decision Making*, Dr Ashok Sharma turns his focus on an increasingly critical sector of India's domestic and foreign policy—energy security—key to its economic ambitions. The country faces several challenges—environmental, equity-based, economic, technological and geopolitical—to meet its energy deficit. They have powerful implications for India's energy policy. Dr Sharma's comprehensive study concludes that the choices the country has made balance the demands of these challenges both effectively and strategically. This timely and scholarly book dealing with the complex issue of energy security is highly recommended for academics, policymakers and anyone seriously interested in India's quest for energy security.

Ambassador Lalit Mansingh,
Former Foreign Secretary of India and
Ambassador to the United States

This remarkable book by Ashok Sharma is, to the best of my knowledge, the first to examine key issues related to India's energy security in a comprehensive manner. Several factors combine to make energy security one of the key challenges facing not just India but the world economy as well. While high GDP growth rates in India are necessary to raise living standards and reduce poverty, such growth rates will also sharply raise energy demand. The source and composition of such energy are also crucial given the twin challenges of climate change as well as the paucity of domestic sources of energy, the latter necessitating India's deep involvement in energy diplomacy to face challenges from China. Sharma's book is an erudite exposition of all these key issues. It will be invaluable to policymakers and students. I recommend it strongly.

Raghbendra Jha,
Professor of Economics and Executive Director,
Australia South Asia Research Centre,
Arndt-Corden Department of Economics,
Crawford School of Public Policy,
The Australian National University,
Canberra, Australia

After his first seminal book *Indian Lobbying and Its Influence on US Decision Making*, Dr Sharma takes a thorough study of India's energy security. Dr Sharma observes that much of the global energy demand will be due to growth from India and China. India, the world's third largest energy consumer, will continue to supersede the growth rate of all major economies. This will increase its oil and gas import dependency. India is also turning to renewables and nuclear sources, signalling a shift to a more environment-friendly energy policy.

Dr Sharma argues that India's energy policy is in need of a multifaceted approach. The domestic measures and reserves are not sufficient. The book is a thorough and meticulous account of India's attempt to diversify its oil and gas imports dependency on the Middle East, and how such a pursuit places India in direct competition with China which is further intensified by a complex great power politics in the Indo-Pacific region. A must-read for both academic and policy professionals and anyone interested in the evolution of India's quest for energy security and the consequent geopolitical ramifications.

Emeritus Professor Dov Bing,
Department of Political Science and Public Policy,
Faculty of Arts and Social Sciences,
University of Waikato, Hamilton, New Zealand

India's Pursuit of
ENERGY SECURITY

India's Pursuit of
ENERGY
SECURITY

Domestic Measures, Foreign Policy and Geopolitics

Ashok Sharma

Los Angeles | London | New Delhi
Singapore | Washington DC | Melbourne

First published in 2019 by

 SAGE Publications India Pvt Ltd
B1/I-1 Mohan Cooperative Industrial Area
Mathura Road, New Delhi 110 044, India
www.sagepub.in

SAGE Publications Inc
2455 Teller Road
Thousand Oaks, California 91320, USA

SAGE Publications Ltd
1 Oliver's Yard, 55 City Road
London EC1Y 1SP, United Kingdom

SAGE Publications Asia-Pacific Pte Ltd
18 Cross Street #10-10/11/12
China Square Central
Singapore 048423

Published by Vivek Mehra for SAGE Publications India Pvt Ltd. Typeset in 10.5/13 pt Adobe Caslon Pro by Zaza Eunice, Hosur, Tamil Nadu, India.

Library of Congress Cataloging-in-Publication Data Available

ISBN: 978-93-532-8539-5 (HB)

SAGE Team: Abhijit Baroi, Guneet Kaur Gulati, Megha Dabral and Rajinder Kaur

*Dedicated to the multitude of Indians
whose human potential was unrealized because of
darkness*

Thank you for choosing a SAGE product!
If you have any comment, observation or feedback,
I would like to personally hear from you.

Please write to me at **contactceo@sagepub.in**

Vivek Mehra, Managing Director and CEO, SAGE India.

Bulk Sales

SAGE India offers special discounts
for purchase of books in bulk.
We also make available special imprints
and excerpts from our books on demand.

For orders and enquiries, write to us at

Marketing Department
SAGE Publications India Pvt Ltd
B1/I-1, Mohan Cooperative Industrial Area
Mathura Road, Post Bag 7
New Delhi 110044, India

E-mail us at **marketing@sagepub.in**

Subscribe to our mailing list
Write to **marketing@sagepub.in**

This book is also available as an e-book.

Contents

List of Abbreviations

ADB	Asian Development Bank
ADMM	ASEAN Defence Ministers Meeting
AERB	Atomic Energy Regulatory Board
AMP	Automotive Mission Plan
AONM	Australian-Obligated Nuclear Material
ARF	ASEAN Regional Forum
AT&C	Aggregate Technical and Commercial
BEE	Bureau of Energy Efficiency
BHAVINI	Bharatiya Nabhikiya Vidyut Nigam
BJP	Bharatiya Janata Party
BPCL	Bharat Petroleum Corporation Limited
BPRL	Bharat Petro Resources Limited
BSE	Bombay Stock Exchange
BTC	Baku–Tbilisi–Ceyhan
CAASTA	Countering America's Adversaries Through Sanctions Act
CAGR	compound annual growth rate
CANDU	Canada Deuterium Uranium
CDM	Clean Development Mechanism
CERC	Central Electricity Regulatory Commission
CIRUS	Canada–India Reactor Utility Services
CNG	compressed natural gas
CNPC	China National Petroleum Corporation
CNSC	Canadian Nuclear Safety Commission
COP	Conference of Parties
COPA	crude oil pipeline agreement
CPEC	China–Pakistan Economic Corridor
CPR	Centre for Policy Research
CSTEP	Center for Study of Science, Technology and Policy
CTBT	Comprehensive Nuclear-Test-Ban Treaty

DAE   Department of Atomic Energy
DGH   Directorate General of Hydrocarbons
E&P   exploration and production
EBP   Ethanol Blended Petrol
EC   European Commission
EPR   European pressurized reactor
EPSA   exploration and production sharing agreement
FAME   Faster Adoption and Manufacturing of Hybrid and
            Electric Vehicles
FDI   foreign direct investment
FEDCO   Feedback Energy Distribution Company
GAIL   Gas Authority of India Limited
GCC   Gulf Cooperation Council
GDP   gross domestic product
GECF   Gas Exporting Countries Forum
GHGs   greenhouse gases
GST   Goods and Services Tax
HBNI   Homi Bhabha National Institute
HCDC   High Committee on Defence Cooperation
HELP   Hydrocarbon Exploration and Licensing Policy
HIRC   House International Relations Committee
HPHT   high pressure–high temperature
IAEA   International Atomic Energy Agency
IAFS   India–Africa Forum Summit
ICC   Indian Chamber of Commerce
IEA   International Energy Agency
IEF   International Energy Foundation
IEP   Integrated Energy Policy
IESS   India Energy Security Scenarios
IEX   Indian Energy Exchange
IGCAR   Indira Gandhi Centre for Atomic Research
IHA   International Hydropower Association
IMF   International Monetary Fund
INDC   Intended Nationally Determined Contribution
INSTC   International North–South Transport Corridor
IOCL   Indian Oil Corporation Limited
IOR   Indian Ocean Region

IPCC	Intergovernmental Panel on Climate Change
IPI	Iran–Pakistan–India
IREDA	Indian Renewable Energy Development Agency
ISA	International Solar Alliance
ITEC	Indian Technical and Economic Cooperation Programme
ITER	International Thermonuclear Experimental Reactor
JCM	Joint Commission Meeting
JNNSM	Jawaharlal Nehru National Solar Mission
JNPP	Jaitapur Nuclear Power Project
KGS	Kaiga Generating Station
KKNPP	Kudankulam Nuclear Power Plant
LAC	Latin America and the Caribbean
LBNL	Lawrence Berkeley National Laboratory
LNG	liquefied natural gas
LOC	lines of credit
IOR	Indian Ocean Region
LPG	liquefied petroleum gas
MDGs	Millennium Development Goals
MEA	Ministry of External Affairs
MMBU	Middle Market Banking Unit
MMT	million metric tons
MNRE	Ministry of New and Renewable Energy
MoU	memorandum of understanding
MSW	municipal solid waste
MTCR	Missile Technology Control Regime
NAM	Non-Aligned Movement
NCA	Nuclear Cooperation Agreement
NDA	National Democratic Alliance
NDCs	Nationally Determined Contributions
NDR	National Data Repository
NEEPCO	North Eastern Electric Power Corporation Limited
NELP	New Exploration Licensing Policy
NEMMP	National Electric Mobility Mission Plan
NEP	National Energy Policy
NHPC	National Hydroelectric Power Corporation
NNPT	Non-Nuclear Proliferation Treaty

NNWS	non-nuclear weapon states
NPCIL	Nuclear Power Corporation of India Limited
NSG	Nuclear Suppliers Group
NSSP	Next Step in Strategic Partnership
NTPC	National Thermal Power Corporation
OALP	Open Acreage Licensing Policy
OBOR	One Belt One Road
OECD	Organisation for Economic Co-operation and Development
OIL	Oil India Limited
ONGC	Oil and Natural Gas Corporation
OPEC	Organization of the Petroleum Exporting Countries
OVL	ONGC Videsh Ltd
PARA	Public Sector Asset Rehabilitation Agency
PHWR	pressurized heavy-water reactors
PLA	People's Liberation Army
PM	prime minister
POK	Pakistan Occupied Kashmir
PPAs	power purchase agreements
PSC	production sharing contracts
PSUs	public sector undertakings
PXI	Power Exchange India
REL	Reliance Energy
RIL	Reliance India Limited
SAARC	South Asian Association for Regional Cooperation
SAGAR	Security and Growth for All in the Region
SCS	South China Sea
SEBs	State Electricity Boards
SECI	Solar Energy Corporation of India
SHS	solar home systems
SIAM	Society of Indian Automobile Manufacturers
SICA	Central American Integration System
SIPRI	Stockholm International Peace Research Institute
SJVNL	Satluj Jal Vidyut Nigam
SLOC	sea lines of communication
SPA	sale and purchase agreement
STP	sewage treatment plant

SUDAPET	Sudan National Petroleum Corporation
TAP	Turkmenistan–Afghanistan–Pakistan
TAPI	Turkmenistan–Afghanistan–Pakistan–India
TERI	The Energy and Resources Institute
TPCL	Tata Power Company Limited
UDAY	Ujwal DISCOM Assurance Yojana
UNCLOS	United Nations Convention on the Law of the Sea
UNFCCC	United Nations Framework Convention on Climate Change
UPA	United Progressive Alliance
WCA	World Coal Association
WISE	World Institute of Sustainable Energy

Foreword

At the outset, I congratulate Dr Ashok Sharma for his comprehensively researched and timely book *India's Pursuit of Energy Security: Domestic Measures, Foreign Policy and Geopolitics*.

To achieve its development goals and to lift millions out of poverty, India needs to continue growing at 8–10 per cent annually. This growth will require more energy. India relies on hydrocarbon sources with coal dominating its energy mix. Given the limited known domestic oil and natural gas reserves, India has no choice but to look overseas for its energy needs.

Over the past two decades, India has applied numerous policy levers to address energy security concerns. The government has introduced a range of economic and regulatory measures to encourage Indian companies to increase oil and gas exploration and to improve productivity. Additionally, the government has taken steps to diversify from hydrocarbons to less-polluting sources by enhancing the share of renewables and alternative sources of energy. However, it is also clear that these domestic measures are not enough and India has to look beyond its borders to satisfy its energy needs.

India's energy crisis, its heavy dependence on imported hydrocarbons, price volatility, geopolitical risk and increasing concerns about economic and environmental consequences have pushed energy security to become one of the country's top foreign policy priorities under Prime Minister Modi. To reduce its over-reliance on the Middle East for energy sources, India has undertaken a concerted effort to pursue oil and gas exploration in Africa, Latin America and the Caribbean, Caspian Basin and in the seas of the Indo-Pacific region. But India's quest for energy security overseas is not free from strategic and geopolitical ramifications. This is visible in India's hydrocarbons explorations abroad, and its bid to become a member of the Nuclear Suppliers Group (NSG) as it looks to enhance the share of nuclear energy in its energy mix.

Since Prime Minister Modi's commitment to reduce carbon emission at the 2015 Paris Climate Summit, India has taken an effective and systematic approach towards energy policy and governance to move the country towards a more sustainable energy future.

Dr Sharma captures these developments with rigorous research and deep analysis. The work is a comprehensive take on India's quest for a fuel mix that provides energy security and helps maintain a high economic growth trajectory amidst global warming concerns. The book critically examines the evolution of India's energy security policy in a broader context of its foreign policy and geopolitics.

This is a welcome book that is highly recommended for academics, diplomats, policymakers and anyone interested in India's quest for energy security. I sincerely hope that the book will prove to be a foundation for other serious studies on this very important subject, and I congratulate and commend Dr Ashok Sharma on this very serious scholarly contribution.

Gen. (Dr) V. K. Singh (Retd)
(PVSM, AVSM, YSM, ADC)
Minister of State for External Affairs,
Government of India, New Delhi, India

Preface

Speaking at a conference in Paris, on 3 June 2017, Prime Minister Narendra Modi vowed that India will go 'above and beyond' the 2015 Paris accord on combating climate change.

In Sydney, 7 October 2017, a crowd of more than one thousand people gathered at Bondi Beach to observe a national day of protest against the proposed Adani's Carmichael coal mine which witnessed thousands of protesters across Australia.

In New Delhi, 26–30 October 2015, India hosted its biggest diplomatic outreach event—India–Africa Forum Summit, which witnessed the largest turnout of African leaders and delegates in India.

Again, in New Delhi, on 11 March 2018, Prime Minister Modi and French President Emmanuel Macron co-chaired the International Solar Alliance (ISA), Prime Minister Narendra Modi's brainchild and an alliance of more than 121 countries, which was formally kicked off with 62 member-countries adopting the 'Delhi Solar Agenda'.

All these events may look different; however, if one digs deeper, each of these disparate events is a manifestation of one underlying theme—India's energy insecurity. These events explain the broad dynamics of India's attempt to address its looming energy crisis, which is the central theme of this book.

Since the advent of the modern world, energy has been key to the widespread economic transformation, global prosperity and well-being of the people. Over the coming decades, energy security will continue to be crucial to the ongoing existence of life in the world. The world's demand for energy grew by 95 per cent in the last 40 years. Global energy consumption is expected to increase by 41 per cent from 2012 to 2035. At present, petroleum accounts for 40 per cent of world energy consumption and the demand is projected to increase eight times by the year 2030. Most of that growth (estimated to be more than 90%) will be accounted for by massive increases in demand from India and China. These resources are finite, and the direct links between energy supply and economic growth have brought the concept of energy security at the centre of policy debate with a considerable impact on the geopolitics.

Energy security is vital for economic development, and in the developing world, its significance is much more than in developed countries. The concept of energy security remarkably has different connotations for different groups of countries. For the countries such as India (and China) energy security is not only about their ability to rapidly fine-tune their new dependence on global markets, a major shift away from their former commitments to self-sufficiency, increase their domestic production, look for new sources of energy abroad, but also about exploring the alternate renewable sources of energy amidst the carbon-controlled environment. Much has been written about the future energy needs of China, the world's largest consumer followed by the United States. There is a need for a comprehensive study of energy security of India, the world's third largest energy consumer.

No more confined to the periphery of the international system, today India is considered as one of the centres of the modern global economic system and is the fastest growing major economy in the world. After more than two decades of economic liberalization, backed by a strong domestic demand, a fast growing middle-class bracket, a young demography of which more than 50 per cent are below the age of 25 with high-tech and English-speaking competency, on path towards better governance and economic reforms leading to an attractive destination for foreign investors, India's economy in all likelihood will continue to supersede the growth rate of all other major economies in the world in the foreseeable future. India needs to sustain a growth rate of 8–10 per cent so as to help eradicate poverty and to meet its economic and social developmental goals. The demand for energy in the coming years will accelerate further as India looks towards modernization and its massive push towards making India a manufacturing hub. Modi government's thrust on 'Make in India' campaign and access to electricity for all (one-fourth of India's 1.2 billion people still have insufficient access to electricity) and the projection of India's growth rate as the fastest growing major economy in the years ahead are all going to create an urgent need for matching the energy demand and supply. More growth means more energy supplies to fuel its economic growth. As a result, energy security has figured as an important policy issue over the past decade and a half, and the subsequent Indian governments have shown an urgency to deal with the looming energy crisis challenges.

In January 2005, in his address to the nation on the eve of the 59th Independence Day, President A. P. J. Abdul Kalam stressed the need for 'energy security' as a transition to total 'energy independence'. The then Prime Minister Manmohan Singh emphasized energy security as the most important security concern, second only to the food security. The present Prime Minister Narendra Modi of BJP-led NDA government, who came to power in 2014 general election on the promise of development agenda, has been taking steps to revive Indian economy and address India's developmental challenges. Cognizant of direct links between energy security and economic development, PM Modi has made energy security one of the top priorities of his government's domestic and foreign policy agenda. It is vital for Modi government's much vaulted 'Make in India' campaign and for developmental goals. It has become one of the top priorities of Modi's foreign policy which is reflected in his record number of high-profile foreign visits during which he has made energy security an important component of bilateral agreements.

According to the New Policy Scenario, the main scenario of World Energy Outlook 2011, India's energy demand will grow by a compound annual growth rate (CAGR) of 3.1 per cent from 2009 to 2035, which is more than double the world's energy demand at a CAGR of 1.3 per cent for the same period. India's share in world energy demand will increase from 5.5 per cent in 2009 to 8.6 per cent in 2035, and the growth would come from all fuels (JooAhn and Graczyk 2012).

There is an increasing gap between energy demand and available reserves. To maintain its projected growth rate until 2031, India would need to increase its primary energy supply and electricity supply. The demand for coal will almost triple from 280 million tonnes of oil equivalent (Mtoe) in 2009 to 618 Mtoe in 2035 at a CAGR of 3.1 per cent. The renewable energy demand is projected to increase from 2 Mtoe in 2009 to 36 Mtoe in 2035. Nuclear energy demand would reach 48 Mtoe in 2035 from 5 Mtoe in 2009 (JooAhn and Graczyk 2012). The demand for oil, natural gas and coal will see the imports of these products continue to rise (Planning Commission 2012).

At the moment, India derives more than 50 per cent of its energy from domestic stocks of coal, mainly in the form of electricity. Coal shall remain India's most important energy source and critical to its growth for the decades ahead. However, it is unlikely that coal will provide half of India's energy in the future.

The alternatives are not obvious, for India, with 17 per cent of the world's population, has just 0.8 per cent of the world's known oil and natural gas resources. Today, oil accounts for 36 per cent of the country's primary energy use. This figure is set to rise both in absolute and in percentage terms. India's domestic production is not sufficient to meet its demand. As a result, India already imports 80 per cent of its crude oil needs. Without new and substantial domestic discoveries, imports are projected to increase to 90 per cent by 2030. By contrast, natural gas currently provides only 7–8 per cent of India's primary energy supply, and most of that gas comes from domestic sources, onshore and offshore. But the position will change significantly if gas utilization rises, as predicted, to 20 per cent by 2025 as India moves towards the world average for the use of natural gas.

To promote oil and gas production at domestic level, the Indian government has been taking several steps which range from encouraging the Indian companies to increase their domestic activities to widening its engagement with multinational companies, broadening opportunities for them to participate in oil and gas exploration in India. Also, to stimulate the investments and development in the exploration of hydrocarbon sources of energy, some of the steps have focused on regulatory changes, a transparent gas pricing policy and redevelopment of uneconomical assets.

The domestic efforts have also seen a concerted focus on exploring various alternative sources of energy which are infinite, renewable and environment-friendly. The government has given a massive push in this regard in the energy production through solar energy, wind power, hydroelectricity power, biomass and nuclear energy.

At the domestic level, the challenges are puzzling and often at odds, such as increasing energy access, building a smart system for drawing investment in energy infrastructure and pricing of energy to facilitate the economic and environmental competence.

Since increased domestic production alone will not be sufficient to meet the projected needs for either oil, gas or coal, India is also expanding its efforts abroad. Eventually, this has made energy security one of the top foreign policy priorities. Despite Modi government's emphasis on domestic production of coal and efforts to diversify the sources of energy, the demand for imported coal is likely to continue. India's energy demand, both for consumer and industry, cannot be met in the near future without the help of imported coal. There is no substantial proof that India can address its energy demand without the use of imported coal to supplement its domestic supply. For example, energy security has been an important aspect of India–Australia relationship in which import of Australian coal remains crucial.

Since two-thirds of India's oil imports come from one single region, that is, the Gulf Cooperation Council (GCC) countries, India is following in the footsteps of other major oil-importing economies and making significant efforts to obtain supplies from sources outside the Gulf. India has taken steps to diversify its hydrocarbon sources outside the domain of Gulf and the Middle East countries such as Latin America, Africa, Caspian Basin, Russia and in the recent years in the waters of Indo-Pacific region. And to increase the gas supply, it has engaged in the construction of gas pipelines with its bordering nations.

Problems of diversification of energy sources for India arise from the political volatility and geopolitics of the region and, above all, it puts India in direct competition with China, particularly in the non-Middle Eastern and non-Gulf regions, as the new destination of energy source which further sparks a geopolitical competition and adds more complexities to an existing competitive and often conflicting India–China relationship.

At present, nuclear energy makes up around 3 per cent of India's energy mix and has the potential to become 20–25 per cent of India's energy mix by 2050 (A. Sharma 2007). India's nuclear energy capability was marred due to lack of enriched uranium, light-water reactors and advanced technology. One of the major landmarks in India's foreign policy since Independence has been the successful conclusion of the civilian nuclear deal with the United States. The nuclear agreement ended India's nuclear isolation and freed India from the technology denial regimes of the Nuclear Non-proliferation Act of 1978 that came after India's first atomic test in 1974. The deal has many benefits, but one of the main advantages is that the deal would help India to address its looming energy crisis by providing access to safeguarded nuclear fuel, advanced light-water reactors and civilian nuclear technology. The nuclear deal is one significant move in the direction as it would reduce India's dependence on hydrocarbon energy source and help in moving towards a cleaner energy.

The nuclear deal brought the energy security at the centre of India's foreign policy debate. After the conclusion of the US–India civilian nuclear deal, India has concluded civilian nuclear deals with many countries such as France, Russia, Canada, Australia and Japan. Given India's non-signatory status of Nuclear Non-Proliferation Treaty, for any country dealing with India in nuclear energy and technology, trade becomes an issue of a mutual trust and confidence.

But the US–India nuclear deal has been seen in the context of the US-designed balance of power strategy in Asia to contain the growing power and influence of China. There was a concerted effort in Capitol Hill in the United States during the passage of US–India civilian nuclear deal by the Chinese and Pakistani lobbies aligning with 'Nuclear Ayatollahs' to block the deal. The Sino-Indian strategic competition resurfaced during India's bid to membership in the Nuclear Suppliers Group (NSG) recently. China used its clout and prerogatives to block India's case for inclusion in the NSG. This has further pushed energy security on India's foreign policy agenda. Nuclear energy is significant for India's efforts to shift to environment-friendly energy sources and alternate sources of energy sources.

Main Arguments and Objectives

Amidst the dynamic international energy setting (Dannreuther 2017) and competition, India is positioning itself to fulfil energy needs for its economic goals, improve the standard of living for its rapidly expanding middle class and provide access to electricity for its 1.2 billion people. In a broad context, India is facing the challenge of finding out a precise fuel mix which will provide both energy security and help it to maintain economic development with a growth rate of around 8 per cent (aimed at double-digit growth in the future) in today's carbon-monitored world.

For a long period, India's energy security has been conceived very narrowly, largely focusing managing supply. This is far too limited. Not only do many new factors need to be accommodated, but they must also be set in a much broader context. Energy security involves geopolitics, bilateral relations with other nations, diversification of sources of supply and diversified types of energy in the form of both renewable and non-renewable energy sources. A new approach is needed.

Today, energy security has evolved from narrow approach and incorporates wider perspectives. The concept of energy security includes a wide range of issues and is conceptualized in the framework of four 'A's—availability, affordability, accessibility and acceptability (Cherp and Jewel 2014). This approach is much more inclusive and takes into account the political, economic, social and environmental issues and concerns under which the energy security policy is being pursued today internationally. India is no exception and is pursuing energy security under the framework of four 'A's to make it accessible to all the sections and sectors at an affordable price in a socially and politically acceptable carbon-controlled environment.

India's energy security measures are not confined to domestic level; they are reflected in a concerted effort at the foreign policy level as well. Rapidly increasing energy demand, heavy dependence on imported hydrocarbons, growing volatility of price, unrelenting political instability, geopolitical risk and increasing concern about economic and environmental consequences have made energy security one of the standout components of India's major policy priority. Today, India's energy security efforts reflect a systematic and inclusive approach.

Given the aforementioned scenario and with new thrust behind efforts to meet its energy needs, this book examines the dynamics of India's energy policy framework in a comprehensive manner in the context of its domestic and international settings. This book aims at an in-depth analysis of India's domestic and foreign policy challenges and measures to meet its fast-growing energy demand in a competitive geopolitical environment as India looks outward and expands to meet its pressing energy security challenges.

In the pursuit of these, this book argues that India's quest for energy security abroad is a multipronged approach and not confined only to the oil and natural gas reserves in the Middle Eastern region. But India's efforts to access the new forms of energy and new geological location for exploration and access to the energy sources is putting it in direct competition with China. In the past, the competition for the energy sources could be seen between the two superpowers during the Cold War period in which energy security formed one of the major components in the US–Soviet Union geopolitical rivalry which catapulted to the military conflicts in the Middle East and Afghanistan. Though India–China energy competition has been mostly confined to commercial- and diplomatic-level tussle, India's quest for energy security is gradually dragging it into a strategic competition with China which may get aggravated due to their security concerns amidst the unfolding strategic scenario, due to the receding US hegemony and militarily aggressive posture of China and the unfolding great power game in the Indo-Pacific region.

To deal with these, the book is divided into six chapters.

Chapter 1 examines the factors and dynamics that explain the concept of energy security. It looks into the varied interpretations of the concept of energy security in economic, social, political and geopolitical contexts. Energy security is on the top of the policy priorities for many countries—ranging from developed, developing to underdeveloped nations. The chapter further deals with the origin and evolution of the energy security concept. Given the dependency on import of energy for meeting the growing need of energy-consuming nations, the chapter further examines the global context of energy security from varied interpretations and priorities of nations ranging from developed, developing to underdeveloped countries. Finally, the chapter examines India's pursuit of energy security in the domestic, international and environmental settings.

Chapter 2, while giving an overview of India's sources of energy, reserves and energy mix, highlights the gap between demand and supply, and failure to tap the potential of its domestic reserves. While making an evaluation of the ongoing debates about the viability of India's energy basket including fossil fuels such as coal and gas, alternate renewable energy sources such as nuclear, and renewables such as wind, waste to energy and solar energy, this chapter concludes that India's energy security policy has been an all-inclusive approach exploring an array of possible energy sources. Moreover, India's energy security quest, policy measures and choices are being pursued in the context of framework of four 'A's—availability, accessibility, affordability and acceptability. Finally, it observes that renewables, especially solar and wind energy in which India is making impressive progress, will have its increased share in India's energy basket in the coming decades. However, coal will continue to lead India's energy mix and light Indian houses, and run Indian industries and businesses for a decade or two.

Chapter 3 makes an evaluation of India's energy security approach in the domestic context. The chapter examines India's pursuit to fulfil the energy needs from domestic sources and the progress made on the exploration and production of the new reserves of hydrocarbon sources, both onshore and offshore, in addition to its renewable potentials. The chapter examines the policy approach of the Government of India at home, including its regulatory measures and steps taken, from the opening of the energy sector to private players to modernization and the technological evolution in the energy sector.

The chapter is divided into five main sections. The first section examines three major energy policy frameworks which have guided India's energy security efforts at the domestic level—Integrated Energy Policy, expert group reports on low-carbon inclusive growth and National Energy Policy. The second section assesses India's move towards liberalizing the energy sector by encouraging private participation. The second section dealt under the heading 'Liberalization of the energy sector: Encouraging private participation' focuses on the Electricity Act 2003, the measures taken by the current government to improve the condition of power distribution through the 'Ujwal DISCOM Assurance Yojana' (UDAY), and the opening of coal mines and the renewable energy sector to private companies. The third section evaluates the Indian government's measures to reduce the burden on hydrocarbons through technology, conservation and energy efficiency. A major focus is on energy efficiency in the automotive sector through the policy of National Electric Mobility Mission Plan (NEMMP) 2020 and the Faster Adoption and Manufacturing of Hybrid and Electric Vehicles (FAME). The fourth section examines the measures for oil and gas exploration and production to enhance the production at the domestic level. This section first delves into the existing policy known as the New Exploration Licensing Policy (NELP) and then the evaluation of new policy introduced under the Modi government which is undertaken mainly on Hydrocarbon Exploration Licensing Policy (HELP), marketing and pricing freedom for new gas production, and the policy for granting extension to the production sharing contracts. An assessment of the overall domestic measures to address energy security is made in the conclusion of this chapter.

Chapter 4 demonstrates that despite all the efforts to diversify the sources of energy such as renewable and alternative sources of energy, hydrocarbons, mainly oil and gas, continue to be significant in India's energy basket. Hydrocarbons constitute the major share of India's energy basket. Despite the recent push to domestic exploration and production activities for oil and gas, India needs to look abroad for these sources. The inadequate domestic reserves of oil and gas and pressing energy demand have pushed India to look for foreign sources.

To explore these, this chapter examines India's attempt to explore and diversify the new sources of hydrocarbons abroad. The first section of this chapter deals with India's NEP and its objectives in the foreign policy context. The second part is about India's energy exploration abroad and its efforts to diversify its energy import sources in which India's energy exploration in Latin America and the Caribbean (LAC), Africa, Caspian Basin and Russia has been discussed in detail. The third section scans India's energy-driven foreign policy and the geopolitics relating to the proposed gas pipeline projects. Before conclusions, the chapter also examines India's new energy sources such as Israel and its energy security effort in neighbouring nations of South Asia, especially Bhutan and Sri Lanka, with a focus on hydropower collaboration.

Chapter 5 observes that despite the doubling of India's nuclear power capacity from 17.7 TWh in 2006 to 35 TWh in 2017, it is not up to its potential. Nuclear energy has the potential to contribute from its present less than 3 per cent to 25 per cent of India's energy mix by 2050. One of the main reasons for the low nuclear power capacity in India has been the international embargo that was imposed on its nuclear power programme after it conducted nuclear explosion for scientific purposes in 1974. The nuclear non-proliferation regime prevented India from obtaining nuclear reactors and enriched uranium. India needed to get rid of the nuclear embargo to enhance its nuclear power generation capacity. This needed a concerted foreign policy effort to dismantle the technological denial regime. Consequently, over the past two decades, India has consistently pursued nuclear issues in its foreign policy and concluded nuclear agreements with several countries which required intense diplomacy, lobbying and negotiations amidst its geopolitics.

In this context, the chapter examines India's nuclear policy, its defiance and dalliance with the international nuclear regime and nuclear weapon states. The chapter gives an account of India's nuclear agreements with various countries, starting with the US–India Civilian Nuclear Agreement and its wider acceptance by the international community reflected in the waiver by the NSG and safeguard agreement with the International Atomic Energy Agency (IAEA). Then the chapter deals with India's nuclear agreements in the foreign policy context with various countries, namely the US, France, Russia, Australia, Canada and UK. India's nuclear agreements with these countries have been significant for a number of reasons including the influence they hold in the international nuclear regime, their legitimate status as nuclear powers, their advanced and high-tech nuclear technology, best nuclear practices and safeguards, and their abundant uranium reserves.

Chapter 6 argues that the energy geopolitics traditionally has been reflected in the hunt for hydrocarbon sources mainly in oil and gas. The energy geopolitics has been concentrated in the Middle East and Caspian Basin for a long time, and lately it could be seen in Africa, Latin America and the Indo-Pacific region, particularly in the waters of the South China Sea and the Indian Ocean. These regions have about 80 per cent of the world's oil and gas, including potential petroleum and natural gas reserves. Both India and China are heavily dependent on these regions for their oil and gas needs. India–China's complex geopolitics rivalry is reflected in their energy and gas exploration as well. Their quest for energy security and pressing energy demand have pushed them towards new destination of energy exploration and import as well. Though both are shifting towards renewable and alternative sources of energy such as nuclear energy, the chase for hydrocarbons abroad has revived their geopolitical and strategic rivalry. Though India–China energy geopolitics largely has been limited to commercial competition, diplomatic tussle and lobbying, it has the potential to catapult to a conflict due to their long adversarial relations.

To deal with these, the chapter examines India's energy security geopolitics in the context of its complex and conflicting relations with China. The chapter aims to illustrate how India's energy security pursuit overseas is putting it in direct competition with China, thus further intensifying their great power competition and long-standing strategic rivalry. The chapter first gives a brief snapshot of India–China bilateral relations and the geopolitical rivalry, an overview of their overlapping oil and gas import sources, their energy security geopolitics in the Middle East, Africa and Latin America, mainly for oil and gas exploration and production, gas pipelines, renewable energy geopolitics, and finally nuclear energy geopolitics in the context of the US–India civilian nuclear deal and India's bid to the nuclear non-proliferation regime.

The concluding chapter observes that India's energy security policy has evolved over the past two decades. Its mounting energy needs have made it a foreign policy priority. This is driven by its need to diversify its hydrocarbon sources from the Middle East to the new import sources. Although it is reflected in the subsequent Indian governments' foreign policy, this is more evident under the Modi government which has taken a concerted and aggressive approach in pursuit of energy security abroad.

India's energy security is being pursued under the conceptual framework of four 'A's—availability, accessibility, affordability and acceptability. Given its scale of demand, the four 'A's will be central in India's energy security concept to maintain its economic growth and the goal of development for all amidst the climate change concern. But it is the affordability and acceptability aspects amidst the environmental concerns that have been the highlight of India's energy policy. Prime Minister Modi has intensely pursued India's energy security amidst the carbon emission concerns and has worked in the direction of his Paris Climate Accord commitment to move to less-polluting fuel.

Finally, India's energy security policy is not devoid of geopolitics. India's quest for energy security abroad has put it in direct competition with China. This is further aggravated by their enduring conflicting and complex relationship, and great power politics in the emerging strategic context in the Indo-Pacific region. The energy security geopolitics between the two Asian great powers is also visible in renewable energy sources and nuclear energy. The US–India civilian nuclear deal and India's waiver at the NSG in 2008, India's bid to the NSG in 2016, and the Chinese opposition to both with failure to stop the nuclear exception in 2008 and success in blocking India's entry to the NSG in 2016, reflect that India–China energy security geopolitics is here to stay and in a more intense form.

Acknowledgements

It is my privilege to show my gratefulness to a number of people who have been encouraging and helpful during the course of the research leading to the publication of this book.

This work emanates from my long-term research interest in the role of Indian lobby in the United States. I noticed that the major issue that galvanized the Indian lobby's effort in the US during the Indo-US nuclear deal was to assist Indian shift to alternate and less carbon-intensive energy sources. But the US–India nuclear deal also faced opposition from China and Pakistan, mainly driven by strategic rivalry between India and China. This was also the time when subsequent Indian governments began to emphasize energy sufficiency as a major security concern for India, next only to food security. What started as a study of the growing influence of Indian lobby in the US during the nuclear deal soon progressed to an evolving interest in India's energy concerns and its interplay with its geopolitical aspirations.

This interest became the focus of numerous articles and presentations, including a journal article 'India and Energy Security' for *Asian Affairs* in 2007 and conference presentations—New Zealand Political Studies Association Conference 2016; Australia–India Business Council, Australia–India Institute and the University of Western Australia at Perth in 2016; and Energy Security Conference at Jawaharlal Nehru University (JNU), New Delhi, in 2017.

At the outset, I would like to acknowledge Professor Daneil Yergin and Professor Benjamin Sovacool whose pioneering works on energy security and geopolitics inspired me to tackle this subject in the Indian context.

I am grateful to General (Dr.) VK Singh for his warm endorsement of the book. I do hope that the work described here generates the kind of discussion that he has hoped for in the forward letter.

I would like to acknowledge the timely support from various ministries, departments and agencies linked to India's energy needs— Ministry of Petroleum and Natural Gas, Ministry of External Affairs and Ministry of New and Renewable Energy. My special gratitude to Shree Dharmendra Pradhan, the Minister of Petroleum and Natural Gas, and his officers Mr Vinay Pradhan (Indian Foreign Service) and Dr Manoj Kumar for facilitating the link with the ministry and Petronet for information, data and a generous grant to support this book project.

I would like to acknowledge the support of Professor Raghbendra Jha at the Australian National University, whose economic insight on India and development projects have always been illuminating and my interaction with him have been helpful in this book project. My sincere thanks to Ambassador Lalit Mansingh for his constant support and encouragement and his insights on Indian foreign policy have always been edifying.

I would like to acknowledge Emeritus Professor Ramesh Thakur, Professor Michael Wesley and Professor Rory Medcalf at the Australian National University for their support and encouragement; their writings and insights on India's nuclear issue and the Indo-Pacific strategic issues have always been enlightening.

My acknowledgement also goes to all my professors at the American Studies and Centre for Political Studies at JNU, and my friends and classmates at JNU. I am also thankful to my former colleagues at the University of Waikato and the University of Auckland, especially Emeritus Professor Dov Bing and Professor Stephen Hoadley for their constant support and encouragement. My gratitude also goes to my colleagues and the Chair Mr Gregory Thwaite at the New Zealand Institute of International Affairs, Auckland Branch, for their encouragement. My special thanks to New Zealand traveller Mr P. Anderson, a historian and geopolitics enthusiast, for his comments on the book.

It is my privilege to acknowledge the support of my colleagues and Professor Shirley Scott, Head of the Department of Humanities and Social Sciences, University of New South Wales, Canberra, at the Australian Defence Force Academy. I am thankful to Associate Professor Christopher Roberts and my colleagues at the Institute for Governance and Policy Analysis at the University of Canberra for their support.

My gratitude also goes to Mr Peter Varghese, the Chancellor of the University of Queensland, for his encouragement. His insights on India and the Indo-Pacific geopolitics have always been illuminating. I am thankful to Mr Jim Varghese, Chair of Australia–India Business Council, for his support and introducing me to various stakeholders in India–Australia energy relations.

My special thanks to Mr Abhijit Baroi at SAGE Publications for facilitating the book from the beginning, Guneet Kaur and the SAGE team for their professionalism in bringing out this book.

My acknowledgement to my brother Distinguished Professor Arun Sharma, Deputy Vice Chancellor, Queensland University of Technology, for his valuable comments and insights on this book project, and for his constant support and encouragement in my academic journey.

In the end, my acknowledgement to all my brothers and sisters, extended family and the new generation of Sharma family of Shree Awadh Koshore Prasad Sharma and Late Girija Devi, my parents, to whom this book is dedicated as well.

CHAPTER 1

Energy Security as a Concept

Thematic Issues in India's Energy Security in Domestic and International Settings

The concept of energy security is one of the most vaguely defined security concepts in the field of security studies. Energy security as a concept in the modern context first came to the forefront after the Middle East oil crisis in the 1970s. Since then there have been attempts by academics, policy practitioners and the businesses involved in the energy sector to define this term. However, interpretations have varied. In fact, it remains difficult to conceptualize energy security to the extent that it could be acceptable to all sectors and sections of society in all countries.

Energy security is the association between national security and the availability of natural resources for energy consumption. Access to cheap energy has become fundamental for the operation of modern nations. However, the unequal and uneven availability and distribution of energy supplies have created major susceptibilities pertaining to energy security issues. Several factors and dynamics help explain the concept of energy security. Nevertheless, fact remains that the concept of energy security has varied interpretations in economic, social, political and geopolitical contexts.

Energy Security: Definition and Varying Interpretations

Energy security is on top of the policy priorities for many countries—ranging from developed, developing and underdeveloped nations. Energy security forms the core of developmental activities and human civilization in the modern world. Many policymakers, scholars and policy practitioners have put forward their versions of energy security. The varying interpretations have led to this vagueness surrounding the concept of energy security.

However, the ambiguity in the standard conceptualization for energy security is becoming outdated. There is unanimity that energy security means different things for individual countries and that it cannot be generalized for all the countries. Any such standardized conceptualization, as astutely observed by Chester, is a folly (Chester 2010). In fact, the concept of energy security is inherently slippery because it is polysemic in nature, capable of holding multiple dimensions and taking on different specificities depending on the country (or continent), time frame or energy source to which it is applied (Chester 2010). For many years, works of scholarship written on the conceptualization of energy security have originated mainly from the lenses of the Western developed nations, primarily the United States and certain European nations. Their attempt at definition was too narrow to define the concept of energy security which has become the core of policy priority for many countries, both in domestic and foreign policy contexts.

Again, for almost more than half a century, the world has been relying on hydrocarbon sources of energy. As resources are finite, the direct link between energy supply and development along with living standards has put the focus on the need for clarity of energy security for a range of countries. During the 20th century, the standard interpretation of energy security simply meant a steady supply of energy.

This definition was primarily framed around oil as the main source of energy.

As the nature of energy-related challenges evolved over time, so did the concept of energy security. New explanations sought to take the energy security debate beyond the main 20th century perspective of an oil-centric steady flow of energy. Today, these ideas are much more multifaceted, embryonic and intensely debated, both at the national domestic policy level and the geopolitical level. Energy security today is inextricably linked with climate change, sustainable development and poverty eradication.

The most distinctive difference in energy security concepts is found between energy importers and exporters, resulting from the emphasis on security of supply for the former and security of demand for the latter. Nonetheless, many energy exporting countries also face domestic energy supply insecurity. Yergin explains that most oil-exporting countries recognize the mutuality of interest and are deeply interested in 'security of demand'—stable commercial relations with their customers, whose purchases often provide a significant part of their national revenues (Yergin 2012). In the same way, Dannreuther considers that the core of energy security is demand security, and this is mainly concerned about stable and secure revenue for development (Dannreuther 2017).

However, the usual definition of energy security is seen mainly in terms of security of energy supply. For example, the United Nations defines energy security as 'the continuous availability of energy in varied forms, in sufficient quantities, and at reasonable prices' (United Nations Development Programme 2000). Similarly, the International Energy Agency (IEA) defines energy security as 'the uninterrupted availability of energy sources at an affordable price'.[1] This is again reflected in IEA's definition of energy security, which encompasses short- and long-term facets. Long-term energy security mainly deals with timely investments to supply energy in line with economic developments and sustainable environmental needs. Short-term energy

[1] https://www.iea.org/topics/energysecurity/

security focuses on the ability of the energy system to react promptly to sudden changes within the supply–demand balance. Lack of energy security is thus linked to the negative economic and social impacts of either the physical unavailability of energy or prices that are not competitive or are overly volatile. These interpretations coincide with stable energy flow, but there has been no agreement on what 'reasonable' prices are for importers and exporters. This conflict of the energy security concept is often criticized by observers. It is not easy, given the multidimensional and evolving nature of energy security, or energy issues in general, for a universally accepted definition to emerge (Energy Charter Secretariat 2015). Another problem with the conceptualization of energy security is that it is either too narrow or too broad. Sovacool and Brown rightly argue that 'notions of energy security are either so narrow that they tell us little about comprehensive energy challenges or so broad that they lack precision and coherence' (Sovacool and Brown 2010).

In fact, several competing definitions of energy security feature in the context of supply security. All of them deal with the idea of avoiding sudden changes in the availability of energy relative to demand. However, the definitions show strong differences in the impact measure that is used for the benefits of increased continuity and the level of discontinuity that is defined as insecure. The scholars can be broadly categorized into three distinct groups. The authors of the first category of energy security are those who place emphasis on the concept of commodity supply continuity. This interpretation of energy security is in consonance with the Department of Energy & Climate Change publication: 'Secure energy means that the risks of interruption to energy supply are low' (Department of Energy & Climate Change 2009). The second group consists of authors who introduce additional security filters. The introduction of these filters is for the purpose of distinguishing and differentiating between secure and insecure levels of continuity of energy. The IEA's interpretation of energy security fits adequately for this category which states: 'Energy security is defined in terms of the physical availability of supplies to satisfy demand at a given price' (International Energy Agency 2001). The third school of thought on energy security broadens the reach of the impact measure beyond the continuity of commodity supplies to the continuity of services, the growth of the economy and impacts on sustainability and safety. A study on gas supply security by Findlater and Noel exemplifies the energy security in the context of continuity of service: 'Security of gas supply (or gas supply security) refers to the ability of a country's energy supply system to meet final contracted energy demand in the event of a gas supply disruption' (Findlater and Noel 2010). In the same category, the energy security definition is expanded to include impact measure of aspects of environmental sustainability. The Asia Pacific Energy Research Centre and European Commission (EC) explain the environmental sustainability aspects of energy security as, 'This study views energy security as the ability of an economy to guarantee the availability of energy resource supply in a sustainable and timely manner with the energy price being at a level that will not adversely affect the economic performance of the economy.' In the contemporary carbon-controlled environment, sustainability as a factor has become an important aspect of energy security policy.

Despite the high importance of energy security in policymaking, several authors have pointed out that the term is not clearly defined. Winzer comes to the conclusion that the multiple interpretations of energy security have made energy security a broad concept which is being used as policy objectives (Winzer 2011). This is explained in a report by Energy Charter Secretariat published in March 2015 which highlights the ambiguity on commonly accepted definitions of energy security. Valentine terms this opaqueness on energy security as 'fuzzy' (Valentine 2010). To Löschel, Moslener and Rübbelke, energy security seems to be a rather blurred concept (Löschel, Moslener, and Rübbelke 2010). In the same way, scholars such as Checchi, Behrens and Egenhofer find that 'there is no common interpretation' of energy security (Checchi, Behrens, and Egenhofer 2009); it is 'elusive' for Kruyt and Mitchell (Kruyt et al. 2009; Mitchell 2002); and it is 'difficult' to define for Chester (Chester 2010; Energy Charter Secretariat 2015; Winzer 2011). From this discussion, it is obvious that the varying interpretations of energy security make it difficult to agree on a universally accepted definition and this often leads to complications in framing a standard policy to ensure energy security.

The Origin and Evolution of the Energy Security Concept

The origin of the concept of energy security goes back to the early years of the 20th century. Energy security began to appear as a policy quandary to ensure the supply of oil to the armed forces of the European powers seeking supremacy at that time, especially during the period of the two world wars.[2] Energy security has been time and again discussed in terms of how Winston Churchill or Georges Clemenceau regarded oil supply security as essential to fuel their armed forces during the First World War. Controlling the oil supply was a major strategic priority for Germany and Japan to invade the USSR and the Dutch East Indies (Indonesia), respectively, during the Second World War (Yergin 1991). During these wars, energy security was equivalent to national security. It was vital to secure oil supplies in order to fuel warships, tanks and aircrafts. In the 1950s and 1960s, world energy demand more than doubled, driven by North America, Western Europe, the Soviet Union and Northeast Asia. In these regions, economic growth, living standard improvement, motorization and electrification pushed energy demand in all sectors. This catapulted a spike in the international energy trade with oil being the dominant commodity which witnessed a quadrupled growth during that phase. Nevertheless, the Western oil entities ruled the international oil supply system with the security of supply by providing cheap oil in a fairly steady way. This circulation of wealth, however, from oil exports incurred the displeasure of the oil-exporting countries. Their discontent led to the formation of the Organization of the Petroleum Exporting Countries (OPEC) in 1960. While security of energy supply was not a high policy priority in developed countries, the majority of the population in many developing countries did not even have access to modern energy, especially electricity (Barnes, Jaffe and Morse 2003; Energy Charter Secretariat 2015; Yergin 1991).

However, the academic interpretations of energy security began to emerge in the wake of the Middle East crisis caused by the

---

[2] For a fascinating and detailed insight into the history of oil, see Yergin, (1991).

unpredictable and unstable situation during the 1970s. This period of energy security studies is considered as classic and continued till the 1980s. Classical scholarly works on energy security reflected the situation of energy security of the 1970s and 1980s. Scholars and policymakers saw energy security mainly from the prism of 1973–1974 energy supply disruption due to the crisis in the Middle East. This was also the time when IEA was formed in 1974 to deal with the energy crisis during the Middle East War crisis in 1973–1974.

This period also saw attempts by scholars, policymakers and policy practitioners to define energy security. The energy security definition that emerged during the 1970s and 1980s utilized the scenarios related to energy security of the environment of that time. As a result, in the 1970s and 1980s, energy security meant a stable supply of cheap oil which was under the threats of embargoes and price manipulations by exporters (Colglazier and Deese 1983). Energy security at the outset was seen as an unwavering supply of energy, primarily oil, as the most vital source of energy, alongside the geopolitical threats arising out of conflicts between nation states and within nation states, particularly in the Middle East.

Now, energy security scholarship has evolved to its current form incorporating broader definitions, alongside wider perspectives. Conventional studies of energy security have intensely evolved over time and include a wide range of issues. It can be seen in the context of four 'A's: availability, accessibility, affordability and acceptability. This approach was first conceptualized by the Tokyo-based Asia Pacific Energy Research Centre in 2007 and has remained an influential concept (Asia Pacific Energy Research Centre 2007). This is much more inclusive and considers the political, social, economic and environmental issues and concerns under which the energy security policy is being pursued by the nations. This framework has been used in the Organisation for Economic Co-operation and Development (OECD) countries but is also applicable to the developing countries and underdeveloped countries.

Energy security has different interpretations and has evolved over a significant period. After the 1970s oil crises, the scholarly interest in energy security diminished in the late 1980s and 1990s because of the stability in the oil prices and receding threats of political embargoes. However, the academic interest in the energy security was revived in the 2000s. This was mainly due to the increasing demand in Asia, disturbances of gas supplies in Europe and the growing pressure on energy security policy to be pursued in a carbon-controlled environment.[3]

Unlike the 'classical' energy security studies of the 1970s and 1980s, the contemporary energy security challenges extend beyond oil supplies and encompass a wider range of issues (Yergin 2006, 69–82). Today, energy security relates to a wider array of energy policy problems including equitable access to modern energy, global warming, sustainable development, poverty eradication, biodiversity and national security. The energy security definition is no longer confined to the classical synopsis, instead it encompasses a wide range of political, economic, environmental and social concerns as seen at different levels of society.

With the evolution of nature of energy-related challenges, the concept of energy security has also transformed. The context of physical availability of energy, primarily oil, in the 1970s moved to the affordability debate in terms of price level, thus redefining the parameters of the discussion on the concept of energy security. However, since the late 1990s, especially since the beginning of the 21st century, the debates surrounding global warming leading to sustainability have surfaced frequently in the energy security debate.

Over the past decade, the clamour for the inclusion of sustainable development in energy security policy is being lauded in an ongoing manner. The United Nations' emphasis on the need to provide adequate, sustainable and environmentally sound supplies of energy to fuel global economic growth has created an imperative for greater international cooperation to increase efficiencies in energy production and use. In recent years, various UN programmes have made a plea for measures that focus on energy efficiency to provide not only economic

---

[3] For detail, see (Cherp and Jewel 2014; Hughes and Lipscy 2013; Yergin 2006).

benefits but also social and environmental advantages (The United Nations, Sustainable Development 2007). Recently, the significance of energy security in attaining the sustainable development goals was emphasized at Astana EXPO on 'Future Energy' by the UN Secretary-General António Guterres, where he made a plea for serious action to make sure that everyone has access to clean, affordable and efficient energy so that they can realize their full potential. Guterres said, 'It reminds us that the world must take urgent action to ensure that everyone has access to clean, efficient and affordable sources of energy […] I hope this EXPO will help us resolve to contribute to a more sustainable world' (Smillie 2017).

The soaring and often unstable energy prices, and now mounting concerns over environmental sustainability, and particularly over the global warming, have accentuated the focus on sustainable development in the energy security debate at the global level. Moreover, this concern is more important to the regions with high economic growth rate such as the Asia-Pacific region. This is a challenge for the fast-growing major economies in this region to pursue their economic growth that has led to a spike in the demand for energy resulting in environmental degradation (United Nations, Economic and Social Commission for the Asia-Pacific 2008).

Similarly, the importance of sustainable development and climate change in framing the energy security policy featured in the 2007 report by the OECD. A study was undertaken by the OECD under the title of *Energy for Sustainable Development* which focused on widening energy access in developing countries, increasing energy research and development and dissemination, promoting energy efficiency and diversity, and reaping the benefits from energy-related climate change policies. The report of the OECD contributed to emphasizing sustainable development, by promoting energy efficiency and diversification, and promoting that energy source be made more environment-friendly, by increasing the share of renewable energy in the energy mix, encouraging measures that are more climate friendly and maximizing the prospects of Clean Development Mechanism (CDM) (OECD 2007). In this regard, Brown and Sovacool add to the debate that energy security and environmental concerns should and can be tackled together instead of separately. This will help in attaining the energy security to sustain economic growth amidst various environmental concerns (Brown and Sovacool 2011).

More recently, energy security is increasingly being linked with the poverty eradication programmes of various governments. While there is no specific Millennium Development Goals (MDGs) with reference to energy, access to energy services is a prerequisite to achieving all eight goals mentioned in MDGs. This was recognized at the World Summit on Sustainable Development in Johannesburg in 2002. Since then, poverty and energy security have become interlinked in the energy security debate (United Nations Development Programme 2005). The sustainable development for the creation of jobs that reduce poverty, governments are facing the daunting challenge of addressing the basic needs of people including getting access to electricity and clean water in a secured condition for economic activities. Today, the focus on energy-related issues has become significant to provide these services as they are inextricably linked to energy security (Hoffman 2004). The basic premise of the inclusion of poverty eradication in energy security is based on the fact that people do not need energy itself but rather the daily activities and services that energy enables. Activities such as heating, cooling, lighting, transportation of people, water and goods, entertainment and a broad range of commercial activities are not possible without energy. This further leads to the argument that governments will look to fulfil these basic needs with the least amount of energy feasible, to reduce energy costs, environmental concerns and national security threat (Hoffman 2004). Today, the lack of access to energy affects nearly half the world's population, but its importance in the eradication of poverty are visible in the Sub-Saharan and South Asian regions. Energy security is not only linked to the developmental matters, but it can aggravate poverty, leading to security problems, including terrorism, social unrest and political violence.[4] To eradicate poverty, these regions will need to grow at a higher rate. To maintain the economic and developmental activities, energy becomes an important factor. Energy is not a goal as such, but

---

[4] https://www.strausscenter.org/energy-and-security/energy-poverty.html

its input is crucial to achieving most of the MDGs and especially the eradication of poverty.

There is a strong correlation between energy consumption and human development. It is very much reflected in the rural poverty in African and Asian countries due to the lack of modern energy sources. Apart from the fact that the energy sector contributes as such to wealth creation, as an input, it is a key factor in economic growth (agriculture and industry), social development (education and health) and communication. Access to modern cooking fuels, motive power and electricity is paramount in improving the livelihood of poor people, both urban and rural. Today the contribution of energy in the eradication of poverty is acknowledged and energy supply has become an important facet of energy security policy.[5] Energy security is a fundamental requirement for achieving the MDGs, which are inherently linked to the eradication of poverty and basic fundamental rights of a human being.

Stressing on significance of the social, economic and political contexts of energy security, evolved over time and place, Dannreuther sees it beyond a concept. For him, energy security is a value that is constantly in dynamic conflict with other core values, namely economic affluence and sustainability.[6]

Energy security has another dimension to look at which differs from the nation/state-centric approach. Unlike the traditional way of defining energy security in the context of nations pursuing their energy security policy at the broader level of domestic and international contexts, there is another dimension to looking at energy security definition. This approach takes into account the status of a person in the context of where a person sits in the society. At the fundamental level, energy security is about the access to basic amounts of energy at affordable prices. In this view, there is an implied supposition that the access to basic energy needs should be unaffected by disturbances

---

[5] https://ec.europa.eu/energy/intelligent/projects/sites/iee-projects/files/projects/documents/e-mindset_the_role_of_energy_in_achieving_mdg1.pdf

[6] For the latest comprehensive study on energy security, see Dannreuther (2017).

in the supply. In addition, there should be reserved energy readily available, reasonably priced and adequate in relation to both amounts available and timely delivery.

From the point of view of the government, responsible for managing the economy and strategic goals, energy security entails energy policies and alternative measures in the event of a supply disruption to provide energy at a price that is considered reasonable for its citizens. This view assumes that the government resorts to actions such as diversification of energy supply, storage of a certain amount of energy reserves, dealing with the major disturbances in supply, price and employment in the energy-related sectors.

The non-nation-centric approach looks at energy security from the perspective of a private citizen. From this angle, energy security is more distinct but does not deviate from the basic definition of energy security—access to readily available resources in adequate amounts at a reasonable price. This individual-focused energy security explanation takes into account microeconomic factors such as small business ventures, farming and local industry. The non-nation state approach of energy security makes a case for the incorporation of these factors for defining energy security and energy policymaking.

In this era of globalization and economic interdependence, energy security is fast becoming an important component of foreign policy objectives of many nation states. This also makes the case for the governance and regulation of energy security in emerging geopolitics. Since the arrival of the concept of energy security in the 1970s, there has been an emergence of some key international organizations dealing with energy security in the international context. Since the emergence of energy security in the 1970s, energy security has been an issue in the global context.

The growing globalization of energy security issues warrants an international system that coordinates and addresses the issues of energy security which are of interest to all nations. The trend in the globalized world is for international economic governance to play an ever-larger role. Given the mutual interdependence of energy producers and consumers, and the importance of this sector not only to growth but also to environmental and other social concerns, strong arguments have been made to move towards an international system which governs energy security with transparent and widely known rules or principles that establish mutually beneficial policies. This would have a much greater chance of achieving energy security than if nations continue to proceed independently (Yueh 2010).

To the IEA, energy security connotes an uninterrupted availability of energy sources at an affordable price. The IEA differentiates energy security in the context of long-term and short-term approaches. The long-term approach of energy security is primarily concerned with the timely investments to supply energy corresponding to economic developments and environmental requirements. The short-term approach of energy security mainly deals with the capability of the energy system to respond on time to abrupt changes in the supply–demand equilibrium.[7] The International Energy Foundation (IEF) was formed in 1979 to oversee energy production, consumption and conservation of energy in the global context. To act as a non-profit foundation fostering world education and scientific research, the IEF aims to promote communication among people interested in energy consumption, production, exploration, conservation, global climate change, standards and learning; disseminate knowledge in energy consumption, production, conservation and global climate change, through symposia, publications and other electronic and print media; and promote and coordinate research and education in energy consumption, production, conservation and global climate change.[8] To ensure that an international perspective on issues related to the growth and development of the foundation is guaranteed, the IEF relies not only on regional headquarters but also on inputs from the International Advisory Committee in more than 165 countries. Similarly, formed in 1991, the IEF, covering all six continents and accounting for around 90 per cent of global supply and demand for oil and gas, is active in ensuring global energy security through open

---

[7] International Energy Agency: https://www.iea.org/topics/energysecurity/
[8] International Energy Foundation: http://www.energy-ief.org/about.html

discussion among the main international energy players in a transparent manner.[9] The International Atomic Energy Agency (IAEA), the first international organization to regulate energy, sought to oversee nuclear energy with the aims to promote the peaceful use of nuclear energy, and to inhibit its use for any military purpose, including nuclear weaponry. Established as an autonomous organization on 29 July 1957, the IAEA could be considered as the first international organization to oversee energy security. Though the IAEA was established independently of the United Nations through its own international treaty, it reports to both the United Nations General Assembly and the Security Council.[10] Although the IAEA was primarily meant for the regulation of nuclear energy, and when it was formed, the number of nations using nuclear energy had not even reached double digits, it could be considered as the first significant and the most effective organization dealing with energy in the international context. Today, nuclear energy is one of the energy sources that are becoming prominent, especially in the major developing countries with huge appetites for energy such as China and India.

In response to the 1973 Middle East crisis, the energy security arrangement surfaced to deal with the oil embargo by Arab states to ensure collaboration and synchronization among the developed nations in the event of a disruption in supply, encourage collaboration on energy policies, avoid bruising scrambles for supplies and deter any future use of 'oil weapon' by exporters (Yergin 2006, 69–82).

In pursuit of developing a plan for achieving the goal of energy security acceptable in the global context, a number of organizations and agencies emerged. These organizations have been on top of the

[9] The IEF constitutes not only the consuming and producing countries of the IEA and OPEC, but also transit states and major players outside of their memberships, including Argentina, China, India, Mexico, Russia and South Africa. Sitting alongside other important developed and developing economies on the 31 strong IEF Executive Board, these important energy players actively support the energy discussion (International Energy Forum: https://www.ief.org/about-ief/ief-overview.aspx)

[10] International Atomic Energy Agency: https://www.iaea.org/about/overview/statute

policymaking, the setting of norms and regulations of energy security at the international level.

The system consisted of organizations such as the Paris-based IEA, whose members were and continue to be industrialized countries, with strategic stockpiles of oil, including the US Strategic Petroleum Reserve, that continued monitoring and the analysis of energy markets and policies, and energy conservation and coordinated emergency sharing of supplies in the event of a disruption.

The challenges are many and certain roadmaps have been suggested. The first step is to build on the existing global organizations that promote dialogue among nations, such as the IEA, IEF, the Energy Charter and IAEA. Second, concrete measures could be undertaken under the leadership of pivotal nations to begin to build the foundations for global cooperation (Yueh 2010, 216–17).

For advancing a new collaborative approach at the international level, the issues that are inherently global in nature could serve as a starting point. These include universal pricing for carbon, which must be agreed in a multilateral setting to be workable. In addition, collective management of strategic oil reserves as well as a multilateral nuclear fuel cycle are further actions that must be operative on an international scale to be successful for similar reasons.

This does not exclude other necessary initiatives that would require international cooperation, such as encouraging higher investment in energy research, development and demonstration linked with wider climate change imperatives. Promoting investment in fuel supply and infrastructure would require countries not to cut back in these areas during an economic downturn. Global cooperation and coordination would then generate the greatest benefit (Yueh 2010, 216–17).

The discussed organizations acknowledge the international context and interdependence in the spectrum of energy security policy formulation. The need for the collaboration and coordination between the nation states is central to facilitate the impartial and frank dialogue on energy security to achieve the universal solution to energy security and advance the mutual understanding among the nation states in their quest for energy security.

The Global Context for Energy Security

Several trends warrant an international cooperative approach to energy security—the fast-growing economies of Asian, African and Latin American countries, the demand from certain poor countries and the continuity of fulfilment for energy security of developed countries. First is that economic growth and population gains will lead to energy demand doubling over the next 30 years, mostly in developing countries. Second, fossil fuels, on current trends, will continue to provide by far the most energy, but these resources are distributed unevenly around the world. This has contributed to a growth in the global trade of energy and is the source of perceived energy insecurity.

Above scenarios have led to confusion between energy security and energy independence. These are two different issues. Energy security, as defined by the World Economic Forum, Global Agenda Council on Energy Security, is the reliable, stable and sustainable supply of energy at affordable prices and at an acceptable social cost. This definition recognizes that environmental and other issues are inextricably linked with those of energy. Energy security, therefore, can only be achieved efficiently through global cooperation and not isolation. The aim, therefore, should be to constitute a much more effective international system for promoting energy security which allays anxieties about the reliability of energy supplies and is rooted in the mutual interdependence of producing and consuming nations. All nations—producers and consumers, rich and poor—must be engaged, while pivotal nations should be called upon to play a leadership role (Yueh 2010, 216–17). This is important given the centrality of politics in the energy security efforts of nation states, as global energy commerce makes producer, transit and consumer countries mutually interdependent. This makes it inseparable from the geopolitics (Dudau and Nedelcu 2016).

Energy security is vital for the development and social and environmental concerns. These have further promoted energy security requiring an international system, with a set of organizations and rules, governing for collective interests and mutually beneficial policies in a transparent manner (Yueh 2010). In the international context of energy trade, Shaffer observes it in the context of international oil trade:

Energy use affects the structure of the international system itself: oil use creates an element of interdependency in the international system. Since oil is a global commodity, each country's demand affects the price and supply availability of oil for all consumers. (Shaffer 2009, 3)

Also, the still unfolding international settings, especially the shifting of economic and military expenditure to the Indo-Pacific region, the uncertainty about the multi-polarity of the world, the emergence and re-emergence of great powers and their assertion in their area of influence and beyond are affecting the energy geopolitics. As the international structure is grappling with these transformations, nation states face the challenge of meeting the domestic demands of development and pursing their national security interests in the evolving international order. Amidst all these, energy security has become a paramount issue more than ever before.

Meanwhile, the global economy has experienced a period of unique transformation. Energy security concerns facing the United States, the world's largest energy consumer, have evolved to encompass oil, natural gas and electricity, and have become significantly more complex. The world's population has grown by almost 20 per cent in the last 15 years alone, while global gross domestic product (GDP) increased by 120 per cent. In many parts of the world, mechanical and analogue systems, traditionally energized by oil–products, are being replaced with automated and networked systems that run on electricity. As a result, the number of devices connected to the Internet worldwide has grown from 400 million in 2001 to 25 billion in 2015. These changes have made electricity and natural gas, in addition to oil, key enablers of many facets of society and have ensured that the modern world is completely dependent on energy (Office of Energy Policy and Systems Analysis 2017).

Across the Asia-Pacific region alone, some 1.7 billion people still rely heavily on traditional biomass for cooking and heating, and almost 1 billion people lack electricity. This has had enormous socio-economic costs—the degradation of the environment, spreading of disease and increasing child mortality and weakening social services. This has also restricted the opportunities for women, who have to gather and use traditional fuels. All of these have had major implications for the MDGs: Without better access to energy services, many of the MDGs may be missed (United Nations, Economic and Social Commission for the Asia-Pacific 2008).

In today's connected world, threats that are intended to disrupt the energy systems and markets in one country can affect multiple countries, regions and the global economy. Thus, energy security concerns now include fuel supply chains; electricity generation, transmission and distribution; the functioning of energy markets; and the ability of the energy system to withstand shocks and disruptions. To effectively ensure energy security, a number of factors must be considered from both domestic and international perspectives, which include ensuring domestic access to energy, securing the electric grid, encouraging the development of global markets and supporting alliances and partnerships that strengthen energy security (Office of Energy Policy and Systems Analysis 2017). Energy security has been defined in the context of, mainly, avoidance of risks which affect the continuity of the energy commodity supply relative to demand, the continuity of service supplies, the continuity of the economy and further impacts on the environment or the society.

Energy Security from Different Nations' Interpretations and Priorities

Despite the growing importance and influence of a broader global context and attempts by the governments of different nations to define energy security, there is no unanimity on what constitutes energy security. This ambiguity is summarized by Luft as where countries stand on energy security depends on where they sit (Luft, Korin, and Martel 2012).

Although in the developed world, the usual definition of energy security is simply the availability of enough supplies at affordable prices, different countries interpret what the concept means for them individually (Yergin 2006). For many nations around the world, the security of supply is the central focus of energy policy. This is very much reflected in the policy goals of the United States and EU visions. The three pillars of the European Union's (EU) energy policy are efficiency, sustainability and security of energy supplies (Commission of the European Communities 2006). A few years after being elected President, Barack Obama said: 'We need a national commitment to energy security, and to emphasize that commitment, we should install a Director of Energy Security to oversee all of our efforts' (Remarks of Senator Barack Obama 2006).

Due to the ambiguities relating to the definition of energy security, the term itself has been often used to cover a wide range of policy goals. This lack of clarity on the definition of energy security makes it difficult to balance against other policy objectives. Paul Joskow expresses this well

There is one thing that has not changed since the early 1970s. If you cannot think of a reasoned rationale for some policy based on standard economic reasoning, then argue that the policy is necessary to promote 'energy security'. (Joskow 2009)

Not only in defining the concept of energy security, the lack of clarity on its definition is also echoed among the different political authorities across the world. For a better understanding, it would be worthwhile to have a look at the way energy security is perceived in different countries. In terms of the meaning of energy security to nations, there is no conformity.

According to the US Congressional Budget Office, energy security in the United States is defined as the ability of households and businesses to accommodate disruptions of supply in energy markets. The US energy security debate is centred on its risk to its economy because of the disruptions in the supply of an energy source (Congressional Budget Office 2012). The ambiguity of the definition of energy security is also visible in political actions. In the United States the focus of energy security has traditionally been on the reduction of vulnerability to political extortion, which has led politicians to call for energy independence and raising the shares of renewable energy (Winzer 2011).

To achieve this energy independence, over the past four decades, energy security measures in the United States have focused on decreasing the nation's dependence on foreign oil. Policies have promoted the production of domestic oil resources, maintenance of the world's largest strategic oil reserve, increased vehicle fuel efficiency standards and a host of other oil-related actions and policies (Office of Energy Policy and Systems Analysis 2017).

As a result, the United States is now the world's largest producer of crude oil and other liquids, and the largest producer of refined petroleum products. A net exporter of refined products, for the first time in decades, the United States now produces more oil than it imports. In addition, the United States has become the world's largest producer of natural gas. Its dramatic growth in gas production has lowered US natural gas prices and allowed the United States to begin exporting liquefied natural gas (LNG), which has increased the competitiveness and transparency of international LNG markets (Office of Energy Policy and Systems Analysis 2017).

The aforementioned report goes well with the observation by Daniel Yergin (2006), the leading authority on energy security, 'To a great extent, the United States goal of "energy independence"—a phrase that became a mantra since it was first articulated by Richard Nixon four weeks after the 1973 embargo was put in place—is increasingly at odds with present reality.' The US energy source is heavily dependent on hydrocarbon sources and more than 80 per cent of the energy consumed in the United States comes from oil, natural gas or coal (Congressional Budget Office 2012). Renewable energy forms only around 8 per cent of the total energy mix in the United States that does not go down well with the global emphasis on ensuring energy security in a carbon-controlled environment.

The push towards renewable and clean energy was emphasized by former US President Barack Obama. Speaking to the community at Georgetown University, Obama reiterated 'energy independence' as America's energy security policy, lamenting the inaction on the move towards clean technology amidst the climate change concerns,

> Unfortunately, some folks want to cut critical investments in clean energy. They want to cut our research and development into new technologies.... So at moments like these, sacrificing these investments in research and development, in supporting clean energy technologies that would weaken our energy economy and make us more dependent on oil. That's not a game plan to win the future. That's a vision to keep us mired in the past. I will not accept that outcome for the United States of America. We are not going to do that. (Obama 2011)

Though in recent years the United States seems to have focused on clean energy technology, it is left behind today in pursuing the clean energy option. The United States used to be the leading producer of wind and solar energy in the 1980s. Today, the United States is left behind by China as the leading wind power energy producer and Germany as a solar power energy producer, respectively.

In the EU context, energy security means the security of supply. But over time, the conceptualization of energy security has undergone a change to incorporate the new factors affecting it. In Europe, the major debate in energy security discussions emphasizes managing the dependence on imported natural gas, and in most countries, apart from France and Finland, whether to build new nuclear power plants and perhaps to return to (clean) coal (Yergin 2006). In the wake of the Fukushima disaster and post Russia–Ukraine conflict, Europe has been exploring renewable energy sources as a serious alternative. In the recent years, the energy security concept in Europe has broadened to take cognizance of the emerging issues in that concept by embracing sustainability along with secure and affordable energy.

For a long time, energy security has become one of the top priorities, mainly due to the mounting concerns about Russia's reliability as a supplier of energy. The EU has been heavily dependent on Russia for its energy, importing a large share of fossil fuels with some members relying nearly 100 per cent on Russian gas. The advent of situations such as the interruption in the gas supply to Ukraine in 2006 and

2009, which had further affected some EU members, and the Crimea situation raised EU's energy security concerns.

These developments brought the energy security in Europe to the forefront and necessitated the formation of a strategy to deal with disruptions in the energy supply. In fact, the EU policy on energy security has been under re-evaluation to incorporate major developments affecting its energy security mainly in the context of the EU–Russia energy relationship, from inter-dependence to dependence and vulnerability (Casier 2011; European Commission 2014). The vulnerability is visible in the EC Energy Security Strategy 2014, 'Many countries are also heavily reliant on a single supplier, including some that rely entirely on Russia for their natural gas needs. This dependence leaves them vulnerable to supply disruptions, whether caused by political or commercial disputes, or infrastructure failure' (European Commission 2014).

In the European context, a study conducted by using a stylized case study of three European countries demonstrated that the ambiguity on the conceptualization of energy security could be ascertained by disentangling the security of supply and other policy goals. The study aimed at achieving clarity on the concept of energy security as the continuity of energy supplies relative to demand. The study observed that if security is defined from the perspective of private utilities, end consumers or public services, the concept could further be reduced to the continuity of specific commodity or service supplies, or the impact of supply discontinuities on the continuity of the economy (Winzer 2011).

In the above context, there has been a serious attempt by the EU to develop a strategic partnership with Russia, a major supplier of oil, gas, coal and uranium to the EU. Russia provides around one-third of EU natural gas imports and sustains approximately a quarter of EU gas consumption. At the same time, the EU accounts for almost 60 per cent of Russia's total gas exports. Especially due to concerns over the stability and predictability of energy relations, these have led to the definition of energy as a legitimate security and political issue, in terms of the strategic nature of EU–Russia energy relations. In addition, the strategic dimension of the EU–Russia relations has led to the justification of the state and EU intervention in the EU–Russia energy trade (Judge, Maltby, and Sharples 2016).

In Brazil, on the other hand, where the vision of energy independence has already become a reality, there were periods when politicians advocated an increasing share of fossil fuel imports and decreasing shares of renewable energy to promote energy security.

For some political authorities, the goal of energy security is the protection of the poor against commodity price volatility. Others highlight the importance of protecting the economy against disruptions of energy service supplies, by allowing the prices of commodities to rise during periods of scarcity. For some people, the goal of energy security is the reliable provision of fuels and the role of nuclear energy is seen as one of enhancing security. For others, energy security is concerned with a reduction of hazards from accidents and proliferation, and the expansion of the nuclear industry is a potential threat to energy security (Winzer 2011).

The energy security interpretation varies between energy importing countries and energy exporting countries. Most of the literature has been focused on the energy security interpretation form energy importing countries. The United Nations' definition also reflects the energy security interpretation from the energy importing nations' perspectives in terms of security of supply. However, for many energy exporting countries, international energy security means a stable energy export flow at a 'reasonable' price that can assure not only new energy investment but also general economic development. Energy exporting countries focus on maintaining the 'security of demand' for their exports, which after all generate the overwhelming share of their government revenues (A. Sharma 2007; Yergin 2006; also see Pant 2004).

It should also be noted that security of domestic energy supply is increasingly becoming an issue in many exporting countries. The demand for inclusion of security of demand for energy supposedly emerged in the wake of the oil price fall in 1986 when the income from oil export began to fall for the oil-exporting countries. In addition, oil-importing nations began to look for other alternatives for energy. Saudi Arabia was the first country to take the lead in this in

1988. Due to the oil price collapses in 1986, 1998, 2009 and most recently in 2014, the oil-exporting nations had to face energy insecurity of demand. The Gas Exporting Countries Forum (GECF) too highlighted the importance of equitable risk sharing among all gas market players to ensure the security of gas supply and demand. Not only Saudi Arabia, but other major energy exporting countries such as Russia, Iran, Canada, Indonesia and Nigeria have also pushed for the energy security demand. Russia is the most vocal nation about security of energy demand, especially in terms of gas demand insecurity. Like Russia, Iran is another country that seeks energy security of demand. A member-country of both OPEC and GECF, Iran has substantial oil and gas reserves and is the third largest oil-exporting country after Saudi Arabia and Russia. The country was one of the first that claimed energy security of demand. Canada is a net exporter of oil, natural gas, coal and electricity, and its energy security features mainly in terms of protecting critical energy infrastructure in the country. Due to the Shale gas revolution in the United States, Canadian gas export has witnessed a fall of around 26 per cent gas exports to the United States since 2007. This has created energy insecurity of gas demand for Canada. Indonesia's concept of energy security increasingly resembles the ones in importing countries, despite its abundant resource base. Another issue with energy policy in Indonesia is energy access. In the case of the energy security of Nigeria, the largest oil and LNG exporter in Africa, and a member of both OPEC and the GECF, the focus is on domestic supply security too (Energy Charter Secretariat 2015).

For Russia, the aim has been to reassert state control over 'strategic resources' and gain primacy over the main pipelines and market channels through which it ships its hydrocarbons to international markets (Yergin 2006). However, Russia's energy security policy needs to be seen during the Soviet era and how it has pursued energy security in the post-Soviet era for over nearly three decades.

The Soviet Union emerged as an energy superpower in the modern age by investing heavily in infrastructure projects including the electrification of vast areas, and the construction and maintenance of natural gas and oil pipelines that stretch out of Russia and into every constituent nation of the USSR (Davies, Harrison, and Wheatcroft

1994). After the demise of the Soviet Union, Russia came out with its first energy strategy known as the Main Directions of Energy Policy and Restructuring of the Fuel and Energy Industry of the Russian Federation for the period up to the year 2010 (Fredholm 2005). This was changed in May 2000 and confirmed by the government on August 2003 under the presidency of Vladimir Putin as the Russian energy strategy to 2020 (Bushuev and Troitskii 2007). The 2010 strategy is now the energy strategy of Russia up to 2030 (Ministry of Energy of the Russian Federation 2010).

Russia's modern energy strategies began forming after the Second World War. With the Soviet Union left standing as one of two global hegemons towering over a divided Europe, Moscow saw no barriers to achieving dominance in the field of global energy. Between the 1950s and 1960s, Soviet oil output doubled, making the Soviet Union the second-largest oil producer in the world and the primary supplier to both Eastern and Western Europe. Revenues from oil exports started to make up nearly half of Soviet export income.

The Energy Strategy of Russia up to 2020 aims at the long-term development of the fuel and energy complex focused on the transition to the path of innovative and energy-efficient development, change in the structure and scale of energy production, development of competitive market environment, and integration into the world energy system. Russia has been able to implement in practice most of the guidelines stated in the Energy Strategy of Russia for the period up to 2020. In particular, the reforms in the electric energy industry, setting up a more favourable tax treatment for the oil and gas industries, promotion of the development of oil refineries and petrochemical plants, and the steps to develop an energy exchange trade, removal of excessive administrative barriers hindering energy companies, and development of infrastructure projects crucial to the development of the domestic energy sector have been taken.

The energy security goal for Russia is to maximize the exploitation of the natural energy resources efficiently, exploit the prospects of the energy sector to maintain economic growth, raise the quality of life of the population and advance the nation's international standing in the global economy. Russia's energy security approach is to find a better quality of fuel and energy mix and thus enhance the competitive ability of Russian energy production and services in the world market. Russia aims at the long-term strategy with the focus on energy safety, energy effectiveness, budget effectiveness and ecological energy security (Ministry of Energy of the Russian Federation 2010). Russia, to achieve its energy security, has focused on reasserting state control over 'strategic resources' to gain primacy over the main pipelines and market channels through which it ships its hydrocarbons to international markets. The nationalization of energy in Russia and the exclusion of Western companies from the Sakhalin-2 projects serve as examples in this regard.

The qualitative results projected for the first phase of the Energy Strategy of Russia for the period up to the 2020 implementation have not been fully achieved. There has been no establishment of a competitive energy market with fair trade principles, no conversion of the related sectors of the economy to a new level of energy efficiency and no transition from the leading role of the fuel and energy complex in the economy of the country to the function of an effective and stable supplier of energy resources for the needs of the country's economy and population. Also, with no significant efforts from the Russian government to reduce carbon dioxide (CO_2) emissions or to produce some of the energy via renewable sources, the chances of things radically changing in the near future are very slim (Millhone, Greene and Vatansever 2010; Ministry of Energy of the Russian Federation 2010).

Russia is also one of the most vocal nations about security of energy demand among energy exporting nations. At a G8 summit in 2006, President Putin stated, 'The energy market must be insured against unpredictability and its level of investment risk must be reduced. In other words, measures taken to ensure reliable supplies must be backed up by measures taken to ensure stable demand' (President of Russia 2006). Russia's 'Energy Strategy up to 2030' considers energy security as one of its main strategic guidelines. The strategy does not demonstrate clarity on the concept of energy security, but it does reflect the Russian government acknowledgement of security of domestic supply and international demand in terms of energy security challenges. With regards to its international energy security of demand, the strategy calls for stable relationships with traditional and new consumers of Russian energy resources. Russia's concern over demand security increased in relation to its gas supply to Europe. EU's gas market liberalization, weak gas demand in Europe, and the instability in demand aftermath of the Ukrainian crisis contribute to the Russian concern (Ministry of Energy of the Russian Federation 2010).

Australia's case of energy security is an interesting one. This could be seen in the context of affordability in term of a reasonable price and acceptability in a carbon-controlled environment. Australia's move from clean energy to the energy guarantee plan is a case. The shift to renewable energy sources was supposed to sustain itself, but that has not been the case in Australia. Australia's energy policy of a national energy guarantee aimed at delivering affordable and reliable electricity for its people is made up of two parts—a reliability guarantee to ensure that energy is always available and an emission guarantee to contribute to Australia's international commitments. The Guarantee builds on Australia's existing energy policy—ensuring energy retailers offer consumers a better deal, delivering more gas for Australians before it is shipped offshore, building Snowy 2.0 to stabilize the system and stopping network companies gaming the system. Combined with its energy policies, including the guarantee, Australia aims to cut the average household energy bill.

The concern for underdeveloped and poor developing countries is how changes in energy prices affect their balance of payments. Whereas for Japan, with no significant domestic reserves of fossil fuel, except coal and import of substantial amounts of crude oil and natural gas, and uranium, energy security means offsetting its stark scarcity of domestic resources through diversification, trade and investment. Japan is facing serious problems in addressing its energy requirements as it had previously relied on nuclear power to meet about 30 per cent of its electricity needs. There has been strong opposition to restart the nuclear reactors (Ryall 2016) since the nuclear reactors were closed down in the wake of the 2011 Fukushima Daiichi nuclear disaster. For China and India, energy security now lies in their ability to rapidly adjust to their new dependence on global markets, which represents a major shift away from their former commitments to self-sufficiency.

India's Energy Settings: Domestic, International and Environmental

India's thrust on energy security mainly emerged in the post-liberalization phase of the 1990s. The Indian government embarked on a programme of liberalization with short-term stabilization combined with a longer-term programme of comprehensive structural reforms to address the severe balance-of-payments crisis that India was facing in 1991.[11] As a result, India entered a phase of economic growth unlike that of the pre-liberalization phase of more than 3 per cent. Within a decade, India's economy began to grow at 5–6 per cent with a projection of annual GDP growth rate of 8–10 per cent.

Sensing this growth projection in its economy, the Indian government began to focus on energy security policy to fuel its economy and improve the standard of living of its people. A cursory look at the energy scenario in 1991 reflects that though India imported petroleum and natural gas, it had abundant coal, hydroelectric power and a burgeoning nuclear power industry. This began to change once India's liberalization policy started to accelerate India's economic growth and modernization process. However, the economic growth in the 1990s could not push energy security to the centre of India's policy priority.

It was not until the late 1990s, or to be more precise, it was during Prime Minister (PM) Atal Bihari Vajpayee-led Bharatiya Janata Party (BJP)/National Democratic Alliance (NDA) government (1998–2004) that energy security began to emerge as a policy priority for India at domestic and foreign policy levels. This was also the period when India had attracted world attention by conducting its nuclear test in 1998 mainly because of what triggered India to go for atomic test. While the security reasons, India's great power ambition, the domestic political context and BJP's revivalist agenda were cited as the main reasons for India's nuclear test, indirectly and unintentionally, it was also seen as a challenge to nuclear apartheid that India was facing

---

[11] *Analysis of GDP of India from 1990–2010.* https://www.scribd.com/doc/50598237/Analysis-of-GDP-of-India-from-1990-2010

since the 1978 after India conducted its first nuclear test in 1974 for peaceful purpose. The nuclear apartheid had restrained India from commerce in high-tech nuclear technology and enriched uranium needed for nuclear energy. Not surprisingly, the Vajpayee government began to focus on mending India's ties with the United States and embarked on a comprehensive India–US strategic partnership that included defence, high technology, and space and energy cooperation as well. This was reflected in the Next Step in Strategic Partnership (NSSP) that was to guide and transform the India–US partnership in the coming years. After the Vajpayee government, parallel efforts regarding high technology transfer and civil space cooperation bore fruit when on 18 July 2005 PM Manmohan Singh signed an agreement to promote cooperation on civilian use of nuclear technology leading to the signing of the US–India Civilian Nuclear Cooperation Agreement (NCA). One of the advantages of this nuclear deal was designed to help India to address its looming energy crisis by diversifying its energy dependency from fossil fuels and enhancing nuclear energy in its energy mix (A. Sharma 2017b).

Vajpayee government's phase could be also considered as India's first major push towards diversification of sources and destination of energy. This was also the phase when India had taken some of the major steps on its economic and foreign policy front. India's economy began to witness a steady growth of 5–6 per cent. This created an urgency to address the looming energy crisis as India was projected to witness a high growth rate of 8–10 per cent. Some of the steps formed the basis of India's engagement with major oil and gas producing countries and included agreements on pipeline deals with important African, Asian and Central Asia countries for the diversification of destinations for the acquisition of energy sources, and the NSSP was a step towards alternate sources of energy.

In the post-Vajpayee government phase, during PM Manmohan Singh-led Congress/United Progressive Alliance (UPA), energy security continued as óne of the significant policy priorities of the government. It was also a period when India's economic growth began to touch double digits and witnessed an average of 8 per cent growth rate. PM Singh emphasized energy security as the most important security concern, second only to food security. In January 2005, in his address to the nation on the eve of the 59th Independence Day, President, A. P. J. Abdul Kalam stressed the need for 'energy security' as a transition to total 'energy independence'. India's energy security began to take shape as reflected in India's Planning Commission vision on energy security in 2006,

> We are energy secure when we can supply lifeline energy to all our citizens irrespective of their ability to pay for it, as well as meet their effective demand for safe and convenient energy to satisfy their various needs at competitive prices, at all times and with a prescribed confidence level considering shocks and disruptions that can be reasonably expected.

The current PM, Narendra Modi of the BJP-led NDA government, who came to power in the 2014 general election on the promise of development agenda, is taking steps to revive the Indian economy and address India's developmental challenges. Cognizant of direct links to energy security with economic development, PM Modi has made energy security one of the top priorities of his government's domestic and foreign policy agenda. It is vital for the Modi government's much highlighted 'Make in India' campaign and for other developmental goals. It has become one of the top priorities of Modi's foreign policy, which is reflected in his record number of high profile foreign visits during which he has given focus to energy security as an important component of bilateral and multilateral agreements.

Amidst the domestic and international scenario, India is in the predicament of pursuing its developmental agenda and fulfilling its energy demand to fuel its growing economy and providing electricity to all its population of which almost 1/4th lacks access to electricity. This needs to be achieved in a carbon-controlled environment by remaining committed to low carbon emission. India is also faced with the dilemma of to what level it should rely only on renewable energy sources to address its mounting energy crisis.

Despite the forward trajectory of enhanced Indian greenhouse gas (GHG) emissions, to date India has contributed less than 3 per cent of cumulative GHG emissions. Per capita emission-wise this is much lower when compared with the developed nations such as the United States which has a per capita emission roughly 10 times greater than that of India (Panagariya 2016). Given this reality and the need to bring electricity not only to those not currently served but also to the tens of millions more who are underserved, further development of coal is going to be a necessary (Palmer 2016). This would make India to look to the international market, as India's coal is considered less clean. India's efforts to achieve its energy security goals and achieving the targeted CO_2 reductions postulated for each electricity fuel source, nuclear, supercritical coal, hydro and perhaps even solar and wind will not be an easy task as envisioned by some policymakers. Consequently, Modi and the Indian economy are unlikely to meet their Intended Nationally Determined Contribution (INDC) pledges (Ebinger 2016). In addition, relying heavily on clean energy such as solar power and wind power may not be enough for India to fuel its economy and consumer demand. Likely, it will lead to a spike in unaffordable energy bills of many households.

The energy scenario in India poses mounting challenges for its future energy policy. With an installed generating capacity of less than 150,000 MW (megawatt) and a per capita consumption of a mere 650 units of electricity per annum, India experienced a huge estimated shortage of nearly 10 per cent in energy terms and almost 17 per cent in terms of peak demand in 2007/2008. India's energy and peak shortages in 2003/2004 were 7.1 per cent and 11.2 per cent, respectively. Since the liberalization of its economy in the 1990s, India has been witnessing a gradual gap in its energy demand and supply. India could only achieve around half of its stated energy capacity addition for the Eighth, Ninth and Tenth Five-Year Plans. The Five-Year Plans failure to achieve the target further pushed the gap between energy demand and supply. This was the situation when over 50 per cent of India's rural population did not even have access to electricity (The Energy and Resource Institute 2009).

In relation to its population, India is poorly endowed with energy resources. Its share in the world population is 17 per cent but the shares in the world gas, oil and coal reserves are only 0.6 per cent, 0.4 per cent and 7 per cent, respectively. This has meant heavy dependence on imports despite a rather low level of energy consumption (NITI Aayog, Government of India 2017b). India's efforts to find new oil reserves had only a little success. However, Indian efforts in establishing natural gas finds have had considerably higher success. The lack of adequate delivery infrastructure has significantly limited the expansion of, and spread of benefits from, this source. The majority of India's population, especially in rural areas, is still dependent on firewood, chips, animal dung and agricultural residues for energy sources.

Until recently estimated to last for another 200 years, at the current reserves-to-production ratio, the government has downgraded its estimation of the useful life of India's coal resources to provide secure access to energy just about 40 years or so—with a peak production level of about 700 million tonnes (mt) per annum. This sector has been plagued by inadequate investment in resource development, inefficiencies in production and inadequate resource allocation for technology development. It has also proven to be one of the most difficult sectors to open up for private participation and the resultant induction of more modern and efficient mining technologies.

Despite subsequent Indian government's several innovative steps to address the major hurdles being faced by its energy sector, notwithstanding some level of accomplishment, India is still facing big challenge in the energy sector. The measures such as the New Exploration Licensing Policy (NELP) introduced in 1997–1998 and implemented in 1998 to provide an equal platform to both public and private sector companies in exploration and production (E&P) of hydrocarbon sources of energy, mainly oil and gas, to increase the production of oil and natural gas and provide level playing fields for both public and private players, to open the coal industry and move it along more market-driven lines, and partnership-based rural electrification programmes have not achieved the desired results (The Energy and Resource Institute 2009). A case of lack of adequate measures in this regard is the lack of steps to explore and tap the full potential of

India's hydrocarbons domain. India has 26 sedimentary basins covering an area of 3.14 million sq km spread over on land, shallow water and deep water. Around 48 per cent of the total sedimentary basin area does not have adequate geo-scientific data which is important as a base to launch future E&P activities, though the present Modi government has taken some major steps in this regard (Ministry of Petroleum and Natural Gas 2017). Conversely, India's quest for energy security is under urgent and intense pressure due to global warming concerns which is basically to move to non-fossil fuels. This means India will need to move towards alternative sources of energy by shifting to renewable and clean technologies which would put a question on the affordability of access to energy because of the elevated cost of fuel, and to shift to more environment-friendly and resourceful ways of commuting which will demand big financial inputs into the existing transport system. These scenarios pose a big challenge to the Indian government's goal of providing universal access to electricity to fuel its fast-growing economy.

The Indian government is facing the uphill task of chalking out strategy, both in the domestic and foreign policy arenas, and adapting to modern high-end technologies to pursue its development goals by considering its commitment to climate change goals. This would require India to operate its energy security in the international context which would be competitive and will need international collaboration at exceptional levels.

The stated concern was very much reflected in PM Modi's unambiguous emphasis at several international meetings, conferences and forums including at the 21st session of the Conference of Parties (COP21) to the United Nations Framework Convention on Climate Change (UNFCCC) held in Paris that India will not be able to achieve its INDC without massive international financial help. While putting the ambitious target of controlling CO_2 emission blamed for global warming at the UN conference on climate change in Paris, attended by the leaders of 150 nations on climate, he asked the developed nations which powered their way to prosperity on fossil fuels to fulfil their duty to shoulder the greater part of the burden of the fight against climate change. He appealed for financial assistance in order for India to achieve these goals, as Modi said, 'Climate change is a major global challenge. But it is not of our making' (*Times of India* 2015).

Economy and energy are dependent on each other. One of the biggest contributors to any modern economy is energy. India is no exception to this. India's growing economy is dependent on its capability to provide energy to its fast-growing economy.

In the pre-liberalization phase of the 1990s, India grew at an average of 3.5 per cent of GDP. Stifling bureaucracy, slovenliness of population and leaning towards socialist economic policy were blamed for this. When adjusted for fertile India's population growth, this was translated into a miserable 1.5 per cent of economic growth a year. While the economies of other formerly poor Asian 'Tigers' such as South Korea, Singapore, Taiwan and Hong Kong took off in the 1950s and 1960s, the sluggish giant India seemed destined to remain forever in the slow lane—with most of its people languishing in poverty. In the 1970s when the countries such as China were liberalizing their economies, India was nationalizing its banks. Nevertheless, this is a thing of the past.

India's Economic Growth

India today is one the leading centres of the global economy. India embarked on a liberalization policy and opened its economy in 1991 and a widening of that opening of economy in the 2000s has seen India advancing towards a free market economy. Since the 2000s, India has witnessed an average growth rate of around 7–8 per cent. India was the only country that was not affected by the global financial crisis. But, prior to Modi's arrival in power in 2014, India's economy went down and its GDP growth almost halved in just three years, falling to around 5 per cent in the 2012–2013 fiscal year down from 9.3 per cent in 2010–2011. PM Modi came to power with a thumping majority on the promise of fixing the economy and putting India's economic progress back on track. Since Modi came to power, there have been positive trends in the growth of the Indian economy. India has been able to attain an annual GDP growth rate of around 7.5 per cent which surpassed China's GDP rate, making India the world's fastest major growing economy. However, due to the 2017 demonetization and implementation of Goods and Services Tax (GST), Indian economic growth has slowed down. However, in the long term, these regulatory measures are meant for the betterment of the economy. The International Monetary Fund (IMF) latest projection for the coming years is that India will retain its fastest growing major economy status in the world by the year 2018 and in all probability continue for a decade and more (*Times of India* 2017). India is also projected to be on the path towards the world's biggest economy in the latter half of the 21st century. The Indian government and policymakers are aware that this growth will need a massive energy input to fuel its economy and fulfil the modern technology-based standard of living aspirations of more than 1.2 billion people. India cannot be complacent in its quest for energy security as expressed by the subsequent Indian governments over the past decade and a half.

India's Young Demography

The youth population of any country, being enthusiastic, vibrant, innovative and dynamic in nature, is the driving force in the development and progress of any country. Their motivation, strong passion and will power constitute the most valuable human resource for fostering the economic, cultural and political development of a nation. A country's ability and potential for growth is determined by the size of its youth population. Their role in building the defence capability of a nation is unquestionably crucial. The energy and passion of youth, if utilized properly, can bring huge positive change to society and progress to the nation. Youth are the creative digital innovators in their communities and participate as active citizens, eager to positively contribute to sustainable development. This section of the population needs to be harnessed. If motivated, skilled and streamlined properly, it can accelerate the progress of a nation.

India is home to the world's youngest populations in the world. India's young demographic dividend is projected to be one of the main drivers of India's growth. According to 'World Population Prospects: The 2015 Revision' Population Database of United Nations

Population Division, India has the world's highest number of 10- to 24-year-olds, with 242 million, despite having a smaller population than China, which has 185 million young people. The regularity and efficiency of census operations in India add credibility to the measurement of youth in India. The decennial enumeration through population census throws up consistent estimates of youth in India. As per India's Census 2011, youth (15–24 years) in India constitutes one-fifth (19.1%) of India's total population.

This demographic dividend also gives India an edge over China. India has the relative advantage at present over other countries in terms of distribution of youth population. As per India's census, the total youth population increased from 168 million in 1971 to 422 million in 2011. India is expected to remain younger longer than China and Indonesia, the two major countries other than India that determine the demographic profile of Asia.

Unlike in China, where the population is ageing rapidly due to the baleful legacy of Deng Xiaoping's one child policy, demography looks likely to provide a powerful tailwind to Indian growth over the coming half century. Around a quarter of all the people entering the global workforce between now and 2025 are projected to be Indian (Chu 2017). India is expected to have a 34.33 per cent share of youth in the total population by 2020. China, in contrast, is seen to have reached the highest share in the year 1990 at 38.28 per cent and is projected to have the share of youth force shrinking to 27.62 per cent by the year 2020, a situation that Japan experienced in around 2000 (The Ministry of Statistics and Programme Implementation 2017).

However, the prospects of this demographic dividend may not last forever. As per the United Nations Population Division's study, the percentage of the 15–34 age group in India's population will reach its peak (64.6%) in 2035 and begin to decline after that. The threats are not merely in terms of wasted educational investment and future contributions to the GDP but also in terms of the burden on pension, insurance and social security schemes (Prasad 2012). If not handled properly, the demographic dividend also may result in social and political upheavals. Proper education to skill the population for the job market and creation of employment opportunities remain an uphill task for the Indian government and of any country to engage the youth population and reap the dividend.

In the 2014 general election, the Modi-led BJP made the Congress-led UPA government's failure to create jobs a major issue. Between 2012 and 2014, jobs creation in India was affected by the economic slowdown and shrinking business opportunities and the unemployment rate increased from 3.5 per cent in 2011 to 3.7 per cent in 2013. A demographic change of 100 million new young voters since the previous election in 2009 (Laughland and Weaver 2014) furthered the necessity of job creation, which the Congress government failed to deliver. This was a promise made by the Modi-led BJP during 2014 parliamentary elections. Modi government was faced with the challenge of creating jobs and had no option but to move to the developmental programme, which not only increases the economic growth but also creates jobs.

Many traditional economies first move from the agricultural to the manufacturing sector. However, in India's case, it has been otherwise. India moved from an agricultural economy to a service sector economy. As a result, India lagged in creating jobs despite its economy growing at a faster rate than many developing and developed countries. In the post-liberalization phase in 1992 though India's industry sector witnessed the growing share of 27.4 per cent in FY 1991 of GDP, it employed only about 9 per cent of the work force in 1991, in the basic industries, mainly textiles, steel and aluminium, fertilizers and petrochemicals, and electronics and motor vehicles. To create jobs, India needs to focus on manufacturing.

PM Narendra Modi has taken certain major initiatives to keep the promise of developing the economy and to provide employment to the young Indians entering the job market every year. One of his initiatives is the much publicized 'Make in India' campaign. India's 'Make in India' campaign has already seen the automotive and defence industries moving in that direction. India has already made moves to tap into its manufacturing potential. Investment from Silicon Valley in Indian manufacturing capacities provides a boost to India's 'Make in India' and 'Digital India' campaigns. Digital India is an initiative to develop internet infrastructure in India with hopes to overcome India's digital divide by delivering electronic governance and universal phone connectivity to 250,000 villages by 2019. This initiative offers enormous potential for large investments in technology manufacturing and job creation.

Another area where India's manufacturing potential can be utilised is the defence industry. The scope for expansion and growth in India's defence industry is massive. India has been the world's largest arms importer every year since 2010, as its defence industry struggles to keep up with its international ambitions. The volume of major weapons imports more than doubled between 2004–2008 and 2009–2013 and India's share of the volume of international arms imports increased from 7 per cent to 14 per cent, according to the reports released by the Stockholm International Peace Research Institute (SIPRI 2014).

A plausible reason for this is that India lacks a defence industry of its own that is sufficient to meet its external challenges and to keep pace with its expanding strategic interests. The need to modernize has indeed been one major reason for India's status as a top spender on arms imports. Nevertheless, there are other factors at work too. India faces the challenges of tackling its adversary nations—Pakistan and China—and what worries its government most is the prospect of its two rivals uniting to form a hostile nuclear and defence partnership. This is not just scaremongering; 54 per cent of Chinese arms exports already go to Pakistan.

In the face of this threat, India feels it still needs to beef up its arms acquisitions to deal with its shortage of fighter jets, submarines, helicopters, howitzers and so on. The prospect of conflict between the three countries is real. India has been at odds with Pakistan since its creation. China fought with India in 1962 and the relationship is still not truly normal, with border stand-offs in the years 2013 and 2017.

India is still struggling to upgrade its arms manufacturing sector, despite this being a priority for over a decade. Continued reliance on foreign suppliers exposes the lack of a clear-cut vision of its defence industry. India has relied on the Soviet Union and now Russia for more than four decades. Since the Cold War, India has tried to increase its arms imports from the United States and Western Europe as part of a broader strategy of diversifying suppliers. This works both ways: The US defence industry is very happy to find new buyers and fund new technologies. American defence giants such as Lockheed Martin and Boeing have been exploring potential business partners in India, attracted by the low-cost, well-educated, English-speaking and technically sound workforce. The United States is now India's second largest arms supplier, accounting for 7 per cent of the total (*Outlook*, n.d.; A. Sharma 2014d).

Relying on imports from first tier arms producing countries has positives and negatives. On the one hand, it has ensured and enhanced India's defence capability, but on the other, it has constrained indigenous research, as technology has been imported rather than developed. Though these arms purchases are done with technology transfer and offset arrangements, India still lags behind the developed countries' defence industry standards. However, another negative impact because of the lack of its own defence industry is the lack of jobs that it would have created otherwise. This is going to change as India looks towards modernizing its defence industry by developing an indigenous defence industry base. India has the capability to give a boost to its defence industry. This will not only ensure the defence manufacturing jobs but will also help in tackling its unemployment problem. The Modi government's massive thrust on developing India's indigenous industry with international and private collaboration needs to be seen in this context as well. This pivot towards India as a global manufacturing hub will further push India's demand for energy to an unprecedented level.

Providing modern amenities to the villages and rural electrification is one of the biggest challenges and priorities for the Indian government. Of the total Indian population, 70 per cent live in villages. To achieve development in the country, it is the foremost responsibility of the government to connect villages with all sorts of modern amenities. One of the pressing challenges that Indian villages are facing is energy insecurity. Lack of access to energy has crippled the development of Indian villages and this has led to a massive migration of people from villages to big cities. This has created an overburden on urban infrastructure.

India's rural areas have not only lacked adequate access to energy but also failed to achieve efficient energy solutions. In 2014, around 600,000 villages needed to be provided with modern amenities. To address the problem of energy access to the rural areas, Dr A. P. J. Abdul Kalam, the Former President of India, emphasized the Provision of Urban Amenities to Rural Areas model as significant to give a fillip to development in the country. Making a plea for energy independence by 2030, he emphasized the need to shift to more alternative sources of energy, mainly solar energy and wind energy (*Hindu* 2014).

In rural area electrification, one of the biggest challenges is to reduce the dependency on biomass fuels used for cooking. This is one of the major causes for air pollution and creates severe health hazards for women. India has to use alternative energy options which are important for health and environmental concerns. NITI Aayog estimates that with the current policy and trend there is going to be a drastic shift from the current level of biomass energy share of cooking which will drop from 22 per cent in 2012 to just 3.3 per cent in 2040 in the baseline case and 3.5 per cent in the ambitious scenario. But India is faced with the challenge of providing environmentally cleaner energy by replacing the present biomass as a source of fuel, which is the major air polluter in the rural areas (NITI Aayog, Government of India 2017b). According to NITI Aayog's future assessment of India's energy demand and supply based on India Energy Security Scenarios or IESS 2047, in per capita terms, annual energy consumption will rise from 670 kgoe in 2015–2016 to 1055–1184 kgoe in 2040. Correspondingly, per capita annual electricity consumption will increase from 1075 KWh in 2015–2016 to 2911–2924 KWh in 2040 (NITI Aayog, Government of India 2017b).

With the GDP composition across different sectors changing with growth, energy shares of different consuming sectors shift. Buildings, industry and transport sectors together are the main gainers in both scenarios. But the gains of individual sectors vary considerably in the two scenarios. The maximum efficiency gains accrue in the transportation sector whose share in the total energy consumption in 2040 turns out to be 23 per cent under the ambitious scenario compared with 25 per cent in the baseline scenario. The above factors when combined with the increase in urbanization will put added pressure on the energy system and bring adverse air quality effects in many towns. City plans must respond to the needs of an efficient energy system in cooking, transport and electricity segments.

India's non-renewable energy reserves are not enough to fuel its fast-growing economy and growing energy appetite of its consumers. India has a limited supply of hydrocarbon sources of energy and is dependent on imports. India's abundant thorium and coal reserves too have some issues. India's coal is not considered as clean as that of other countries and India lacks the high-end technology and best practices of mining which can enhance its quality and lessen its impact on environment. As a result, India is dependent on imports of clean coal which come mainly from countries such as Australia. Again, for the best mining practices, India has to collaborate with other major coal exporting countries with better mining practices, such as Australia. India's thorium reserves are the second largest in the world, but the country has not been able to attain the processing technology to convert it to enriched uranium for its nuclear reactors for producing electricity. India's Non-Nuclear Proliferation Treaty (NNPT) status constrained it to acquire nuclear reactors and enriched uranium for moving to nuclear energy for electricity production. India's major source of oil and gas imports has been the Middle East region, which has huge oil and gas reserves. India has maintained its ties with the OPEC which is the leading supplier of oil to the world with its share of around 35 per cent of the total exports. India is no exception and is heavily dependent on import from this bloc for its oil and gas needs. This foreign dependency on these major sources of energy and renewable and other alternate sources is still not sufficient to keep pace with the energy needs of its accelerating economy and the widening gap between energy demand and supply. India's energy security policy has acquired an international context, eventually forcing it to become a major foreign policy priority.

In addition, the disruption and threat to its traditional destination for the sources of energy because of security concerns, unseen disruptions because of political instability, religious extremism and terrorist organizations, and threats to the supply lines have pushed India to look for other foreign destination for the hydrocarbons sources of energy.

Given this scenario, it is clear that quest for energy security is not going to be an easy task for India. It will need to be pursued in a manner that makes energy accessible to all the sections and sectors at an affordable price in a socially and politically acceptable carbon-controlled environment. The four 'A's is going to be the conceptual framework for the Indian energy policy and possibly of other debates which may be included in the future. But in the case of India, given its scale of demand, the four 'A's will be significant to sustain targeted levels of economic growth and spread of all-round and inclusive social development.

However, given the increasing energy demand and the debates surrounding energy security in a carbon-constrained environment, India will have no option but to continue to be dependent on energy imports, but it should also look towards major alternative sources of energy on land, offshore and sea. This will require a massive domestic and foreign policy effort. Modi's record number of visits to foreign nations has included energy security discussions but pursuing energy security is not always in a spirit of cooperation and resulting collaboration. It also involves geopolitical dimensions.

CHAPTER 2

India's Energy Reserves and Energy Mix

Addressing the Demand in a Carbon-Controlled Environment

Energy use is fundamental to modern nations, being vital to the economic development of any country. Its importance and demand are more critical in developing economies as they grapple with the gigantic task of addressing their growing economies' energy needs which require massive investment to meet those needs. Households and businesses use energy from oil, natural gas, coal, nuclear power and renewable sources (such as wind and the sun) to generate electricity, provide transportation, and heat and cool buildings. Energy consumption also constitutes a significant portion of a nation's GDP. Disruption in the supply of energy sources usually results in increased fuel prices, impacting households, consumers, businesses and overall economic activities of the nation.

India is no exception to the dynamics of energy supply and demand. As a fast-developing economy, India is heavily dependent on the supply of energy. Any disruption in the supply of energy can reduce India's national economic output, slow down the development, push down manufacturing and stymie growth in the industry and, at an individual level, reduce the income and standard of living of its people.

Since India's economic liberalization phase in the 1990s, the economic development, over the past decade and a half, has aggravated the need for energy security and energy policy to be taken seriously. India has taken a comprehensive approach to meet its growing energy demand. There is a big gap between India's energy demand and supply despite subsequent Indian governments' serious focus to address India's looming energy crisis over the past decade and a half.

This chapter, while giving an overview of India's sources of energy, reserves and energy mix, highlights the gap between demand and supply, and failure to tap the potential of its domestic reserves. While making an evaluation of the ongoing debates about the viability of India's energy basket including fossil fuels such as coal and gas, alternate renewable energy sources such as nuclear and renewables such as wind, waste-to-energy and solar energy, this chapter concludes that India's energy security policy has been an all-inclusive approach exploring an array of possible energy sources. Moreover, India's energy security quest, policy measures and choices have been pursued in the context of four 'A's: availability, affordability, accessibility and acceptability. Finally, it observes that renewables, especially solar and wind energy in which India is making impressive progress, will have their increased share in India's energy basket in the coming decades. However, coal will continue to lead India's energy mix and light Indian houses, and run Indian industries and businesses.

India's Energy Scenario: Gap Between Demand and Supply

According to the New Policy Scenario, the main scenario of World Energy Outlook 2011, India's energy demand is projected to grow by a compound annual growth rate (CAGR) of 3.1 per cent from 2009 to 2035, which is more than double the world's energy demand at a CAGR of 1.3 per cent for the same period. India's share in world energy demand will increase from 5.5 per cent in 2009 to 8.6 per cent in 2035, and the growth would come from all fuels (Ahn and Graczyk 2012).

There is a widening gap between energy demand and available reserves. To maintain its projected growth rate until 2031, India would need to increase its primary energy supply and electricity supply. The demand for coal will almost triple from 280 Mtoe in 2009 to 618 Mtoe in 2035 at a CAGR of 3.1 per cent. The renewable energy demand was projected to increase from 2 Mtoe in 2009 to 36 Mtoe in 2035. Nuclear energy demand would reach 48 Mtoe in 2035 from 5 Mtoe in 2009 (Ahn and Graczyk 2012). The demand for oil, natural gas and coal will necessitate the import of these products (Planning Commission 2012).

India has just 0.8 per cent of the world's known oil and natural gas resources. Today, oil and gas account for around 36 per cent of the country's primary energy use. This figure is set to rise both in absolute and in percentage terms. India's domestic production is not sufficient to meet its demand. India already imports more than 70 per cent of its crude oil needs. Without new and substantial domestic discoveries, imports are projected to increase further by 2030. By contrast, natural gas currently provides around 6–8 per cent (its share was at the peak at 12 per cent in 2011 and then fell by half in 2016 in India's energy mix) of India's primary energy supply, and most of that gas came from domestic sources, onshore and offshore. However, the position will change significantly if gas utilization rises as India moves towards the world average for the use of natural gas.

Coal continues to be the main source of energy for fuelling the Indian economy. India derives more than 50 per cent of its energy from domestic stocks of coal, mainly in the form of electricity. Coal shall remain India's most important energy source and critical to its growth for decades ahead. However, it is unlikely that coal will ever provide more than half of India's energy in the long term.

India's Coal Policy in a Carbon-Constrained Environment: India's Coal Reserves, Domestic Production and the Future Scenario

India continues to be excessively reliant on coal for its energy sources. As per the 2016 *BP Statistical Review of World Energy*, the global coal consumption declined by 1.8 per cent in 2015, which was below the 10-year average annual growth of 2.1 per cent. All of the net declines were accounted for the US at –12.7 per cent, the world's largest volumetric decline, and China –1.5 per cent. But, India's consumption of coal continued with +4.8 per cent, a modest increase when compared to Indonesia at +15 per cent. Global coal production fell by 4 per cent, with large declines in the US (–10.4%), Indonesia (–14.4%) and China (–2%). Coal's share of global primary energy consumption fell to 29.2 per cent, the lowest share since 2005 (BP 2016). Though the global trend of coal consumption as an energy source has seen its biggest decline in the past 10 years, India continues to rely heavily on coal for energy needs.

Of the major fuel sources, coal (55%) currently meets the largest share of India's energy requirement. The comprehensive policy document, 'Integrated Energy Policy' (IEP; effective in 2006 and ended in 2017), in its first report in 2006, stated: 'Coal shall remain India's most important energy source until 2031–2032 and possibly beyond'. For the 10 supply options projected by IEP, coal's share remains between 42 and 65 per cent, indicating its criticality to growth (Planning Commission, Government of India 2005). After more than a decade, coal continues to be a major source of India's energy mix. Coal accounted for 58 per cent of India's energy mix in 2015. By 2047, however, coal's share could shrink to 42–48 per cent owing to the growing share of other sources of hydrocarbon and renewables (*Reuters* 2017).

India ranks fifth in terms of coal world reserves. The Geographical Survey of India estimates India's coal reserves at 308.801 billion tonnes as on 1 April 2016 (Ministry of Coal 2017). The known reserves of coal increased 0.7 per cent over the year 2015, with the discovery of an estimated 2.20 billion metric tons (2.43 billion short tons). The estimated total reserves of lignite coal as on 31 March 2016 was 44.59 billion metric tons (Ministry of Statistics and Programme Implementation 2017). The energy derived from coal in India is about twice that of the energy derived from oil, whereas worldwide, energy derived from coal is about 30 per cent less than energy derived from oil.

Coal remains India's main provider of energy and tops its energy mix. India is the second largest producer of coal after China. However, India's coal production is not sufficient enough to meet its demand for coal. The data in Figure 2.1 shows the gap in the demand and supply of coal for the period till 2013–2014.

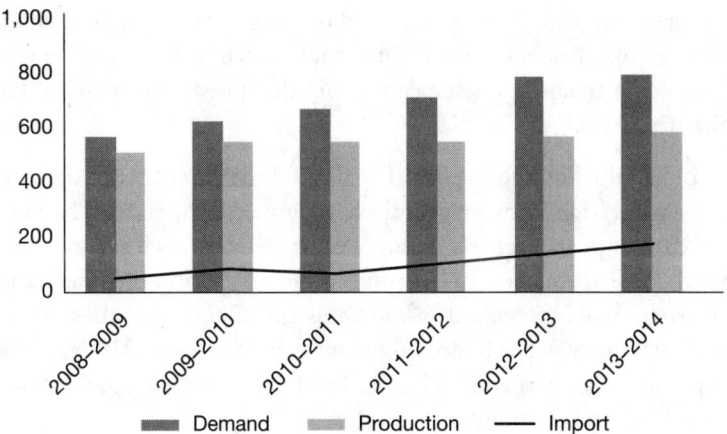

Figure 2.1 *Demand, Production and Import of Coal*
Source: Saritha S. Vishwanathan, Amit Garg, Vineet Tiwari (2018). Coal transition in India. Assessing India's energy transition options. IDDRI and Climate Strategies, available at https://coaltransitions.files.wordpress.com/2018/09/coaltransitions_finalreport_india_20181.pdf

Although in the year 2015–2016, India's coal production stood at 639.23 million metric tons (MMT; 704.63 million tons short) witnessing a growth of 4.93 per cent over the previous year, the big gap between the demand and supply in the graph could be explained by factors such as the poor quality of coal and inadequate availability of coking coal in India. India's coal sector is beset with major inefficiencies, productivity stagnation and obsolete technology. Environmental issues such as excessive pollution, ash-handling and disposal are additional constraints.

It is estimated that production from the large existing opencast mines will start declining by 2030. Furthermore, efforts to bridge the current gap between demand and supply by producing more coal from existing mines will have an adverse impact on coal availability in the longer run. A decade ago, India had set up Joint Working Groups on coal with France, Germany, Russia, Canada, Australia and China. These collaborations aimed at bringing in new technologies in both the underground and the opencast sectors, promoting efficient management, skill development and training, seeking bilateral funds for the import of equipment not manufactured in India and accessing foreign financial assistance to meet the investment required (A. Sharma 2007).[1]

India relied on the import of coal to fill the gap between demand and supply. Mainly, two types of coal are imported by India. The thermal coal for power-generation supplied by countries such as Indonesia, South Africa, Russia and Colombia. The other form of India's coal imports is coking coal, also known as metallurgical coal, used for steel-making which is imported mainly from Australia. Although the importation of thermal coal has declined in recent years, the demand for coking coal continues.

In recent years, India's coal policy has focused on quality over quantity as evident in its coal import trends. Indonesia that has been the leading supplier of thermal coal to India has witnessed this decline. The year 2016 coal import data shows a drop in the import from Indonesia losing much of its Indian coal market to its competitors such as South Africa, Colombia and Russia. The argument, based on the data and certain assumptions about the energy, or calorific value of coal from various countries show that the calorific value of India's coal imports is not declining nearly as fast as the actual tonnage (*Reuters* 2016). Though there is a decline in India's coal imports, the energy value is declining at a far slower pace than the physical volumes as India shifts to cleaner fuel reflected in its coal import trends.

For the first five months of 2016, India's coal imports were 82.57 mt, a drop of 5.4 per cent over the same period for the year 2015. Breaking the data down shows that while Indonesia, with 36.72 mt, remained the top supplier, it went down almost 20 per cent from the 46.9 mt shipped in the first five months of 2015. This benefitted South Africa that exported 16.58 mt coal, an increase of 26 per cent, to India in the first five months of the year 2016; Russia, which exported 1.8 mt to India over the period versus 1.45 mt coal a year ago and Colombia with 981,000 tonnes compared with just 46,000 shipped in the same period in 2015. The imports from Australia in

[1] Ministry of Coal, Government of India: http://www.coal.nic.in/.

the first five months were 18.74 mt higher than the 18.53 mt in the same period (*Reuters* 2016).

Indonesian thermal coal is generally of lower quality of a calorific value of 4,500 kilocalories per kilogram (kcal/kg), compared with South African and Colombian coal with their energy value of 6,000 kcal/kg, and Russian coal at 5,600 kcal/kg. Taking this into consideration, the energy value of India's imports from these four nations in January–May 2016 period is 5.9 per cent below what it was for the same period in 2015. However, the physical volume of coal from the four nations is 9 per cent lower; meaning the decline in tonnes was far more pronounced than the drop in energy value.

The declining trend in the import of coal continued in the year 2017. Coal imports declined by 24 per cent on a yearly basis. Coal import (all types of coals) in August 2017 stood at 14.97 mt (provisional), against 19.75 mt in August 2016. According to the official data, the coal import fell 6.37 per cent to 191.95 mt in 2016–2017 from 203.95 mt in 2015–2016. The lower volume of coal and coke imports in August (down 24.2% year on year) is mainly due to a 2.4 mt decline in non-coking coal imports during the month under review. The main reason for the decline in the import of non-coking coal is due to higher production by Coal India moving towards a position of surplus coal with an availability of almost 20 days' stock (*Economic Times* 15 September 2017; *Times of India* 2016).

The trend shows that India's coal imports will continue to decline in volumes with a trend towards higher-quality fuels. But Australia's coking coal which is mainly used for India's steel industry continues to be in demand. India's non-coking coal importers are taking the view that it's better to pay more for higher grade cargoes than to merely take the cheapest on offer. This fuel mix strategy of importing lower volumes of higher-quality coal is in line with the Indian government's stated aim of stopping the import of thermal coal by the year 2022. The dictates from India's Ministry of Coal to the government-owned and operated thermal power producers to stop all imports should be seen in the context of the availability of surplus coal from the state miner Coal India.

The Ministry of Coal in India, responsible for the exploration of coking and non-coking coal, aims at securing the availability of coal to meet the demand of various sectors of the economy in an eco-friendly, sustainable and cost-effective manner. To achieve this goal, the Indian government is focusing on increasing coal production through government companies as well as the mining route by adopting state-of-the-art and clean coal technologies with a view to improving productivity, safety, quality and ecology; augmenting the resource base by enhancing exploration efforts with the main thrust on increasing proven resources; and facilitating development of necessary infrastructure for the prompt mining of coal.[2]

The stated goal of the Ministry of Coal and India's move towards higher-quality coal is in line with the Indian PM Modi's commitment at the Paris Climate Summit to pursue India's energy security policy in a carbon-controlled environment. While ensuring the energy available to its economy and consumption, carbon reduction is pushing India to explore the option of renewable energy. However, renewable energy alone cannot meet India's massive energy demand. India's energy mix is still dominated by coal and it is around 55 per cent. So any increase in renewable energy will not be enough to keep pace with India's fast-growing economy, its move towards the manufacturing sector to create more jobs for its young population and the increasing demand of its rapidly expanding middle-class consumption. Coal is likely to be a major source of India's energy mix.

In recent years, India has made progress on domestic coal production. Due to the increased productivity of domestic mines, cheaper renewables and lower than expected energy demand, India now has surplus coal. On the increase in domestic coal production, India's erstwhile Energy Minister Piyush Goyal made a statement that India has sufficient coal capacity to power itself without foreign coal. However, despite this progress on coal form indigenous mines, India's reliance on foreign coal is likely to continue in the coming years.

---

[2] Ministry of Coal, Government of India. 'Vision': https://www.coal.nic.in/content/vision.

India's coal import sources are moving towards those countries which have the higher grade of coal. Australia figures prominently in the scenario. India has many power plants which are still dependent on foreign coal as these are power plants which are designed to run only on imported coal. Too many Indian power plants had been designed to run on foreign coal. Despite the trends of reduced coal imports since the Modi government's arrival, to stop coal import totally is not possible. Many recently built coal power plants were designed to process more efficient, higher-calorific value coal which is not suitable for using the coal produced by the mines in India. The erstwhile Indian government's Energy Minister Piyush Goyal had been critical of the previous government's lack of farsightedness and undervaluing of the potential of Indian coalmines, 'Previous governments did not ever imagine that India could be self-sufficient and depend solely on its own coal' (Shafi 2017). As a result, 83,100 MW of coal-based thermal power capacity has been set up in India, partially or fully reliant on imported coal. India's total electricity capacity is currently 220 GW. In the year 2016, for the first time, India had a 69 m-tonne coal surplus.

In a fast-changing move from fossil fuel to a cleaner energy, India cancelled coal power plants projects up to around 14 GW of electricity capacity and was about to close down 37 coal mines in May 2017. In the first instance, this could be seen in the context of availability of surplus of coal. However, in the larger term it is India's goal to move towards cleaner energy (Johnston 2017).

India is the world's fastest growing polluter with CO_2 emissions volume growth at 8 per cent and 5 per cent in 2014 and 2015, respectively. To reduce dependence on fossil fuels and its commitment towards cleaner energy, the Indian government has an ambitious plan to increase India's installed capacity of renewable energy to overtake that of thermal coal by 2022. India has set an ambitious target of reaching 175 GW of installed renewable capacity by the year 2022.

The Indian energy minister's view about India's use of a reduced amount of coal in the future to fuel its fast-growing economy over the next decade has been backed by the recent report published by the New Delhi-based research think tank The Energy and Resources Institute

(TERI). The TERI report highlights that India could get rid of coal power plants by 2050 if the right policies were adopted. If the prices for both batteries and renewable energy sources continue to fall at the current rate, India can reduce its CO_2 emissions by up to 10 per cent or 600 mt after 2030, should renewable energy and batteries become less costly than coal within a decade. The TERI research report further reveals that if India adopts a strategy that emboldens the possibility of an electricity grid that essentially operates on renewable sources, it can get rid of all coal-power plants by 2050.

China—the biggest emitter of CO_2 at 29 per cent—witnessed CO_2 emissions reduction by only 0.7 per cent in 2015, compared to a growth of more than 5 per cent per year in the previous decade. In the US, the second biggest emitter of CO_2 at 15 per cent, emission decreased by 2.6 per cent in 2014. India contributed 6.3 per cent of all global CO_2 emissions in 2015. India's average emission of CO_2 is far below that of China and the US (*Indian Express* 2016).

India is pursuing its coal policy in a carbon-controlled environment and in an affordable manner. However, the assumption of the Indian government that India will not need coal is an overambitious proclamation. Given India's fast-growing economy and its commitment to providing regular electricity to all its citizens, and push towards providing employment to its people, coal will remain a paramount source of energy in the coming decades. Furthermore, India's fast-growing energy consumer population and its booming economy have made it the world's fastest growing CO_2 emitter. India's carbon emissions volume grew by over 8 per cent in 2014 and 5 per cent in 2015, respectively. India targets to reduce carbon emissions by 20–25 per cent of 2005 intensity by 2020 (Froome 2014).

The electricity in India is heavily dependent on non-coking coal. The electricity generated by burning coal power plants releases fly ash. This highly polluting fly ash contains extremely polluting substances such as radioactive isotopes and heavy metals including mercury. These highly toxic releases pollute the air and the water around the power plants, and the pollutants are being found in the food chain as well. As a result, coal-based electricity generation through power plants has received widespread opposition (Sarma 2017).

Given the widespread debate over the use of coal and its impact on global warming, India committed to reduce the carbon emission through its INDCs compliance to the UNFCCC; India's commitment to reduce the emission intensity of its GDP significantly by realizing about 175,000 MW (40% of the cumulative electricity generation capacity) from non-fossil energy resources by 2022 is aimed at achieving the UN Climate Change Commitment. This means that the share of coal will decline in the coming years in India's electricity generation mix with the Indian government's priority to efficiency improvements in the electricity supply chain and progressive shift towards decentralized renewable energy systems (Sarma 2017).

However, India's image as a fast-growing polluter contradicts its commitment of limiting its CO_2 emission to the Paris Climate Summit. India's move towards cleaner energy complements its energy policy in a carbon-monitored environment. However, on the front of making energy accessible to its citizens and to fuel its booming economy, it may not be possible without coal. India will need coal and the option should be to get better coal with modern and upgraded technology. In response to the report by TERI on India pursuing its energy need without any new coal power plant after 2027, the World Coal Association (WCA) questions its viability. According to WCA, at any rate of India's investment and progress in renewable sources of energy, it is not plausible for India to attain its goal of universal energy access to its citizens' consumption and grow its economy without coal in the coming years.

According to WCA CEO, Benjamin Sporton,

> India's energy needs are too huge for any suggestion that it will not need coal in the future. In a country where 244 million have no electricity and no access to clean cooking facilities, it is impossible to find a solution without coal being part of the energy mix—coal is essential to global efforts to achieve universal energy access. (Lock 2017)

The WCA argument is that in the case of India, both fossil fuels and renewables are needed. Though renewable energy will have a major role in India's energy policy amidst the concern about climate change, coal will continue to be the main energy source for the electrification and industrialization in India. The IEA also makes its assessment that coal will continue to make the largest contribution to electricity generation in India through to 2040.

The argument for investments in new coal power plants in India for the near future is based on economic and developmental grounds. India has huge developmental and energy challenges, and the government is clear that all sources of energy will be needed to power up the economy, including coal. Although the competitiveness of renewables and gas-fired technology is likely to improve over time, coal is expected to remain the most affordable option through to 2035 (Lock 2017).

The assumption of meeting these energy needs without coal ignores many important factors affecting India's energy needs. This may lead to an energy policy based on false assumptions. The world would have welcomed this assumption given the urgency of reducing CO_2 emission, but it is not achievable, especially in India's case. India's Paris Climate Summit commitments in all likelihood go with the right mix of energy sources in which coal is likely to continue for a very long time to come. This has become more relevant today than ever, especially considering the Indian government's ambitious goal of providing universal access to energy to its citizens and fuelling its economy which will require steady, unwavering and reasonably priced energy, in a carbon-controlled environment.

From this debate, it is clear that India will continue to rely on coal even if it makes progress in the shift to renewables. In this context, the self-reliant argument of ending coal import may not hold ground. This is in opposition to the Queensland Carmichael mines Indian owner, the Adani Group, and several Australian government ministers, who have pitched the $16 billion project as critical to alleviating energy poverty in India. According to Adani's top executive in Australia, Jeyakumar Janakaraj, the mine and rail project is a way to address power poverty for hundreds of millions in India.

One-fourth of India's population is still denied access to regular electricity. India is a country with a surplus of energy but still has the maximum number of people living without electricity. India has set a goal of generating 40 per cent of its electricity needs from non-fossil fuels. India has 250 million people who have no access to electricity and yet surprisingly India produces a surplus of energy.

Supplies of coal in a cost-efficient manner and foreign coal-run power stations have the capacity to provide electricity to 100 million homes or more. The coal from the Queensland Carmichael mine of Adani—the first coal exports—are expected to reach Indian power plants by 2020. As a result, India's coal import destination is moving towards the countries which have a higher grade of coal. Australia figures prominently in this scenario.

In fact, it is important that implementation of the Paris Climate Agreement integrates environmental imperatives with the aims of universal access to energy, energy security and socio-economic development. Rightly so, the Indian government will need to rely on modern technologies to renovate, upgrade and extend the lifespan of coal power stations in use. But that still would not be sufficient as India's many recently built coal power plants are designed to run imported coal. Australian coal has a high calorific value and is considered as cleaner coal of high quality. Despite the shift to renewable energy sources, coal will continue to be in demand. Australian coal will be relevant on two grounds. First is the environmental concern in which Australian coal will be fit for India's ambitious target on CO_2 emission. Second is that the price of Australian coal rather than India's energy policies would determine the level of those imports. This is in line with India's power ministry which, when asked if India will import coal from Australia, replied that provided it is imported at an affordable price.

In the year 2017, according to the Central Electricity Authority's *Load Generation Balance Report*, India will produce 8.8 per cent surplus electricity over and above that required for consumption. This falls in line with PM Modi's most keenly followed promise over the last four years—his promise to light every home in India. PM Modi, in his India's Independence Day speech on 15 August 2015 had promised that all the remaining 18,452 un-electrified villages will be electrified within the next three years. But according to the progress data on the government's website, by March 2016, 74 per cent (13,640) of the 18,452 villages marked for electrification have been electrified. By that benchmark, it may seem a success, but in reality, few are benefiting. The government's statistics show that only 8 per cent of villages have every home connected. A village is considered as electrified if public places in the village and 10 per cent of its households have electricity (Bansal 2016). By this benchmark of village electrification, not every citizen of India can be considered as connected. This gives an inflated number of villages electrified, as the presence of electrical infrastructure does not automatically translate into electrification. That is why village electrification is continuing even in the villages that have been declared electrified under the scheme of 'intensive electrification' which targets all households and not just 10 per cent.

Many remote villages without road access have been connected through solar energy panels and micro-grid systems. But the contribution of renewables under PM Modi's much-vaunted flagship programme of rural electrification, Deen Dayal Upadhyaya Gram Jyoti Yojana, is being probed on the ground of effectiveness and dependability and the extent to which renewables can form India's energy mix. Further, demand is going to be created by further urbanization including the pressure of urbanization caused by the creation of smart cities. Again, in India, the unpredictability of wind is greater than most regions and changes significantly with the seasons. It is also contended that solar energy for large-scale solar projects is not available in more than 90 per cent of the country.

The Modi government, unlike the previous government, has taken a massive step in regards to supplying electricity for every citizen of India, and has made an impressive leap forward especially in connecting to the villages. But in reality, the demand for energy for electrification in India is underestimated. The Indian government's ambitious goal of producing 40 per cent of the required electricity without coal by 2030, though a welcome vision in the climate-conscious world, is overestimated. Not surprisingly many experts from different corners have difficulty in agreeing with the government's claim and future projections. In the words of Arunabha Ghosh, the CEO of India's Council on Energy, Environment and Water, Coal would remain 'vital' to India's power production for another two decades. The current projection indicates coal consumption for power generation rising from roughly 700 mt today to 1.5 billion over the coming 20 years and this energy dividend certainly cannot be driven by non-fossil fuel alone; hence, coal will remain important, and India would continue to import coal from countries such as Australia, which has a higher calorific and energy value, and a low ash content (Bennett 2017a).

Similarly, while questioning the optimistic time frames embarked by the Indian government in its view of the ambitious and extensive task that India has set of increasing access to grid-connected electricity and reducing fossil fuel-based power at the same time, the *Coal in India*, a report by the Australian government's Office of the Chief Economist, presents of the following view:

> India is likely to continue to rely on imports. The expansion in India's coal use presents some opportunities for the Australian industry, which is not currently a large supplier of thermal coal to India. Australia has large deposits of high-energy, low ash coal that is suitable for use in advanced coal-generation technologies. The roll out of advanced coal generation technologies in India presents a significant long term opportunity for coal producers (Froome and Dargaville 2015).

India is presently going through a defining shift in its energy mix from fossil fuels to renewable resources. In the coming decades, it is highly unlikely that the country will be a large consumer of coal for electricity generation or a major coal importer (Sarma 2017). India's shift to renewables has been remarkably transformative and it is expected that by 2030, 175,000 MW of renewables will come on stream against an electricity capacity of just over 300,000 MW (Bennett 2017a). This elevates India to a country pursuing one of the most forceful goals for renewable energy in the world. This should be seen in the context of unanimity across India's political spectrum on the issue of climate change and global warming, and its quest for energy security in a socially acceptable milieu.

However, the appetite for electricity and other forms of energy will increase dramatically, so the challenge of meeting these demands with local fossil fuel will be significant. India has demonstrated a rapid uptake of renewable energy technologies, but to continue that expansion at required rates is also a massive challenge (Froome and Dargaville 2015).

The government aims to protect the environment as well as make India's power supply more self-sufficient. This is a rather optimistic objective, considering coal still makes up nearly 70 per cent of India's current electricity generation capacity and demand is still 10 per cent greater than the capacity to supply. The requirement for electricity has been gradually increasing. Given the Indian electricity accessibility scenario in which about one-fourth of its population is still not connected to the electricity grid, more than half the population is still not getting a regular supply of electricity, and given the current government's commitment of providing electricity to all, the demand of electricity is not going to stop.

India is aiming for 175 GW of renewable to be built by 2022 which is equivalent to the building of four times Australia's energy infrastructure. Given the scenario that India is all set to surpass China's population and become the most populated nation in the world by 2030, its extraordinarily expanding middle class population and its growing economy which is set to become the world's second largest by 2050 and the largest in the second half of the this century, energy demand will grow dramatically and seems irreversible.

Despite progress on renewables, coal is indispensable for India's electricity production. The Indian government aims to boost coal production from its indigenous coal mines instead of relying on imports. This is cost-effective as coal from the domestic mines costs around US$24 a tonne, compared to US$62 a tonne for imported Australian coal (Froome and Dargaville 2015). But the claim of progress made by the Indian government on India's reliance on coal from indigenous mines being a reason to stop importing coal in the coming years is an overly optimistic and unrealistic goal for India. The time frame set by the Indian government seems very optimistic, given the substantial task it has set for itself of increasing access to grid-connected electricity and simultaneously reducing fossil fuel-based power.

India's Hydrocarbons Sources: Oil and Gas

India in 2006 stood as the sixth largest oil consumer in the world and the third largest consumer in Asia. Oil accounted for 36 per cent of the country's primary energy use. India's proven reserves of crude oil are 732 MMT. Domestic production is some 32 mt of oil per annum, but demand is 113 mt. As a result, India needed to import 70 per cent of its crude oil needs (A. Sharma 2007). The gap between its demand and supply continues to this day. In the year 2016, as per the BP Energy Outlook, India relies on oil for 29 per cent of its total energy needs (*BP* 2018). India today is the third largest oil consumer in the world at the end of year 2016 with 212.7 mt, China second with 578.7 mt and the US as the leading consumer with 863.1 mt. According to the IEA estimates, India will be the centre of global oil demand growth until 2030. India's consumption of gasoline and diesel is expected to double by 2030 despite the aspiration to sell only electric vehicles in 2030. The two fuels account for more than half of India's oil usage. As per the Ministry of Petroleum and Natural Gas's Petroleum Planning and Analysis Cell, India's oil consumption growth rate was fastest in 14 months from 15.3 million in 2017 to 16.9 mt in 2018 (Sundria and Chakraborty 2018).

India's total proven oil reserves at the end of the year 2016 are estimated at 0.6 thousand mt which is 0.3 per cent of the world share. Obviously, India is totally dependent on imports for its oil needs. India's annual growth rate of consumption of oil in 2016 was 8.3 per cent, China's at 2.7 per cent, the US at 0.5 per cent with the world growth rate being 1.5 per cent. The oil consumption of the annual growth rate share in the world in the year 2016 was recorded with India at 4.8 per cent, China 13.1 per cent and the US 19.5 per cent. India's oil consumption average growth rate during 2005–2015 was 4.9 per cent, China's 5.5 per cent and the US –0.9 per cent, whereas the world average was 1 per cent (*BP* 2017). This growth is attributed to improved road freight transport following the stabilization of the new nationwide sales tax and the growing use of automobiles. This is not going to stop given India's expanding middle class, and urbanization and industrialization with the focus on making India a manufacturing power with an expected growth rate of 8–10 per cent in the coming years for at least two decades.

Obviously, India's proven reserves cannot fulfil its oil needs. It is heavily dependent on imported oil. India's indigenous oil exploration has not been fulfilling, and India has thus been intensely involved in oil exploration abroad and seeking to broaden its supply sources over the past 15 years. (India's energy oil exploration abroad and consequences of its geopolitics is dealt with in Chapters 4 and 6.)

Natural Gas

According to the 2016 *BP Statistical Review of World Energy*, the world natural gas consumption grew by 1.7 per cent in 2015, a significant increase from the very weak growth (+0.6%) seen in 2014 but still below the 10-year average of 2.3 per cent. As with oil, consumption growth was below average outside the OECD (+1.9%, accounting for 53.5% of global consumption) but above average in the OECD countries (+1.5%). Among emerging economies, Iran (+6.2%) and China (+4.7%) recorded the largest increments in consumption, even though growth in China was sluggish compared with the 10-year average growth of 15.1 per cent. Meanwhile, Russia (–5%) recorded the largest volumetric decline, followed by Ukraine (–21.8%). Among OECD countries, the US (+3%) accounted for the largest growth increment, while EU consumption (+4.6%) rebounded after a large decline in 2014. Globally, natural gas accounted for 23.8 per cent of primary energy consumption. Global natural gas production grew by 2.2 per cent, more rapidly than consumption but below its 10-year average of 2.4 per cent. As with consumption, the US (+5.4%) recorded the largest growth increment, with Iran (+5.7%) and Norway (+7.7%) also recording significant increases in production. Growth was above average in North America, Africa and the Asia-Pacific. The EU production once again fell sharply (–8%), with the Netherlands (–22.8%) recording the world's largest decline (BP 2016).

The latest global energy consumption scenario indicates that although India is not one of the leading gas consumption countries, it is expected to be so in the coming years. India, the world's third largest energy consumer, is striving to expand its natural gas use and shift away from oil and coal for power and industry. Though India is the world's fourth largest importer of LNG—behind Japan, South Korea and China—and is seen as a market on the rise, growth in consumption of the fuel is still far outpaced by oil.

In 2006, natural gas provided only 8 per cent of India's primary energy supply and most of that gas came from domestic sources, onshore and offshore. It rose to a high of 12 per cent in 2011 but went down by half in 2016 (*Economic Times* 6 December 2016). The main reason for this low share of gas in India's energy mix is often cited as India's lack of infrastructure to support a gas-based economy. As a result, coal and liquid fuels continue to dominate India's overall energy mix. Impetus is being given to increase the share of natural gas in India's energy mix by the current government. Acknowledging the environmental acceptability and economic affordability feature of gas, Indian PM Narendra Modi stated that natural gas was the 'next-generation fossil fuel, cheaper and less polluting' (*Economic Times* 5 December 2016).

There is an understanding among the policymakers and the government that India will need to enhance its gas import infrastructure to meet the demand of gas and explore the natural gas reserves in India. The Modi government is focusing on developing the infrastructure and pumping money to uplift the gas infrastructure. India's oil ministry is planning to invest up to US$15 billion in laying pipelines and setting up LNG terminals over the next five years. The government has planned to extend piped gas to 10 million houses over the next five years; double the length of the national pipeline network to 30,000 km and build a new gas line in the underdeveloped eastern region.

Both private and public sectors are getting involved in this task. The state-owned consultancy Engineers India Ltd Bombay Stock Exchange (BSE) –0.53 per cent recently entered into an agreement with Gazprom to study the construction of a gas pipeline to India from Russia. India is keen on tapping foreign expertise and know-how. India's LNG imports from RasGas represent around 10 per cent of the Middle Eastern nations' overall annual output. RasGas has a big presence in India through its 25-year contract with the top importer of the fuel, Petronet LNG BSE –0.52 per cent, and provides more than 50 per cent of the country's LNG shipments. The Russian entity, Gazprom, aims at tapping India's fast-increasing energy demand and India's move to increase the gas share in the energy mix. Gazprom has a long-term supply agreement with state-controlled GAIL India Ltd BSE –0.26 per cent and intends to enhance its stake in India's gas market and compete with RasGas. India renegotiated a long-term LNG supply deal with RasGas last year, nearly halving the price and avoiding a $1.5 billion penalty fee for lifting less gas than agreed as customers there preferred cheaper spot supplies (*Economic Times* 6 December 2016).

Shifting to Alternative and New Forms of Energy

The issues of affordability in terms of high and fluctuating costs of oil, accessibility of oil due to instability in oil-producing region and acceptability amidst environmental concerns are driving India to reduce dependency on hydrocarbon sources of energy and explore alternative sources of energy.

The fluctuating price of oil peaked at $70 a barrel in the summer of 2006, with projections of $100 a barrel in the near future, which was alarming for India which imported 70 per cent of all oil used. Diversifying oil supply sources may help with security of supply, but it does not alter the economic burden of a high oil price. Thus, it is a policy necessity to reduce dependence on oil and explore alternative and renewable sources of sustainable energy.

Nuclear Energy: Its Potential, Security Concern and the Future Scenario

The world energy scenario of nuclear energy use shows that the global nuclear output grew by 1.3 per cent, with China (+28.9%) accounting for virtually all of the increase. China has passed South Korea to become the fourth largest producer of nuclear power. Elsewhere, increases in Russia (+8%) and South Korea (+5.3%) offset declines in Sweden (–12.6%) and Belgium (–22.6%). The EU output (–2.2%) fell to the lowest level since 1992. Nuclear power accounted for 4.4 per cent of global primary energy consumption (*BP* 2016).

According to the BP survey, nuclear energy accounted for 6.1 per cent of the total global primary energy consumption, with much higher shares for advanced countries, namely 38.6 per cent for France, 12.6 per cent for Japan, 11.4 per cent for Germany and 8 per cent for the United States (*BP* 2005, 38). The US 2005 National Energy Policy (NEP) Act encourages development of advanced nuclear reactors, promising financial assistance to relevant contractors and subcontractors. Its 2006, National Security Strategy suggests 'a global nuclear energy partnership for common endeavour toward applying sophisticated nuclear cycle and reactor technologies for giant strides in the expansion of safe and clean nuclear energy so as to meet the soaring demand worldwide' (Kurian and Vinodan 2013; The White House 2016).

Nuclear power plants are characterized by high construction costs and relatively low operating costs. Nuclear power plants involve massive facilities with high construction costs. The sheer scale of commercial-sized nuclear reactors entails that most components must be particularly designed and constructed, often with few potential suppliers worldwide. These components are then assembled on site and structures are constructed to house the assembled components. All stages of design, construction, assembly and testing require highly skilled, highly specialized engineers and differences in reactor design and site-specific factors have historically meant that there was little scope for spreading design and production costs across multiple projects (Davis 2012). In addition to the high costs involved, the production of GHGs during production; decommissioning the nuclear plants; safety and security of maintaining and keeping the nuclear establishments; nuclear wastage and above all the threat of nuclear terrorism and norms against nuclear weapons have a combined holistic effect on the prospects of nuclear energy gaining prominence.

In the US, the leading energy consumer for a long time, the use of nuclear energy has been a major source of electricity generation and energy security. The demand for nuclear power plants and the use of nuclear energy got a massive boom in the 1960s and 1970s in the United States. By the year 1974, the United States had 54 operating atomic reactors and 197 reactors were on order. This boom era and enthusiasm for nuclear power was due to many reasons. The US power consumption trend projection of a strong growth trajectory in the demand for electricity did not coincide well with the prices of coal in the US which was then at its peak in real terms. This gave nuclear power primacy and its growth witnessed a massive upward trajectory. This positive trend could be seen in the US Atomic Energy Commission (1974) report that predicted that the nuclear power would constitute 50 per cent of all US electricity generation in the US by the end of the 20th century (Davis 2012).

However, this trend could not continue for long. No new reactors were ordered and to make things worse, companies began to suspend construction on existing orders. In the 1970s, the safety argument and environmental concerns about nuclear power plants became prominent.[3] Other forms of hydrocarbon sources such as natural gas were not significant mainly because of the lack of use of extensive modern combined cycle technology and due to deficiencies related to federal price controls on natural gas which restricted the availability of it for electricity production. The safety and security of nuclear power, the handling of radioactive materials, and the storage and disposal of spent fuels became the issues on which the various communities started challenging nuclear power projects in the US courts both at state and federal levels. This resulted in protracted construction interruptions and it affected the public outlook about nuclear power, thus, further damaging its prospects. The boom period of nuclear power plants in the 1950s and 1960s was also supported by the steady fall in electricity prices due to economies of scale, decreasing commodity costs and relatively low inflation. But in the 1970s, firms began to increase prices which led to the scrutiny of utilities' capital expenditures, and in

---

[3] For the trends on nuclear energy, see Joskow (1982) and McCallion (1995).

specific, investments in atomic plants. The prospects of nuclear power further dimmed after the incident at the Three Mile Island plant in Pennsylvania which suffered a partial core meltdown in 1979. This accident did not cause any injury but aggravated public concerns about nuclear safety. The combination of severe public concern about the risk of nuclear accidents and escalating construction costs put nuclear projects in an extremely vulnerable position. The nuclear industry was ramshackle by the time the Chernobyl disaster happened in April 1986. The number of nuclear plants in the United States is now around 104 at 65 sites, accounting for 20 per cent of US electricity generation. They all were ordered prior to 1974. A similar trend about the use of nuclear power could be seen in other nuclear technology countries such as the UK, France, Germany and Japan in the 1960s and 1970s. As per the US Department of Energy, 2011 data, US net generation of electricity in 2010 included coal (45%), natural gas (24%), nuclear (20%), hydroelectric (7%) and wind and other renewables (4%) (Davis 2012).

The prospects for revival of nuclear power were dim even before the partial reactor meltdowns at the Fukushima nuclear plant. Nuclear power has long been controversial because of concerns about nuclear accidents, proliferation risk and the storage of spent fuel. These concerns are real and important. In addition, however, a key challenge for nuclear power has been the high cost of construction for nuclear plants. Construction costs are so high that it becomes difficult to make an economic argument for nuclear power, even before incorporating these external costs. This is particularly true in countries like the United States where recent technological advances have dramatically increased the availability of natural gas. The 2016 US energy by energy source data table shows that the possibility of revival of nuclear energy as a major component of the US energy mix is minimal as nuclear electric power constitutes 9 per cent of the total electricity generation, reduced by more than half since 2010. In the same year, natural gas increased to 29 per cent and renewables more than doubled to around 10 per cent (US Energy Administration 2017).

But worldwide, there has been a steady increase in the nuclear power capacity. The future projection is that the proportion of nuclear power generation will increase steadily as envisaged by OECD's IEA.

In its 2017 edition of its World Energy Outlook, the IEA taking the current situation and concerns of controlling global warming by reducing carbon reduction, in its 'New Policies Scenarios', envisages the installed nuclear capacity growth of over 25 per cent from 2015 (about 404 GWe) to 2040 (about 516 GWe). The scenario envisions a total generating capacity of 11,960 GWe by 2040, with the increase concentrated heavily in Asia, and in particular China (33% of the total). In this scenario, nuclear power contribution to global power generation will be about 14 per cent of the total (International Energy Agency 2017).

The report states,

> In the Sustainable Development Scenario, low-carbon sources double their share in the energy mix to 40% in 2040, all avenues to improve efficiency are pursued, coal demand goes into an imme-diate decline and oil consumption peaks soon thereafter. Power generation is all but decarbonised, relying by 2040 on generation from renewables (over 60%), nuclear power (15%) as well as a contribution from carbon capture and storage (6%)—a technology that plays an equally significant role in cutting emissions from the industry sector (International Energy Agency 2017).

This sustainable development argument makes nuclear energy a viable option for countries such as China and India. These two develop-ing nations are faced with the challenge of development as well as controlling their carbon emissions and moving towards less polluting sources of energy.

Not surprisingly, China is leading in the construction of nuclear power reactors. But China has been witnessing a steep rise in nuclear power for the generation of electricity over the past decade which accounts for the increase in the building of nuclear reactors, and leads in their construction in the world (Davis 2012). Nuclear power plant construction is currently being witnessed in the Asian region, especially China and India with fast-growing economies and with a rapidly rising electricity demand. In all, about 160 power reactors with a total gross capacity of some 160,000 MWe are on order or planned, and over 300 more are proposed. At present, there are around 440 nuclear power reactors operating in 30 countries plus Taiwan, with a combined capacity of over 390 GWe. In 2015, these provided 2,571 billion kWh, about 11 per cent of the world's electricity. About 50 power reactors are currently being constructed in 13 countries, notably China, India, UAE and Russia.[4]

It is noteworthy that in the 1980s, 218 power reactors started up, an average of one every 17 days. These included 47 in the USA, 42 in France and 18 in Japan. These were fairly large—the average rated power was 923.5 MWe. With China and India's nuclear sectors growing, it is not hard to imagine a similar rate of reactor construction in the years ahead. Energy security concerns and greenhouse constraints on fossil fuel burning have combined with basic economics to put nuclear power back on the agenda for projected new capacity in many countries. At the end of 2017, 57 power reactors were under construction around the world. Many countries with existing nuclear power programmes either have plans to or are building new power reactors.[5]

Coming to India, the concerns about the safety and accidents at nuclear power plants sites figured in the debate in recent years, especially after the Fukushima accident in 2011. After that Japanese nuclear disaster, India witnessed mass protests by inhabitants, surrounding the designated nuclear power plant sites. Protesters challenged the argument about nuclear energy being clean and environment-friendly and pointed to the damage that atomic energy can cause in accidents and raise safety concerns. The nuclear power plant projects became the target of mass protests which were witnessed at the French-backed 9,900 MW Jaitapur Nuclear Power Project (JNPP) in Maharashtra (Mishra 2017b) and the Russian-backed 2,000 MW Kudankulam Nuclear Power Plant (KKNPP) in Tamil Nadu which is a showcase step in India's ambitious plan to bring a total of 57 reactors on line to power its economic rise (Levy 2016). The West Bengal government also sided with the protesters and denied approval to a recommended

---

[4] World Nuclear Association: http://www.world-nuclear.org/information-library/current-and-future-generation/plans-for-new-reactors-worldwide.aspx
[5] Ibid.

6,000 MW facility near the town of Haripur that intended to host six Russian reactors (*Hindustan Times* 2017).

But in India, where the demand for energy is so massive, these concerns seem not to have hampered the construction of nuclear power plants. Since the conclusion of the US–India civilian nuclear deal, India's nuclear power plants have seen gradual progress. The nuclear deal opened the international market for India to trade and commerce for nuclear technologies and fuel for civilian proposes, mainly for electricity generation to address its mounting electricity demand. Nuclear energy is also considered to be economically viable for India. Another argument for the use of nuclear energy is that it is less polluting and environment-friendly. However, the critics of nuclear energy in India question nuclear energy because of safety concern related to the nuclear power plants which in case of accidents can cause damage which would be devastating and more massive than any other forms of energy-related accidents.

For India, nuclear energy is an important option. Nuclear power is the fourth largest source of electricity in India after thermal, hydro-electric and renewable sources of electricity. As of 2016, India has 22 nuclear reactors in operation in 8 nuclear power plants, having a total installed capacity of 6,780 MW. Nuclear power produced a total of around 35,000 GW of electricity in 2016 with a percentage share of 3.38 per cent in total electricity production. Six more reactors are under construction with a combined generation capacity of 4,300 MW.[6] At present, India has an installed nuclear capacity of 3,310 MW and within the next three to four years this should double to 6,730 MW. More distant, and more ambitious, targets are 20,000 MW of installed capacity by 2020 and 64,000 by 2032. Increased nuclear power would reduce India's dependency on hydrocarbons and be environment-friendly too. It is potentially an infinite source of energy.

Since 2004, subsequent Indian governments have been committed to achieve this target for nuclear power generation in India's energy mix. In 2004, the target for nuclear power was to provide 20 GW by

---

[6] International Atomic Energy Agency: https://www.iaea.org/PRIS/CountryStatistics/CountryDetails.aspx?current=IN.

2020, but in 2007, PM Manmohan Singh referred to this as 'modest' and capable of being 'doubled with the opening up to international cooperation'. Then in June 2009, Nuclear Power Corporation of India Limited (NPCIL) set the target of 60 GW nuclear by 2032, including 40 GW of PWR capacity and 7 GW of new pressurized heavy-water reactors (PHWR) capacity, all fuelled by imported uranium. In 2009, India aimed to increase the contribution of nuclear power to overall electricity generation capacity from 2.8 per cent to 9 per cent within 25 years. In 2009, India ranked ninth in terms of the number of operational nuclear power reactors. The 2032 target was reiterated late in 2010 and increased to 63 GW in 2011. With the arrival of the Modi government, India's nuclear power goal increased further in July 2014. The Department of Atomic Energy (DAE) was asked to triple the nuclear power capacity to 17 GW by 2024. PM Modi applauded the Indian scientists for developing the indigenous nuclear programme, making India self-reliant in the nuclear fuel cycle. He stressed that it is significant that India's nuclear power programme maintains the commercial viability and affordability compared with other clean energy sources. In March 2017, parliament was told that the 14.6 GW target of nuclear capacity by 2024 was maintained, relative to 6.7 GW (gross) grid-connected (World Nuclear Association 2017). India's nuclear power programme is flourishing and mainly home-grown. India's aim is to have 14.6 GW nuclear capacity on line by 2024 and 63 GW by 2032. The target for nuclear power in India's energy mix is 25 per cent by 2050.

India's non-signatory status of the Nuclear Non-Proliferation Treaty, which India considered biased in the favour of the five permanent members of the United Nations Security Council (UNSC), excluded it from trade in nuclear reactors, technology and enriched uranium. This hampered the progress of India's civil nuclear energy up until 2009. India's nuclear power programme was denied the benefits that it could have had because of international collaboration. Despite more than three decades of nuclear apartheid and isolation from the international nuclear trade because of India's atomic explosion in 1974, which was lifted after the conclusion of the US–India nuclear deal, India's nuclear power programme has developed and advanced within the context of its indigenous capability. The absence of sufficient indigenous uranium reserves and the ban on the import of uranium from the Nuclear Suppliers Group (NSG) nations compelled India to focus on processing its thorium reserves. India has been developing a nuclear fuel cycle to exploit these reserves of thorium, which are abundant and ranked second in the world after Australia. There have been issues surrounding international cooperation, especially because of the basic incongruity between India's civil liability law and international conventions which have slowed and in some bilateral dealings limited the foreign technology provisions.

To achieve the target, India needs international cooperation and especially the enriched uranium. It is evident that even the 20 GW target would require substantial uranium imports and an acceleration of nuclear power plant construction. In this context, the landmark US–India civilian nuclear deal is of considerable significance. Nuclear energy can be developed using uranium, plutonium and thorium, which is abundant in India. But until it develops the thorium route, India has to import uranium. To meet its uranium needs in the long run, India needs uranium from the NSG of countries. However, NSG nations sell uranium only to those countries whose nuclear facilities are under IAEA safeguards. According to the agreement done under the US–India nuclear deal, India placed 14 of its civilian nuclear reactors under IAEA safeguards, and in return, the US agreed to sell nuclear technology and provide light-water reactors to meet the energy demands of India. The US is also committed to urging other NSG nations to supply uranium to India's civil nuclear reactors under IAEA safeguards. India's membership in the NSG group was blocked by China which used its UNSC veto power on the ground of India's non-signatory status to the Nuclear Non-Proliferation Treaty. But this Chinese veto on India's bid for NSG membership is seen in the context of geopolitics which this book deals with in the last two chapters.

As far as the economic viability of nuclear power is concerned, it is reliably estimated that for power supply at locations far away from coal reserves, nuclear power would be cheaper in terms of long-run marginal cost, particularly if hydel sources are also not available in these areas. According to a French government report comparing the cost of electricity from gas, coal and nuclear power in the period 2007–2015 and factoring in the cost of limiting carbon emissions, 'the cost of nuclear production is more stable than that of coal and much more than that of gas' (A. Sharma 2007).[7] In spite of nuclear power plants being capital-intensive with long gestation periods, their operational costs are relatively much less. Their cost-effectiveness lies in the fact that India has now overcome the difficulties faced during the indigenization of the technology up to the mid-1980s and has excellent R&D centres, a nuclear design setup, engineering test tube labs and a large construction and operating work force. It has also mastered the complete fuel cycle from uranium mining and fuel fabrication to spent fuel reprocessing and radioactive waste disposal. Nuclear energy's viability is further enhanced due to its being environment-friendly and a carbon-free electricity supply (Sethi 2004). India, unlike other big energy-consuming nations such as the United States and Russia, does not have ample natural gas reserves or any other low carbon emitting sources of energy. Furthermore, India's increasing and impressive record of renewable energy is not sufficient to fulfil its mounting energy demands and hence it makes sense for the Indian government to be ambitious to increase the nuclear energy share in India's energy mix from around 3 per cent to 25 per cent by 2050.

Earlier constraints on India's nuclear power expansion were lack of adequate financing, technological denial regimes, continued non-availability of uranium at low cost, the limitations of the Indian manufacturing industry and negative public perceptions about nuclear energy. But now, after the signing of the Indo-US nuclear deal and further debate within the country, nuclear energy is emerging as one of the viable options for dealing with India's need for more energy to sustain economic development, though safety argument remains alive.

After the US–India nuclear deal, India began using imported enriched uranium for light-water reactors that are currently under IAEA safeguards, but it has also developed other components of the nuclear fuel cycle to support its reactors. Development of select technologies has been strongly affected by limited imports. India has

---

[7] NucNet News No. 333/03, http://www.industrie.Gouv.fr as quoted in WNA Newsletter (January/February 2004).

made use of heavy-water reactors as it allows uranium to be burnt with little to no enrichment capabilities. India has also done a great amount of work in the development of a thorium-centred fuel cycle. While uranium deposits in the nation are limited, there are much greater reserves of thorium and it could provide hundreds of times the energy with the same mass of fuel.

Despite opposition, the capacity factor of Indian reactors was at 79 per cent in the year 2011–2012 compared to 71 per cent in 2010–2011. A total of 9 out of 20 Indian reactors recorded an unprecedented 97 per cent capacity factor during 2011–2012. With the imported uranium from France, the 220 MW Kakrapar 2 PHWR reactors recorded 99 per cent capacity factor during 2011–2012. The availability factor for the year 2011–2012 was at 89 per cent.

India has been making advances in the field of 'thorium'-based fuels, working to design and develop a prototype for an atomic reactor using thorium and low-'enriched uranium', a key part of 'India's three-stage nuclear power programme'. A prototype reactor that would burn uranium–plutonium fuel while irradiating a thorium blanket is under construction at Kalpakkam by Bharatiya Nabhikiya Vidyut Nigam (BHAVINI).

India's Renewable Energy Sources

Renewable energy is energy obtained from sources that are essentially inexhaustible. Examples of renewable resources include wind power, solar power, geothermal energy, tidal power and hydroelectric power. The most important feature of renewable energy is that it can be harnessed without the release of harmful pollutants and it is available in abundance. The world has been shifting towards renewable energy. Renewable energy is acceptable in all the forms of 'A's-accessibility, availability, affordability and acceptability.

According to BP statistical Survey 2016, renewable energy sources in power generation continued to increase in 2015, reaching 2.8 per cent of global energy consumption, up from 0.8 per cent a decade ago. Renewable energy used in power generation grew by 15.2 per cent, slightly below the 10-year average growth of 15.9 per cent but a record increment (+213 terawatt-hours), which was roughly equal to all of the increase in global power generation. Renewables accounted for 6.7 per cent of global power generation. China (+20.9%) and Germany (+23.5%) recorded the largest increments in renewables in power generation. Globally, wind energy (+17.4%) remains the largest source of renewable electricity (52.2% of renewable generation), with Germany (+53.4%) recording the largest growth increment. Solar power generation grew by 32.6 per cent, with China (+69.7%), the US (+41.8%) and Japan (+58.6%) accounting for the largest increases. China overtook Germany and the US to become the world's top generator of solar energy. Global biofuels production grew by just 0.9 per cent, well below the 10-year average of 14.3 per cent: Brazil (+6.8%) and the US (+2.9%) accounted for essentially all of the net increase, partly offset by large declines in Indonesia (–46.9%) and Argentina (–23.9%) (*BP* 2016). According to the 2017 *BP Statistical Review of World Energy*, China continued to dominate renewables growth, contributing over 40 per cent of global growth—more than the entire OECD—and surpassing the US to become the largest producer of renewable power (*BP* 2017). India has yet to tap the full potential of its renewable energy sources, though a start has been made. The figure of India's energy mix in 2014 (Figure 2.2) clearly reflects the renewable sources of energy in India's energy basket.

Over the past decade and a half, India has focused on renewable energy. In recent years since the arrival of the Modi government, India has made excellent progress on the renewable energy front. The Ministry of New and Renewable Energy (MNRE) has taken several steps to achieve PM Narendra Modi's goal of a clean energy future for the 'New India'. India is pursuing the largest renewable energy capacity expansion programme in the world. The government is aiming to increase the share of clean energy through a massive thrust in renewables. A capacity addition of 27.07 GW of renewable energy has been reported during the last three and half years under grid-connected renewable power, which includes 12.87 GW from solar power, 11.70 GW from wind power, 0.59 from small hydropower and 0.79 from bio-power. This is a leap forward. Not surprisingly, the progress in the

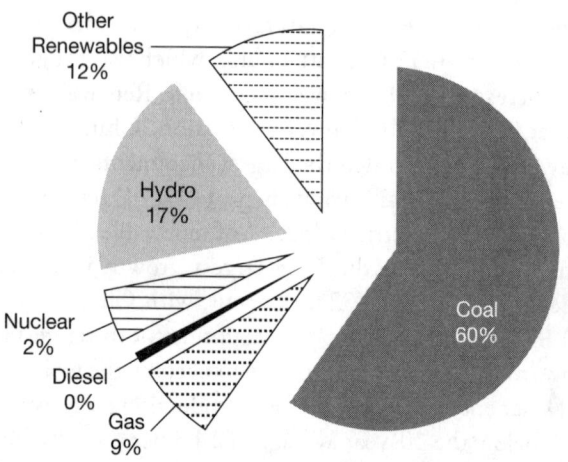

Figure 2.2 *India's Generation Mix as on 31 March 2014*

Source: Central Electricity Authority (2014), http://cea.nic.in/; Craig Froome, "India's energy future: Australian coal or renewable revolution?" *The Conversation*, 6 June 2014, available at https://theconversation.com/indias-energy-future-australian-coal-or-renewable-revolution-26569

renewable energy sector has encouraged the Indian government, leading to its commitment to the United Nations Framework Convention on Climate Change on INDC that India will achieve 40 per cent cumulative electric power capacity from non-fossil fuel-based energy resources by 2030 with help from the transfer of technology and low-cost international finance, including from the Green Climate Fund (Ministry of New and Renewable Energy, Government of India 2017).

The leading renewable sectors are as follows.

Wind Power Energy: Advancing Towards Tapping Full Potential

Worldwide, wind power has now been established as the most promising renewable energy technology for generating electricity. Wind power, which produces electricity from the naturally occurring power of the wind, is considered as the most economical source of large-scale renewable energy. In 2004, wind power constituted 0.6 per cent of global electricity. Total worldwide wind power capacity is some 48,000 MW from about 75,000 wind turbines installed in over 60 countries. China ranks number one in the world in terms of wind energy. By the end of 2004, China had produced 200,000 off-grid wind turbine generators (Feller n.d.), and is expected to produce 15 percent of all electricity from renewable resources by 2020.

According to the 'world wind assessment report', by the end of 2013, the US had more than 61.1 GW of installed wind power capacity. But China, the largest global energy consumer, overtook the US as the largest net importer of petroleum and other liquid fuels in 2014 and emerged as the leading wind farmed country with more than 91 GW capacity. Europe, India, China and US together account for 93 per cent of the total wind power installed capacity. India accounted for about 21 GW of cumulative installed capacity of wind power by the end of March 2014 (WWEA 2014).

India has made significant progress in harnessing wind energy resources for grid power generation. Installed capacity is 4435 MW, which is about 65 per cent of the total installed renewable grid-connected electricity generation capacity. This was because of the attention given to wind energy in the early 1980s. Initiatives were taken under a comprehensive Wind Resource Assessment Programme which focused on wind monitoring, wind mapping and complex terrain projects. Under the initiatives, a thousand stations were installed, which at that time was touted as one of the world's largest wind resource assessment efforts. Private investment has contributed significantly to the wind programme in India. This has not only helped them to reduce the dependence on conventional hydrocarbon sources of power but also given them a business opportunity to exploit India's lucrative energy market.

At present, all major financial and commercial institutions, including the Indian Renewable Energy Development Agency (IREDA), are financing wind power projects and 16 states have announced policies to encourage commercial development of the renewable energy sector, including wind power generation. Hindustan Petroleum Corporation

Ltd has identified around seven sites in Maharashtra and Karnataka for its own wind projects and aims to generate around 500 MW of energy within the next three or four years. Other energy conglomerates have already announced plans to invest in wind energy. Oil and Natural Gas Corporation (ONGC) is developing projects across six states in the country, while Reliance Energy (REL) is planning projects in coastal states including Karnataka (ORF Energy News Monitor 2006, 23).

Wind energy is a renewable, available in abundance and is accessible. It is a non-polluting and environment-friendly source of energy which enhances its acceptability. The Intergovernmental Panel on Climate Change (IPCC) assessment shows that wind energy offers significant potential for near-term (2020) and long-term (2050) GHGs emission reductions. The need to reduce GHG emissions is yet another major driver and a compelling reason for countries to set up wind power plants. The countries and regions that face energy security issues also face the challenge of environmental sustainability in planning their future energy mix (WWEA 2014). Not surprisingly, wind energy is fast becoming an important component of the energy mix of China and India which are hard-pressed to pursue their energy policy in a carbon-constrained environment. On the cost-effectiveness front too, the focus on wind power is a viable option for a country such as India which is dependent on imports to fulfil its energy needs. The initial investment in producing wind energy is costlier than fossil-fuelled generators but it runs at the lowest operating costs making wind energy affordable.

In the past, India's wind power energy programme development has been mainly privately driven. As per the data in the year 2016, India made impressive progress under the Modi government reaching an all-time high in wind power capacity, as the government focused on exploring all sources of energy. The capacity addition of Indian wind energy producing capacity reached a record high of 5,400 (MW). In the year 2018, the expectation was to hit 60 GW, though in the first quarter of the year 2017 wind power production could not keep pace with the year 2016 trend. This was due to the uncertainty surrounding the low tariff, the excessive supply of thermal power, prices going down at power exchanges such as Indian Energy Exchange (IEX) or

Power Exchange India (PXI)—similar to the BSE or NSE, where companies can buy or sell electricity, and the state government utilities' preference to buy power only if they need it rather than get into long-term agreements. The Indian wind power energy sector has been able to attract global giants such as Enercon and Vestas. Also there has been a change in the previous way of establishing the tariff in the form of power purchase agreements (PPAs) which ensured the long-term payments at a predetermined price signed between the various state governments' electricity regulatory authorities and private wind power producing company. According to the new provision, auctions would determine tariffs, which would typically be lower than those set earlier (Sushma 2017). But contrary to the expectation, during the second auction held in October 2017, the wind power tariff fell to an all-time low of ₹2.64 per kWh in the second wind auction of 1,000 Mw conducted by the Solar Energy Corporation of India (SECI). The tariff was much lower than the first wind auction which concluded at ₹3.46 per kWh in February 2017 (*Indian Express* 6 October 2017). In a move to bring transparency and lessen the risk factor to the wind power companies, the Indian government came out with the applicable guidelines for tariff-based wind power auctions in the December 2017.

India's energy security measures have focused on harnessing the abundant source of wind. India's focus on generating electricity through wind power over a decade and a half has begun to show results. In recent years, India's wind power producing capability has increased conspicuously. India's total installed wind power capacity was 32.72 GW. By the end of 2015, India stood fourth largest in installed wind power capacity globally. India produced 11.70 GW wind power capacity in three and half years in the period May 2014–November 2017. India's ambitious bidding trajectory for wind power capacity is 60 GW over the next three years starting from 2017 (Ministry of New and Renewable Energy, Government of India 2017).

India has huge potential in wind power both onshore and offshore. Data showed in an evaluation of India's wind power potential conducted in 2011 that the potential was considerably higher than what was assumed (Hossain, Sinha, and Kishore 2011). This assessment has subsequently been confirmed by a number of independent studies undertaken by different institutions including Lawrence Berkeley National Laboratory (LBNL)—44, World Institute of Sustainable Energy (WISE)—45, TERI—46, and Center for Study of Science, Technology and Policy (CSTEP). In another assessment, conducted on offshore wind power potential, a figure of 966 GW was realized. A committee was constituted by the MNRE, Government of India, to come up with a national assessment figure of wind power potential to have a comprehensive, accurate and a better understanding of wind power from the energy security policy perspective (WWEA 2014).

Solar Energy: India's Move Towards a Rightful Place Under the Sun

In the 21st century, amidst the new goal for sustainable development and the tackling of global warming, fossil fuels, which sustained the modern industrial revolution in Europe, are no longer acceptable in the long term. Today renewable energy is taking the lead in which solar energy is all set to be a leading energy source. Solar energy is the energy which is abundant and fundamentally infinite. The nations that are between the Tropics of Cancer and Capricorn mainly from Asia, Africa, South America and the Pacific get abundant sunlight. These countries are also the fast-growing economies, essentially developing nations with large populations, but with inadequate reserves of energy sources. In the long run, solar energy will become the main viable, ecological, affordable and easily accessible energy solution for these countries.

Scientifically also, there is no substitute for solar energy potential. The massive amount of energy that the sun can deliver in competition with any other renewable resource is simply unmatched. Sunlight has by far the highest theoretical potential of the earth's renewable energy sources. The solar constant (the solar flux intercepted by the earth) is 1.37 kW/m. This theoretical potential is 89 terawatts (TW), more energy striking the earth's surface in one and a half hours (480 EJ) more than worldwide energy consumption in the year 2001 from all sources combined (430 EJ) (Tsao, Lewis, and Crabtree 2016; van Zyl 2017). The consumption of energy by the earth is at a rate of about 17.7 TW that would reach 30 TW by the second half of this century supposing the world population grows at a similar pace. The solar energy irradiating the surface of the Earth is almost four orders of magnitude larger than the rate at which the world can consume it. This is obviously more than sufficient if harnessed properly (van Zyl 2017).

This advancement in new technology can further revolutionize the use of solar energy. Black solar photovoltaic (PV) panels are the most familiar to generate electricity. The ground-breaking leap in the solar panels advancement in the future is likely to be an innovative technology where black solar PV panels are translucent. In this case, the new transparent solar panels will replace regular glass wherever suitable. For example, the vertical glass panels on buildings will become the source of energy that will power the building. Likewise, the glass screen on the mobile phone will charge the phone and eventually make batteries redundant (van Zyl 2017). Solar energy is the world's most plentiful and ubiquitous energy source, and researchers around the world are developing ways to convert sunlight into a useful form of energy. One of the main goals is to elevate the efficiency of solar energy to a higher level by turning sunlight into liquid fuels. Research in solar-derived liquid fuels, or solar fuels, aims to make a range of products that are compatible with our energy infrastructure today, such as gasoline, jet fuel and hydrogen. The goal is to store sunlight in liquid form, conveniently overcoming the transient nature of sunlight. Once it becomes a reality, solar energy can fuel cars and planes in the future (Scheffe 2015).

Solar energy PV technology converts sunlight directly into electricity. PV technology makes use of the abundant energy in the sun and has little impact on the environment. It can be used in a wide range of products, from small consumer items to large commercial solar electric systems. Its energy source—sunlight—is free and abundant; PV systems can guarantee access to electric power as well.[8]

Many countries including Germany, Japan, the US, China and Korea have announced policies and funding programmes to support

[8] Energy Information Administration, Official Energy Statistics from the US Government: http://www.eia.doe.gov/.

the deployment of solar PV systems. There are important emerging PV programmes in China and India. China is pushing the use of renewable energy sources such as wind and solar. On 1 January 2006, the Chinese government passed its first law on renewable energies. China aims to increase renewable consumption in the energy mix from the current 7 per cent to 15 per cent by 2020 and to achieve 180 MW of PV deployments per year by 2010 (*China Daily* 2006).

India has been developing solar PV energy programmes for the last three decades. In India, there are about 300 clear sunny days in a year and solar energy is widely available in most parts of the country. As a domestic source of electricity, it has the potential to contribute to India's energy security. It is especially useful in rural areas where PV technology offers a unique decentralized option for providing electricity locally at the point of use. It can also provide viable power sources for applications such as lighting, water pumping and telecommunications, and as a relatively new, high-tech industry, it can create job opportunities and strengthen the economy.

The Ministry of New and Renewable Energy is focusing on R&D in PV. In 2005–2006, financial support was provided for PV development programmes: R&D, industrial development and quality assurance including tests centres, capacity building through training, organization of seminars, workshops and symposia, and initial demonstration of solar PV devices systems in this country.[9]

By December 2005, PV systems of about 245 MWp aggregate capacity (about 1,300,000 systems) were installed for various applications. Also under the PV programme, about 1 million systems have been installed. This includes 560,000 solar lanterns, 342,000 solar home lighting systems, 54,000 street lighting systems, 7002 water-pumping systems and about 4.75 MWp aggregate capacity

---

[9] 'Administrative approval of Renewable Energy Programmes', Ministry of Non-conventional Energy Sources, Government of India: http://mnes.nic.in/frame.htm?majorprog.htm.

of standalone and grid interactive power plants/packs (A. Sharma 2007).[10] In November 2009, the Union Government of India aimed to achieve grid parity (electricity delivered at the same cost and quality as that delivered on the grid) by 2020. Achieving this target would establish India as a global leader in solar power generation. It also proposed to launch the Jawaharlal Nehru National Solar Mission (JNNSM) also known as the National Solar Mission under the National Action Plan on Climate Change. The scheme inaugurated by former PM Manmohan Singh on 11 January 2010 aimed to generate 1,000 MW of power by 2013 and up to 20,000 MW grid-based solar power, 2,000 MW of off-grid solar power by the end of the final phase of the mission in 2020. The JNNSM aimed to achieve grid parity by 2020 (Sethi 2009). The target now stands at 100,000 MW under the PM Narendra Modi government as revised in the 2015 Union Budget of India (Government of India 2015).

In May 2014, India's solar power generation capability grew by almost five times to 12,288.83 MW in 31 March 2017 from 2,650 MW on 26 May 2014. In addition, India's solar power capacity accounted for an addition of 5,525.98 MW in 2016–2017, which was the best so far (*Economic Times* 6 April 2017). India's solar energy capacity has expanded by a record 5,525.98 MW in 2016–2017, according to the latest figures provided by the MNRE. In comparison, India had added 3,010 MW of solar capacity in 2015–2016, which shows that growth nearly doubled in one year. Cumulative solar capacity currently stands at 12,288.83 MW, against 6,762.85 MW at the end of March 2016. India aims to continue this impressive growth by committing to a cumulative target of 20,000 MW by adding another 7,750 MW in 2017–2018 (*Economic Times* 6 April 2017).

India with its dense population and with large amounts of solar radiation reaching it makes for a perfect blend of consuming solar energy. Many parts of India do not have an electrical grid, so one of the first applications of solar power has been for water pumping, to begin

[10] 'Solar Energy Programmes at a Glance', Ministry of Non-conventional Energy Sources, Government of India: http://mnes.nic.in/frame.htm?majorprog. htm.

replacing India's four to five million diesel-powered water pumps, each consuming about 3.5 kilowatts and off-grid lighting. Some large projects have been proposed, and a 35,000 km² (14,000 sq mi) area of the Thar Desert has been set aside for solar power projects. This will be adequate to produce 700–2,100 GW. India's consistent focus on increasing the solar power in its energy mix has resulted in a growth of 113 per cent (Kenning 2016).

India's solar power initiatives have received international accolades as well. One of the successful solar programme initiatives is the United Nations Environment Programme backed Indian Solar Loan Programme which was awarded the esteemed Energy Globe World Award for Sustainability for forming a consumer financing programme for solar home systems (SHS). First introduced in 2003 in India's southern state of Karnataka, the Indian Solar Loan Programme in order to bring SHS to rural households was supported by two of India's largest banks, Canara Bank and Syndicate Bank, along with their eight associate regional rural banks (Grameen Banks), partnered with UNEP to establish and run a loan programme through their branch offices across Karnataka and part of the neighbouring state of Kerala. Within three years, the solar power projects programme initiatives resulted in 2,000 bank branches financing more than 16,000 SHS, especially in the southern Indian villages where the electricity grid did not exist. The programme still exists in Karnataka, and is being replicated in other adjacent states such as Kerala and Andhra Pradesh (United Nations Environment Programme 2005). Another example in India's progress on solar energy progress in recent years has been that India's Cochin International Airport made India the first country to establish an airport fully operational by solar energy. This airport on the southwest coast of India is a sustainable project and this extraordinary feat is expected to save 300,000 tonnes of carbon emissions over the next 25 years (*Science Alert* 2015), which is said to be the equivalent of planting three million trees (Kim 2015).

To achieve the target set at the Paris Climate Summit, the Modi government has given a renewed focus on the expansion of solar and wind power which will help exceed the Paris Climate Summit targets by almost half and two–three years ahead of what was planned.

According to the Modi government's draft of the 10-year energy blueprint projection, Indian will exceed the Paris Climate Accord target of electricity from non-fossil fuels by 2027. The Paris Climate Accord set a target of 40 per cent of electricity generation through renewable energy by 2030. But the Modi government's Draft on National Electricity Plan envisions that instead of the target set at the Paris Climate Summit, 57 per cent of India's electricity will be generated through renewable energy by the year 2027, implying that India will not create any new coal-fired power stations by 2030 (Central Electricity Authority, Ministry of Power, Government of India 2016).

The Modi government's renewed focus on renewable energy and streamlining of energy policy have attracted private sector investors, both domestic and overseas, in the renewable energy sector. The Indian government has also appealed to the rich nations to help India exceed the Paris Climate Summit target. Despite the government's focus on the renewable energy sector, there has been no major state investment. In fact, the year 2015 trend shows that it is the inflow of investments from the domestic and overseas private sectors that have helped the solar energy sector to achieve its impressive growth. One of the major private sector investment commitments in the Indian solar energy sector is $20 billion from Japan's Softbank in conjunction with Taiwan's Foxconn and India's Bharti Enterprises: the domestic and overseas collaboration. The French state-owned energy company also joined India's solar energy market to tap its gigantic potential in wind and solar ray availability by committing to invest $2 billion (Shafi 2016).

The Indian energy giants have also shown serious interests in helping India to achieve its Paris Climate Accord CO_2 emission, committing to invest in the solar energy market. While Adani opened a solar plant in Tamil Nadu, the world's largest solar plant in Tamil Nadu, Tata proclaimed its goal of producing around 40 per cent of its energy from renewable sources by 2025. India's ambitious target in solar energy is also derived from its excellent technological progress which resulted in the falling of solar energy price by 80 per cent between the years 2011 and 2016. The big energy conglomerates committing to India's shift to non-fossil fuel is because it is economically viable and commercially justifiable without subsidies (Shafi 10 May 2017).

In the 2027 forecasts, India's goal is to generate 275 GW of total renewable energy, in addition to 72 GW of hydro energy and 15 GW of nuclear energy. Nearly 100 GW would come from 'other zero emission' sources, with advancements in energy efficiency expected to reduce the need for capacity increases by 40 GW over 10 years. Acknowledging the Modi government's efforts to increase the renewable energy share in India's energy mix, Buckley is of the view that the fast development in the solar energy sector could not have been imagined before 2014 (Shafi 10 May 2017).

India's unrelenting efforts towards embracing solar energy have culminated in India leading the global efforts towards solar energy revolution. On 11 March 2018, India found its rightful place under the sun as PM Narendra Modi's brainchild, the International Solar Alliance (ISA), formally kicked off with 62 member-countries adopting the 'Delhi Solar Agenda', seeking to increase the solar power in their energy mix with a goal to address the global warming concern and make clean and affordable electricity available to the underprivileged. Modi and French President Emmanuel Macron co-chaired the ISA founding conference in the presence of heads of state of 23 countries and ministerial representatives from 10 other nations, indicating an acceptance of India's leadership role in the global fight against climate change. This is important as ISA, the world's single largest renewable energy programme, will be headquartered in Gurgaon in Delhi's suburbs, the first time that an international treaty body will have its secretariat in India (*Times of India* 12 March 2018). The ISA, an alliance of more than 121 countries, aimed at energy security, envisions to mobilize more than $1 trillion of funding by 2030 for achieving the target of over 1 TW of solar generation capacity; making solar energy available at affordable rates, creating solar grids and establishing a solar credit mechanism; reducing the cost of finance and technology; and addressing the hurdles that may reduce the pace of progress of solar energy generation in the ISA member-nations.

The ISA not only puts India on the driving seat for solar energy but also shows India's serious commitment to a developmental path through green and climate-friendly energy. India's energy security is being pursued in a more systematic manner with a clear vision. India's solar energy policy has all the tenets of the four 'A's: availability, accessibility, affordability and acceptability. India's energy security policy will continue to do so in the near future.

Waste-to-Energy: Future Potential

The high volatility in fuel prices in the recent past and the resulting turbulence in energy markets has compelled many countries to look for alternative sources of energy, for both economic and environmental reasons. With growing public awareness about sanitation and with increasing pressure on the government and urban local bodies to manage waste more efficiently, the Indian waste to the energy sector is poised to grow at a rapid pace in the years to come. The dual pressing needs of waste management and reliable renewable energy source are creating attractive opportunities for investors and project developers in this waste-to-energy sector.[11]

The logic behind the waste-to-energy could be seen in the fact that most wastes are eventually released into land and water without appropriate treatment, finally polluting water and air. But the waste-to-energy concept can alleviate the harm that the solid and liquid waste does to the environment by taking on climate-friendly waste treatment methods and know-how before disposal. The advantage that waste-to-energy brings to address the energy crisis and environment is gripping. It is also a better option than just releasing the waste into the water and land or even for the landfill purpose. Waste-to-energy has immense potential in a country like India which is fast urbanizing and industrializing with a population of 1.2 billion people. The waste-to-energy is renewable energy which is abundant and easily accessible, affordable in the long run after high cost at the initial stage of establishment of waste disposal treatment plants; and above all the acceptability of waste-to-energy lies in its attributes of climate-friendly and clean energy which meet the GHG emissions concern.

---

[11] EAI, *India Waste to Energy–Concepts*: http://www.eai.in/ref/ae/wte/wte.html.

The Government of India is aware of these attributes of the waste-to-energy and has been constantly working in this direction, though the country's potential is hardly tapped yet. The focus would lessen the quantity of wastes, which would help PM Modi's Swachh Bharat Abhiyan (Clean India Mission), produce a sizeable amount of energy and help India achieve its Paris Climate Agreement goal of reducing the GHG emission.

Every year, about 55 mt of municipal solid waste (MSW) and 38 billion litres of sewage are generated in the urban India. In addition, large quantities of solid and liquid wastes are generated by industry. Waste generation in India is expected to increase rapidly in the future. As more people migrate to urban areas and as incomes increase, consumption levels are likely to rise, as are rates of waste generation. It is estimated that the amount of waste generated in India will increase at a per capita rate of approximately 1–1.33 per cent annually. This has significant impacts on the amount of land that is and will be needed for disposal, economic costs of collecting and transporting waste, and the environmental consequences of increased MSW generation levels (Emmanual 2012; Kumar, Subbaiah and Rao 2014).

India has had a long involvement with anaerobic digestion and biogas technologies. Wastewater treatment plants (WWTP) in the country have been established which produce renewable energy from sewage gas. However, there is still significant untapped potential (Clarke Energy 2014). Also, wastes from the distillery sector are on some sites converted into biogas to run a gas engine to generate onsite power. Companies such as A2Z Group of companies, Hanjer Biotech Energies, Ramky Enviro Engineers Ltd, Hitachi Zosen India Pvt Limited and Clarke Energy are the leading players in this sector of energy (Emmanual 2012).

There are various type of waste in India which can be converted for energy purposes—Urban waste such as MSW, sewage and faecal sludge, industrial waste classified as hazardous industrial waste and non-hazardous industrial waste, biomass waste, and biomedical waste.[12] The Ministry of Environment and Forests has revised Solid Waste Management Rules by introducing responsibility on generators to segregate waste into three categories: wet, dry and hazardous waste. India, with its ambitious target of adding 175 GW of renewable energy to the grid by 2022, plans to include 10 GW of electricity generation capacity from bio-power. To achieve the target, the Modi government has taken certain steps to boost the waste-to-energy sector. The new policy has assigned responsibility on local bodies having 1 million or more people to set up waste processing facilities by 2019 (*Business Wire* 2017). Further, the Tariff Policy 2016 mandates power distributors to buy all the electricity generated from waste-to-energy plants in a state. The remunerative tariff set for it by the Central Electricity Regulatory Commission (CERC) has helped raise investors' interest in this sector. Appropriate funding to the municipal corporations, the setting up of a $1.25 billion fund, backed by the state-owned Power Finance Corp. Ltd and Rural Electrification Corp. Ltd along with some private investment over the next three years are some of the steps taken by the Modi government towards the Swachh Bharat Abhiyan for cleanliness, waste management and waste-to-energy projects.

Waste-to-Energy potential in India lies in the technologies used for the electricity generation from waste. Energy can be recovered from the organic fraction of waste (biodegradable as well as non-biodegradables) through thermal, thermochemical, biochemical and electrochemical methods. However, out of these four technologies for the generation of energy from waste, thermal conversion is being questioned because of the use of incineration that has high emission characteristics. Electrochemical conversion requires broad assessment studies on bulk scale liquid waste treatments and stands at a budding level in India as well as worldwide.

In India, multiple biomass materials are also being used for generating energy. Rice husk, wood and other agricultural residues already play a significant role. Indeed, more than 70 per cent of the population still depends on biomass for its energy needs. If biomass

---

[12] EAI, India Waste to Energy: http://www.eai.in/ref/ae/wte/wte.html

used in cooking were to be substituted by liquefied petroleum gas (LPG) or other petroleum products, India would need to import an additional 30 mt of such products, at a cost of $8 billion per year. Biomass could potentially also play a role in petroleum substitution. Bio-fuels such as sugarcane-based gasoline substitutes may help from a security of supply perspective. Some say that the high water use implicit in sugarcane production as well as its heavy dependence on fertilizers makes the economics of such substitution debatable. However, India can look to Brazil as an example. In Brazil 70 per cent of fuel consumed comes from non-petrol sources, three-quarters of vehicles run on ethanol, produced from sugarcane, molasses or corn. Ethanol can also be made from grass, wood chips, any husk or agricultural waste (ORF Energy News Monitor 2006, 24). In 2006, although it was estimated that around 11 per cent of world primary energy need was met by biomass, only 1 per cent of electricity generation was based on biomass, which accounted for about 10,000 MW of electricity generating capacity worldwide. The US alone had 7,500 MW. In India, bagasse, the fibrous residue of sugarcane that is left after juice extraction, is generally the sole energy source for all sugar mills. The Indian government has been keen to develop its use, and as a result, a full-fledged programme for the promotion of optimum bagasse cogeneration in Indian sugar mills was initiated in 1993–1994. In 2006, installed captive cogeneration capacity was 2,000 MW. Interestingly, although Brazil is the largest sugarcane producer in the world, it does not use sugarcane residuals for power generation due to the cheap electricity available from its large hydropower station (A. Sharma 2007).

The Modi government's agenda on sanitation and sewage treatment is focused on improving the socio-economic aspects of India. Financial schemes from the central government to build sewage treatment plants (STPs), river cleaning programmes and providing clean sanitation facility for all are indicators of enterprising solutions from the Modi government (Clarke Energy 2014).

An all-inclusive approach to sanitation and renewable energy generation at the micro level can contribute towards energy security.

At community levels, a centralized collection and waste treatment from toilets cannot lead to cleanliness but its scientific treatment can create renewable energy. Biogas generated from the waste treatment can generate electricity. At many public toilets across India, the biogas-to-power concept is being implemented. There are success stories of safe and hygienic on-site human waste disposal technology; a new concept of maintenance and construction of pay-and-use public toilets, popularly known as Sulabh Complexes with bath, laundry and urinal facilities being used by about 10 million people every day generates biogas used for generating electricity. Many countries are adopting the concept of waste treatment and renewable energy production (*Hindu* 2008). With the rapid urbanization and increasing population of Indian cities, it becomes paramount to treat the waste generated. The Indian government's announcement to develop STP through project funding is aimed at removing financial hurdles. The government's fast-track implementation of developing sewer networks and connecting to a STP (or WWTP) at community levels or at a central location is targeted at the rapid implementation of proven ways of sewage collection and disposal (Clarke Energy 2014).

According to the 2011 estimate by the MNRE, there exists a potential of about 1,460 MW from MSW and 226 MW from sewage a total of close to 1,700 MW of power. However, out of that only about 24 MW was exploited which was less than 1.5 per cent of the total potential. The IREDA has so far realized only about 2 per cent of its waste-to-energy potential. This lack of growth is attributed to multiple factors, waste-to-energy is still a new concept in the country; most of the proven and commercial technologies in respect of urban wastes have to be imported; the costs of the projects especially those based on bio-methanation technology are high as critical equipment for a project have to be imported. In view of low level of compliance of MSW Rules 2000 by the municipal corporations/urban local bodies, segregated MSW is generally not available at the plant site, which may lead to non-availability of waste-to-energy plants and lack of financial resources with municipal corporations/urban local bodies. Again the lack of conducive policy guidelines from state governments in respect of land allotments, supply of garbage and power purchase and the evacuation facilities contribute to this lack of growth.[13]

The waste-to-energy concept has multifaceted advantages and is not confined only to energy production. In fact, waste-to-energy is economically viable and can bring considerable financial benefits. By using the proper technology with the best possible processes, entire waste can fetch good value and be a potentially profitable business venture. With the support of government incentives, business attractiveness can get enhanced further. To support waste-to-energy projects, the Indian government grants capital subsidies and favourable tariffs. In addition, related opportunities can be created by replicating the successful MSW management to other waste such as sewage waste, industrial waste and hazardous waste. The success of the Indian waste-to-energy model could be exceeded internationally, especially to African and other Asian countries.[14] This could further fetch profits for the waste-to-energy business enterprises. Above all, in the context of India's commitment to reducing the CO_2 emissions and PM Modi's much vaunted Clean India Mission, the waste-to-energy concept is both affordable and acceptable, and India's newly formed ministry exclusively for drinking water and sanitation is another reinforcing factor for the argument that India's quest for energy security is in line with the concept of four 'A's—affordability, accessibility, availability and acceptability.

Hydro-electricity Power

China has done extremely well on the hydroelectricity front. According to the International Hydropower Association (IHA), China ranked number 1 in hydroelectric power and dominated the market for new development and total installed capacity in hydropower, adding 19.4 GW of new capacity within its borders, including 1.2 GW of pumped storage during the year 2015.

[13] EAI, *India Waste to Energy—Concepts*: http://www.eai.in/ref/ae/wte/wte.html.

[14] Ibid.

At the end of 2015, the world's total installed hydropower capacity reached 1,211 GW, including 145 GW of pumped storage. During 2015, 33 GW of new hydropower capacity was commissioned world-wide, with China accounting for close to 60 per cent of this new capacity. By 2015, Chinese total hydropower capacity had reached 320 GW. The United States had 102 GW of installed capacity by 2015, followed by Brazil at 91 GW, Canada at 78 GW and India and Russia with 51 GW each, with Japan just a tad behind at 50 GW. India witnessed one-tenth of Chinese hydropower capacity added during the year 2015. However, India ranked among the top five nations with new capacity addition. Despite severe water shortages in recent years mainly because of drought and inadequate monsoon, India was nevertheless at number 5 in terms of total installed hydropower capacity globally (India Water Review 2016).

In the year 2016, as per the *BP Statistical Review of World Energy*, global hydroelectric output grew by a below average 1 per cent, compared with a 10-year average of 3 per cent. China (+5%) remains by far the world's largest producer of hydroelectricity; as with nuclear power, China accounted for all of the net global increase, even though growth in percentage terms was less than half the recent historical average. Elsewhere, growth in Turkey (+64.6%, following a very weak 2014) and Scandinavia was offset by drought conditions in Italy, Spain and Portugal (−28.6% combined) and Brazil (−3.3%). Hydroelectric output accounted for 6.8 per cent of global primary energy consumption (*BP* 2016).

India is ranked third worldwide in the total number of dams. As of 31 March 2016, India's installed utility scale hydroelectric capacity was 42,783 MW or 14.35 per cent of its total utility power generation capacity (Central Electricity Authority, Ministry of Power, Government of India 2017).[15] Additional smaller hydroelectric power units with a cumulative capacity of 4,274 MW have been installed. India's hydroelectric power potential is estimated at 84,000 MW at 60 per cent load factor (Central Electricity Authority, Ministry of

[15] http://www.cea.nic.in/monthlyhpi.html

Power, Government of India 2017). India's hydroelectric power plants established at Darjeeling and Shivanasamudram in 1898 and 1902, respectively, were among the first in Asia and India has been a strong player in global hydroelectric power development.

In a move to boost the hydropower sector, the Modi government adopted a three-pronged strategy in 2016. The approach includes increasing the sphere of small hydro projects from 25 Mw to 100 Mw. This move is to achieve the renewable energy targets of states and also bring a large number of projects under the net of government subsidy and other tax benefits. This is considered as an important step given the stagnation in the installed capacity of hydropower projects which remained 40,000 Mw for the years 2013–2015, even as that of the renewable energy sector saw a significant jump (Jai 2016).

Though India has made good progress in hydropower electricity generation, it has yet to exploit the hydropower potential from the Godavari, Mahanadi, Nagavali, Vamsadhara and Narmada river basins. Various state governments have not been able to address the concerns and opposition from tribal population. The hydropower sector is overwhelmingly dominated by the public sector responsible for 92.5 per cent of India's hydroelectric power generation. Public sector undertaking (PSU) such as the National Hydroelectric Power Corporation (NHPC), North Eastern Electric Power Corporation Limited (NEEPCO), Satluj Jal Vidyut Nigam (SJVNL), THDC and NTPC-Hydro are the leading players in India's hydropower generation. The scope for the private sector lies in the development of hydroelectric energy in the Himalayan mountain ranges and in the northeastern region of India.[16] The private sector growth in this sector, especially in the northeastern region, is also significant strategically. While balancing the electricity grid in this region, the focus is on hydropower projects to advance economic growth in states that border India's neighbouring countries, China and Pakistan. It is strategically important given the Chinese claim to a part of the northeastern state

[16] Hydro Electric Potential in India: http://www.cea.nic.in/monthlyhpi.html

of Arunachal Pradesh, which is among Indian states with the greatest hydroelectricity potential (*Financial Express* 2018). India also imports surplus hydroelectric power from Bhutan, and Indian companies have ventured into hydropower projects with its neighbouring nations such as Nepal, Bhutan and Afghanistan. However, India's hydropower development will be constrained domestically by real and perceived concerns about dislocation and environmental damage. Internationally, water sharing has remained historically a point of contention between India and Nepal, and India and Bangladesh and a heavier reliance on hydropower could become a part of China's India-constraining strategy.

India's hydroelectricity potentials are still not tapped fully. It is estimated that around 100 GW of electricity generation potential in India's rivers is lying untapped. One of the main reasons is the high tariff. The Modi government is likely to take action in this regard by lowering the tariff. As India targets to add 175 GW of renewable energy, it will need to focus on hydropower electricity projects, often located in remote regions, to stabilize the grid (*Financial Express* 2018).

The hydropower energy sector is witnessing an impressive growth around the world. Given the emphasis in the Paris Agreement on climate change and new sustainable development goals, hydropower is significant in meeting these energy, water and climate challenges. As per IHA chief executive Richard Taylor, 'Through its ability to support clean energy systems and provide multiple water services, hydropower can be the key to realizing the ambitious global targets outlined at COP21' (*India Water Review* 2016).

India is not averse to this trend and the Indian government has given substantial focus in recent years to the hydropower sector. Given its huge appetite for electricity, the steps for increasing the share of hydropower for electricity generation in India's renewable energy in particular and overall energy mix, in general, will be the right move as India treads towards the dependable, economically affordable and environmentally acceptable source of energy.

Hydrogen, another alternative fuel for motor vehicles may also be an option for India. This requires a considerable amount of planning and investment. Even if the technical problem in hydrogen-powered vehicles is fully solved, there is bound to be a difficult phase involving the development of production and distribution. Further, the difficulties still experienced in hydrogen-based vehicles need to be tackled at a practical level with a view to emphasizing user-friendliness, as now seen in petrol and diesel automobiles.

Summing up, there is a big gap between energy demand and supply. Nevertheless, India in recent years has made an impressive progress on meeting its energy demand and has explored both fossil and non-fossil sources of energy. One of the striking features of India's energy quest is its push towards renewable energy. Renewables are going to stay in the coming years, and trends indicate that technological drawbacks that hampered the development of renewables for electricity generation are being overcome. Experts are of the view that technological advancement is making it easier than expected to include more and more renewables in the electric grid, and to handle periods when demand is high but the wind is not blowing and/or the sun is not shining. According to Morgan Bazilian, the leading energy specialist at the World Bank, who tracked the issue for 20 years, the very high levels of variable renewable energy can be accommodated both technically and at low cost. The *Technology Review* article on 'Smart Wind and Solar Power' published in 2014 explains that the big data and artificial intelligence are producing ultra-accurate forecasts that will make it feasible to integrate much more renewable energy into the grid (Romm 2016).

There is a wide spectrum of renewable energy technologies. They are currently at widely different degrees of maturity, with wind and biomass energy already commercial, and solar PV well on the way to becoming so. Though India will continue to rely on fossil fuels such as coal and oil and gas for the near future, renewable energy will become a significant source of India's energy mix in the coming decades. India has made excellent progress over the past three years, especially on the renewable energy front. One of the major reasons for the push towards the renewable is global warming and India's challenge to meet the CO_2 emission reduction committed at Paris Climate Summit. However, India's focus on renewables is not only environmental acceptability but because of their abundance and accessibility and financially affordability also.

CHAPTER 3

India's Domestic Measures to Energy Security

Exploration and Production, Privatization and Energy Efficiency

For years, India relied heavily on hydrocarbon sources of energy. Coal along with oil has played a significant role in fuelling India's economy, lighting Indian houses, and running cars and industries. For coal, India has relied on its own indigenous reserves, and it has sought to fulfil the demand by importing coal, especially less polluting and cleaner coal for its industrial needs. For oil and gas, India is heavily dependent on the foreign supply—import from the Gulf countries. The renewable energy has been an insignificant constituent of India's energy mix, and alternatives such as nuclear energy due to international apartheid for a long time have been inconsequential. India's approach to energy security has evolved over time from the regular accessibility and availability framework to an economic affordability and acceptability framework. India's energy security is not only about fulfilling its energy demand. The quality too matters in its pursuit of energy security. India's mounting energy concerns amidst the growing awareness of climate change and global warming have pushed it to adopt a multi-pronged strategy, both in the domestic and foreign milieu.

To promote oil and gas production at the domestic level, the Indian government has taken several steps which range from encouraging the Indian companies to increase their domestic activities to widening its engagement with multinational companies, broadening opportunities for them to participate in oil and gas exploration in India. In addition, to stimulate the investments and development in the exploration of hydrocarbon sources of energy, some of these steps have focused on regulatory changes, a transparent gas pricing policy and redevelopment of perceived uneconomical assets.

Domestic efforts have also seen a concerted focus on exploring various alternative sources of energy which are renewable and environment-friendly. The government has given a massive push in this regard in the energy production through solar power, wind power, hydroelectricity power, biomass and nuclear energy. At the domestic level, the challenges range from increasing energy access to building a smart system for drawing investment in energy infrastructure and pricing of energy to facilitate economic and environmental competence.

The main objective of this chapter is to make an evaluation of India's energy security approach in the domestic context. The chapter examines India's pursuit to fulfil the energy needs from domestic sources and the progress made on the E&P of new reserves of hydrocarbon sources, both onshore and offshore, in addition to its renewable potentials. The chapter examines the policy approach of the Government of India at home including its regulatory measures and steps taken from the opening of the energy sector to private players to modernization and technological evolution in the energy sector.

This chapter is divided into five main sections. The first section examines three major energy policy frameworks which have guided India's energy security efforts at the domestic level—IEP, expert group reports on low-carbon inclusive growth and NEP. The second section assesses India's move towards liberalizing the energy sector by encouraging private participation. The second section dealt under the heading 'Liberalization of the energy sector: Encouraging private participation' focuses on the Electricity Act 2003, the measures taken by the current government to improve the condition of power distribution through the Ujwal DISCOM Assurance Yojana (UDAY) scheme, the opening of coal mines and the renewable energy sector to private companies. The third section evaluates the Indian government's measures to reduce the burden on hydrocarbons through technology, conservation and energy efficiency. A major focus is on energy efficiency in the automotive sector through the policy of National Electric Mobility Mission Plan (NEMMP) 2020 and the Faster Adoption and Manufacturing of Hybrid and Electric Vehicles (FAME). The fourth section examines the measures for oil and gas E&P to enhance the production at the domestic level. This section first delves into the existing policy known as the NELP and then the evaluation of new policy introduced under the Modi government which is undertaken mainly on Hydrocarbon Exploration and Licensing Policy (HELP), marketing and pricing freedom for new gas production, and the policy for granting extension to the production sharing contracts (PSCs). An assessment of overall domestic measures to address energy security is made in the conclusion of this chapter.

India's Energy Security Policy Framework: Towards a Comprehensive Policy Approach

India's energy security-related policy has been mainly the responsibility of various ministries and government bodies: Ministry of Power, Ministry of Coal, Ministry of Petroleum and Natural Gas, and Ministry of New and Renewable Energy are the main ministries responsible for exploration, production, transmission, regulation and management. Governmental bodies such as the Atomic Energy Commission and Planning Commission (till 2014) are the regulating bodies for energy sources. However, despite a number of ministries involved in the overseeing of energy, India's energy policy has lacked coherent direction, and this is often because of the overlapping jurisdiction of various ministries and the lack of an integrated policy approach. The report submitted by Centre for Policy Research (CPR) for NEP also highlighted this shortcoming in its 2017 report.

In fact, India's energy policy exists in a complementary and distinct manner, but there has been no single inclusive and integrated energy approach which addresses energy security in a holistic manner. All the ministries have their independent vision statements. For instance, 'The Hydrocarbon 2025 Report', issued by the Ministry of Petroleum and Natural Gas, focused only on the oil and gas (Madan 2006). This lack of coherency in the energy security approach is due to the absence of an agency or ministry with an all-encompassing vision.

India's failure to tap its energy potential is described by Charles K. Ebinger as a story of lost opportunities. The numerous agencies/ministries dealing with different energy resources use different sets of energy data thus hampering the adoption of a cohesive and coordinated approach to energy security.[1] The policies on pricing, energy production and tax subsidies all differ across the different energy sources. In addition, energy security space lacked a coordinated and cohesive response from a number of non-energy-related ministries such as the Ministry of External Affairs (MEA), the Ministry of Science and Technology for R&D, the Ministry of Road Transport and Highways, the Ministry of Railways, and Ministry of Environment, Forest and Climate Change (Kumar 2012). India's energy security efforts have suffered because of this lack of successful implementation of policies in a cohesive, coordinated and timely manner.

IEP

However, over the past two decades, there has been a move in the direction of an integrated approach to India's energy policy. At the domestic level, the first major integrated policy initiative could be seen in India's Planning Commission's Twelfth Five-Year Plan which set out an all-encompassing and inclusive policy approach in the form of IEP. The IEP which was supposed to be effective from 2012–2017 aimed at a concerted approach to frame India's energy policy. The IEP should be considered as the first guided approach to energy security.

[1] For detailed, deep insight on the subject, see Ebinger (2011) and Planning Commission, Government of India (2011, 18).

It sprang from an expert committee report established in 2004 by the 10th Planning Commission at the behest of the Indian PM and the Deputy Chair to frame a comprehensive energy policy in the wake of looming energy demand in India. As a result, the IEP came into effect in the year 2006 and remained effective till the year 2017.

One of the striking features of IEP was its focus on renewable energy. An insight into the IEP's recommendations reflects that it sought to address the four 'A's. The IEP sought to address India's energy security issues focusing on accessibility, availability, affordability and acceptability amidst environmental concerns. The IEP laid down several policy choices for the government in its quest for energy security. These options consisted mainly of the following strategies.

The focus of IEP was on energy diversification and efficiency. Given the projected increase in energy demand, the IEP envisioned coal a significant and major source of energy provider in India beyond 2030. The IEP also proposed the energy demand be met with other traditional hydrocarbon sources such as petroleum, liquid natural gas and recommended alternative sources of energy including atomic and renewable energy. The IEP advocated R&D in technological innovation by funding educational institutions, increasing investment in R&D and recommended the creation of a national energy fund to finance commercially viable technologies for enhancing efficiency and renewable energy. It proposed that energy research be focused on all possible renewable energy sources and not just nuclear energy research (Badrinarayana 2010).[2]

To achieve its energy diversification and efficiency goals, the IEP recommended reforms in the energy sector with an emphasis on regulatory intervention, increasing market competitiveness, and pricing. While advocating the liberalization of certain energy sectors, the IEP proposed the government facilitate investment in energy diversification by adopting a blend of strategies which included regulatory intervention, market competitiveness, energy pricing changes and financial assistance (Badrinarayana 2010). It recommended opening the coal sector to market-based production contracts, liberalizing international

---

[2] For details, see Planning Commission, Government of India (2006).

coal trade and easing the government's control over coal by disinvesting from the government-owned Coal India Limited. Accordingly, the IEP proposed regulatory measures to enhance spending in energy generation. This was to be realized by making federal laws allowing the states, private stakeholders and people to invest in central government energy projects. A number of steps to give a boost to investment in energy production, supply and access by changing pricing methods were also prescribed by the IEP. Though the IEP recommended the lessening of the regulatory responsibility from the government in the energy sector to encourage independent regulation and facilitate the market competitiveness and fair pricing, it did not propose complete deregulation in preference for market-based pricing.[3]

The IEP acknowledged the supply and availability aspects of energy security in the foreign policy context. It emphasized the need for the acquisition of better technology to exploit indigenous sources, the vulnerability to the supply of energy, especially oil and gas because of volatile political situations in India's energy source countries, and the need for strengthening of diplomacy to pursue energy security in the context of foreign policy and geopolitical risks.

Taking cognizance of the growing environmental and global warming concerns, the IEP recommended the government to take measures to reduce CO_2 emissions by inducting the CDM as per the norms acknowledged and prescribed under the Kyoto Protocol. The IEP also advocated increasing sector-wide energy efficiency, promotion of mass transportation, development of renewable energy, expeditious nuclear energy development and clean coal technology development as key measures to mitigate climate change (Badrinarayana 2010).

The IEP's climate change and global warming concerns were reflected in its recommendation for renewable energy. It proposed the following steps.

- Phasing out of capital subsidies by the end of the 10th Five-Year Plan linked to the creation of renewable grid power capacity;

[3] For details, see Planning Commission, Government of India (2011).

- Requiring power regulators to seek alternative incentive structures that encourage utilities to integrate wind, small hydro, cogeneration and so on into their systems, and the linking of all such incentives to energy generated as opposed to capacity created;
- Enhancing the contribution of renewable energy in India's energy mix with the focus on certain renewable energy sources.

The broad objectives of the IEP, as summarized, emphasized on the following points (Planning Commission, Government of India 2005):

- Condense energy intensity by 20 per cent.
- Increase average gross efficiency of power generation up to 34 per cent (from 30.5%).
- Address constantly the energy demand of all sectors, including the energy needs of vulnerable households in India.
- Ensure the availability of safe and convenient energy at the lowest cost in a technically efficient, economically feasible and environmentally sustainable means.
- Lastly, the IEP envisaged coal to remain the main source of power generation in India and proposed a combination of strategies to address India's energy security problem, such as exploiting the hydro and nuclear potential, rationalization of fuel prices, lesser energy intensity, energy efficiency, promotion of urban transport, enhancing the domestic resources and addressing the environmental concerns by moving towards environmentally suitable alternative fuels and employing an innovative tax system to discourage the environmentally untenable practice of energy.

However, the IEP was not free from criticism. The IEP had shortcomings both at the policy and implementation level. At the implementation level, the IEP faced many challenges which the Indian government could not address. As a result, the desired goals set out in the IEP recommendations were not achieved. One of the main shortcomings was the slow and tardy approach towards power sector reforms. The impediments to power sector reforms were mainly due to the lack of cooperation between the union and the state governments which resulted in an obdurate power-sharing arrangement.

The government bodies, responsible for the generation, distribution and regulation of electricity, such as the National Thermal Power Corporation (NTPC), NHPC at the central government level and State Electricity Boards (SEBs), were slow on the power-sharing reforms. The IEP recommendations on the privatization front was initiated in the 1990s liberalization process in the power sector reforms in view of the increasing gap between demand and supply of electricity particularly because of the poor economic conditions of SEBs. But the progress made was not up to the mark and could not achieve the desired target. Furthermore, the negative perception among the general public against private reforms in the power sector, lack of particulars regarding power sharing permits, grid facility, rules dealing with the commercial aspects of power, tariff validation, and tackling larceny and corruption in the power sector impeded the IEP recommendations, preventing goal achievement. This was very much reflected in the concern expressed by the former PM Manmohan Singh as he indicted the inability of India's energy portfolio to meet the energy demand of its economic aspiration and pointed to the need for a formal process to assess and evaluate policy steps undertaken to achieve the goals of the IEP (Badrinarayana 2010; iGovernment 2009).

The broad objectives discussed earlier reflect that the IEP also advocated for energy security at an affordable price in a sustainable manner and acknowledged the growing environmental concerns. The broad objective of the IEP shows that India's energy policy has been pursued mainly under the conceptual framework of four 'A's—accessibility, availability, affordability and acceptability. For a long time, flawed subsidy policy, lack of proper implementation of reforms, inexact energy costing and insufficient spending in the renewable energy sector impeded India's energy security drive. On top of that, the Indian government needed to pursue its energy security goal in a way that would allow India to shift to a low-carbon economy.

Overall, the IEP was the first all-encompassing outlook on IESS by the Planning Commission of India, which recognized the manifold energy challenges that the Indian government needed to address, such as meeting the energy demands, security of supply, alleviating the environmental concerns, and moving towards the renewable and alternative sources of energy. The IEP could be considered as futuristic in its approach, aimed at addressing India's energy security trapped between economic growth and the climate change concerns. The policy recommendations of the IEP were carried till the year 2017 under the 12th Five-Year Plan till 2014 and continued under the NITI Aayog from 2014 to 2017.

Expert Group Reports on Low-Carbon Inclusive Growth

The Planning Commission of India set up the Expert Group on 'Low-Carbon Strategies for Inclusive Growth' to put forward an Indian economy on low carbon emission consistent with an inclusive growth. The interim report was submitted by the Expert Group in May 2011, but the final report was presented in April 2014.

Despite India's ranking as one of the lowest emitters of GHGs in the world on per capita basis at less than one-third of world average, less than one-fourth that of China and one-twelfth that of the US, it is one of the most exposed countries to global warming and environmental extremes. India has been at the forefront of global efforts to mitigate the effects of climate change. Since the 2009 Kyoto Protocol to the 2015 Paris Climate Summit, India has been showing serious intentions to act on the global efforts to address this challenge which is reflected in its pledge to keep its per capita emissions below the per capita emissions of industrialized countries. The interim report of the Expert Group, which provided a set of choices, pointed to India's ability to lessen its carbon emission intensity by 20–25 per cent over 2005 levels by the year 2020. The final report gave a more comprehensive and long-term evaluation of these choices and their overall economic and welfare outcomes (Planning Commission, Government of India 2014).

This significant policy step by the Government of India in its pursuit of energy security, focused on environmental concerns and could be considered as a major policy statement by the leading experts on how to pursue economic growth under the growing awareness and norms of the environment and climate change.

The low-carbon strategy gave top precedence to inclusive growth. The strategy focused on energy efficiencies and low carbon supply technologies, such as solar and wind power, and the use of public transport and non-motorized transport. It also recommended a dramatic reduction in the cost of renewable energy and more financial support for the expansion of the renewable energy sector. The low-carbon strategy aimed at reducing pollution and minimizing the reliance on imported hydrocarbon sources of energy (Planning Commission, Government of India 2014).

The recommendations by the Expert Group on 'Low Carbon Strategies for Inclusive Growth' focused on the following features (Planning Commission, Government of India 2014):

- Energy pricing: An important policy instrument that advances the efficient use of energy resulting in a selection of proper technology and energy constituents.
- Energy efficiency in the household use of electricity, energy-efficient buildings and industry: Energy efficiency has a transcendental role in any modern economy. For example, the labelling and star rating have contributed significantly to encouraging the use of more energy efficient appliances.
- Policy for transport sector: metro, rapid and high-speed rail services, and improvement in the public transport sector.
- Policy for renewable energy, nuclear power, advanced technologies in the coal sector, and the construction of smart grid.

The low-carbon strategy advocated for a multi-sectoral, multi-departmental and multidimensional approach given the difficulties in pursuing growth, inclusion and sustainability as independent imperatives—a unified framework covering all the policy goals for rapid, inclusive and sustainable growth to be implemented in a well chalked out strategy (Planning Commission, Government of India 2014). A cursory look at the central focus of the aforementioned policy recommendation demonstrates that the low-carbon strategy was the energy security goals to be pursued under the four 'A's in which the main focus was on affordability and acceptability concerns.

NEP

The Modi government came into power in 2014 General Election with a developmental agenda. The government found that the lack of energy was a major impediment to India's overall progress. The Planning Commission has shaped India's energy security policy for a long time, in addition to the Indian government's various ministries. In a first major policy initiative, the Modi government dissolved the Planning Commission to be replaced by the National Institution for Transforming India known as NITI Aayog. A cursory look at India's energy policy under the Modi government would suggest that there has been a continuation of the IEP recommendation but with much-grounded realities, speedy implementation and visionary aspect in its outlook.

One of the important features of NEP is its focus on energy efficiency. This not only addresses the availability and affordability aspects but also reduces energy consumption and promotes conservation, eventually making it environmentally suitable. Aiming at two near-term goals to be realized in 2022 and the medium-term goal to be achieved in 2040, the NEP takes a comprehensive approach to reassess India's energy policy system. NEP's recommendations on energy efficiency have mainly focused on the following four broad aspects:

- Electricity at an affordable price: This aims at ensuring 100 per cent electrification nationwide and 24-hours supply every day by 2022. It aims at providing clean cooking fuel and technologies to all within a realistic time frame. Till 2017, around 500 million people were still using biomass for cooking which is considered dangerous to health and environmentally unacceptable as well.
- Improve energy security by energy independence: Enhancing energy security by diversification of energy sources both by reducing the import dependency mainly of oil and gas and by increasing the share of other sources of energy in India's energy mix. India has yet to tap its own energy potential in both the renewables and non-renewables sector and it has substantial hydrocarbon reserves

of oil, gas and coal, and immense prospects in renewable energy such as solar and wind energy.

- Greater sustainability: This is in the view of unseen and sudden disruptions in the energy supply which may occur because of political disturbance, terrorist attack, blackmailing by the energy provider nation or because of natural calamities that are likely to increase as a result of global warming. The NEP also focuses on the concerns due to the bad air quality that is a serious problem in big cities in India mainly because of the use of fossil fuel. According to the estimates available in 2017, 90 per cent of commercial primary energy demand was fulfilled by fossil fuel. To meet the demand for fossil fuel, India relies heavily on imports and it is likely to be increased in the near future. Hence, the NEP emphasizes the need for decarbonization to attain the goals of both energy security and sustainability.

- Boosting economic growth: India continues to grow fast, and this cannot be sustained if there is a disruption in energy supply. The NEP aims at securing competitive prices and growth in the energy sector as a source for foreign investments (refining/distribution of petroleum products received significant investments over the last couple of years).

The NEP sets seven specific targets and guidelines on how to meet these four key objectives (International Energy Agency 2018; NITI Aayog, Government of India 2017b). Energy efficiency is identified as a common area for all of the objectives and its importance is, therefore, highlighted throughout the policy document. Seven areas of intervention where the objectives are to be applied stated in the policy are given as follows (Created by LBNL, AEEE and FICCI with inputs from stakeholders 2015):

- Energy consumption by businesses, households, transportation and agriculture
- Energy efficiency/decarbonization measures on the demand side
- Production and distribution of coal
- Electricity generation, transmission and distribution
- Augmenting the supply of oil and gas, both by domestic E&P

- Refining and distribution of oil and gas
- Installation, generation and distribution of renewable energy

India's energy policy watchers have evaluated the practicability of NEP. One view is that the NEP gives the possible way out for existing and future energy challenges confronting India and provides an outline for the future strategy. This is reflected in NEP's assessment of demand and supply, and evaluation of the structure dealing with the regulatory, infrastructural, technical capabilities and international collaboration issues. This assessment points to the lack of a coherent set of prioritized strategies. Despite encompassing a broad range of energy sector issues, the NEP is ambiguous about evaluating priorities, preparedness or uncertainty. The NEP is focused broadly on supply planning and targets, but it fails to spot the factors and issues affecting the individual demand sectors, and the future strategy to deal with them (Tongia and Ali 2017). The NEP is a fresh insight which deals comprehensively with the matters of the supply side, including energy security, access, affordability and sustainability. However, the assessment points out that the lack of supply would not be an issue in the future; the important thing is ensuring energy security cleanly and inclusively (Tongia and Ali 2017).

The aforementioned discussion shows that the energy sector is not static. In fact, it is speculation-driven and in a state of constant change due to the interplay of various actors and factors. This makes it difficult for policymakers to frame a long-lasting energy policy. The government and agencies involved with energy policy are faced with the challenge of incorporating the dynamism and hypothetical possibilities of the energy sector. The Indian energy sector is not an exception in this regard. The Indian energy policy has transformed from a very narrow to a more inclusive comprehensive approach. This is reflected in subsequent governments' various policy initiatives. The IEP and the Expert Group reports on low-carbon inclusive growth focused on the comprehensive energy challenges which were not confined to one ministry, department, business or consumer segment. The NEP has sought not only to address the challenges identified in the previous energy policy recommendations, but it is in its own right a further stride in achieving India's energy goals.

Liberalization of the Energy Sector: Encouraging Private Participation

One of the important steps in India's energy policy is its move towards opening the energy sector to the private players. The steps have been taken by subsequent governments to enhance the stakes of the private stakeholders in the energy space since the 1900s when India opened its economy to the wider world. But the bottlenecks such as the habit of free access and people's suspicion towards the privatization have not been able to let the things rolling on this front. Under the Modi government, the privatization of the energy sector has received a considerable push.

Electricity Act 2003

The passage of the Electricity Act of 2003 could be considered as a significant move in regards to the privatization of the power sector. This Act reflected government's intention to a total shift from the earlier path of negotiated memoranda of understanding (MoU) with investors to a market-driven situation. The Act aimed to encourage competition between the investors for energy production and trans-mission contracts. Only two Indian states, Orissa and Delhi, had this in place before the passage of the Electricity Act of 2003. The Act provided for the appointment of any person (franchisee) to undertake distribution and supply on behalf of the licensee (state distribution utility) within the licensee's area of supply. The introduction of reforms and competition under the Electricity Act of 2003 brought a significant private sector response in energy generation, a limited but respectable response in transmission (because not many transmission lines were tendered in the first place), but a very limited response in distribution (Mukherjee 2014).

The energy generation section of the power sector value chain has witnessed the maximum private participation, with mainly Indian companies. Private companies operating in the power and infrastruc-ture sectors and the cash-rich non-infrastructure sector as well showed the interests. But in the transmission and distribution sectors, the companies were owned and operated by central and state government-owned entities, respectively (Mukherjee 2014).

The Act had positive results within three years of the period since its emergence in 2004. During FY 2006/2007, the Act showed impressive outcomes in energy generation capacity. However, the distribution part was adversely affected by weak financial conditions. The revenues collected from customers were not adequate to cover all costs incurred along the value chain before final delivery to the end user. This is mainly attributed to power theft, the poor condition of power grids due to negligence in maintenance, technical losses due to the poor physical condition of energy networks hindering supply of purchased power to the customer, and the habit of providing free and the unmetered power to agricultural users. These drawbacks compelled the DISCOMs to be reliant on state subsidies. Moreover, the lack of commercial culture, and lack of personal accountability and operational inefficiency in the bureaucratic setup further handicapped the state-run power sector. The record shows that India's losses taken together, technical and commercial (Aggregate Technical and Commercial [AT&C]), are the top in the world. Over one-third (35%) of the volume of power purchased by the distribution, the utility is on average lost and never billed to the final customer (Mukherjee 2014).

The distribution segment needed urgent commercial attention and management practices that come through private participation. The half-hearted attempts at private participation in the distribution sector could not yield the desired results as it continued to remain in the stranglehold of state-owned utilities with compliant management that accommodated the dictates of politicians. However, the privatization attempts brought a significant change in some urban areas where it was implemented. Distribution segment finances continued to worsen considerably to a level that it was termed at times as 'India's subprime crisis'. The data shows that from 2008 to 2013, the problem aggravated to a massive level with the annual financial gap of US$20 billion before subsidies. The lack of an adequate financing system resulted in poor quality of supply and inadequate capacity utilization in generating stations as the distribution companies lacked the purchasing power to buy enough power from generating stations. The distribution segment's losses put the power and financial sector at high risk. This was evident in data of nonperforming assets which multiplied nearly 10 times between September 2011 and September 2012 (Mukherjee 2014).

India is also moving in the direction of exploring shale reserves in its oil and gas blocks by changing the definition of petroleum in the guidelines to include the shale resource and by opening the sector to private players. India recognized the potential of shale reserves in 2010. Several global agencies made assessments regarding the potential of shale resources in the Indian sedimentary basins. On 14 October 2013, the Government of India launched a set of policy guidelines in exploration and exploitation of shale gas and oil. The law permitted state-controlled ONGC and Oil India Limited (OIL) to explore shale resources on land blocks that were allotted to them on a nomination basis before the advent of the NELP. But private entities were excluded from the exploration activities of shale reserves in India.

So far, India has not commercially exploited shale gas but the two public sector companies (ONGC and OIL) are working on test wells. ONGC has started exploration in nearly 50 blocks, while OIL is working on three blocks. India's move to open the shale reserves to the private companies is significant as the shale reserves were clearly excluded under the Petroleum and Natural Gas Rules, 1959, and the PSCs governing conventional oil, gas and coal bed methane fields.

As per the US Energy Information Administration, India is estimated to have 96 Tcf of recoverable shale gas reserves. In January 2011, Schlumberger made an initial gas-in-place estimate of 300 Tcf to 2,100 Tcf and ONGC, in 2013, estimated the shale gas reserves to 187.5 Tcf in five basins, namely, Cambay, KG, Cauvery, Ganga and Assam, and Assam-Arakan. India's shale gas potential is in six basins, which include the Cambay (Gujarat), Assam-Arakan (Northeast India), Gondwana (Central India), KG onshore (Andhra Pradesh), Cauvery onshore and Indo-Gangetic basins (Malhotra 2017).

According to the information provided by India's Petroleum and Natural Gas Minister Dharmendra Pradhan on 27 March 2017 in the Lok Sabha (the lower house of the Indian Parliament) the ONGC completed the drilling of a total of 21 wells in 18 blocks for shale gas and oil. OIL completed geological and geophysical studies and geochemical analysis in its identified areas worldwide. In recent years, shale reserves exploration has become an alternative energy source to conventional oil and gas for addressing energy demands. The US, Canada and China have already given major focus on shale reserves exploration but India has yet to make significant progress. The US has been looking at joining India's shale reserves exploration. In 2010 under the joint shale exploration under the MoU, the State Department agreed to cooperate with the Indian petroleum ministry on shale gas resource assessment, technical studies, regulatory framework consultations, training and investment promotion through the exchange of experiences and best practices, and study tours. India's Reliance India Ltd (RIL) has significant joint venture stakes in the US shale reserves sector. Other significant private players such as Essar Oil, Vedanta Resources and Cairn Energy have been looking for opportunities and the government's assent to exploit the shale reserves potential in India (Malhotra 2017).

UDAY

UDAY, which aimed to improve the financial health and operation turnaround of power distribution companies (DISCOMs), is a significant step in the energy sector reforms. The bad financial situation of the distribution segment got the attention of the new union government formed after the 2014 general election. As a result, the Ministry of Power, Government of India, launched UDAY which was approved by Union Cabinet on 5 November 2015. The UDAY is seen as a game changer for the private players in the electricity distribution. In total 32 states and union territories of which 27 states, except Odisha, West Bengal and Nagaland, have joined the centrally-sponsored UDAY (Ministry of Power, Government of India 2018). This scheme works by state governments taking over three-fourths of DISCOMs' debts, leaving the DISCOMs with more funds for operations, and putting through incentives-backed measures to ginger up operations.

According to the memorandum, the scheme is only applicable to State-owned DISCOMs. For this purpose DISCOMs may include combined generation, transmission, and distribution undertakings. Participating states would need to improve their operational efficiency by actions such as compulsory feeder and distribution transformer metering by States, Consumer indexing, upgrading or change of transformers and meters, smart metering of all consumers above 200 units a month (Ministry of Power, Government of India 2018). Both the Government of India and the State governments will take steps to reduce the cost of power. The union government expected to focus on increasing the supply of domestic coal and the coal sector-related measures, and the State governments will need to focus on prospective power purchase through transparent competitive bidding by DISCOMs and improvement in the state generating units.

The noticeable benefits of UDAY began to surface within a very short period. The two crucial measures of performance of DISCOMs are clearly visible. The first benefit could be seen in the lessening of the gap between the average cost of electricity supplied and the average revenue realization. The second is the decline in AT&C losses, which is the loss of energy due to the inefficiency of equipment and theft of electricity. It should be noted that the electricity theft has been a major problem in India over the years. The data shows that the scheme has done well in certain states. The gap between cost and realization decreased by 15 paise and the AT&C losses came down by 4.22 per cent in the first nine months of 2016–2017 over the FY 2015–2016, but that is an average figure. In the first nine months of 2016–2017, 12 states saw higher AT&C losses than in the previous year, showing that the result is not positive in all states (Ramesh 2017). The progress has shown an encouraging trend. An evaluation of UDAY made by India Ratings agency showed that both the financial outcome and operational efficiency of reduction in AT&C losses have improved at an aggregate level. This initial trend is attributed to the reduction in interest cost which has benefitted DISCOMs' finance (*Economic Times* 31 July 2017).

But some assessment taken on India's energy policy doubts the continuity of the positive trends in the distribution segment. According to the KPMG evaluation, the reforms to uplift the condition of DISCOMs may follow the track of previous efforts in this direction. Unless electricity comes out with the 'public goods' perception, the expected benefits may not last long. The assessment suggests that it is imperative to look for a substantial reduction in losses and an increase in tariffs. Power procurement practices at DISCOMs continue to be driven by lower short-term prices in power exchanges, and no major new long-term power sale agreements were executed leaving a larger power generation capacity high and dry (KPMG 2017). Even the India energy watchers who think that UDAY provides big opportunities for the private sector participation in electricity distribution point to similar issues which will need to be addressed. For the long-term success of UDAY, the government will need to legislate structural changes to bring down the AT&C losses. The focus will need to be on improving the efficiency in billing and collection which would require smart meters and ensuring payment from electricity consumers. This is an important step given the trends and statistics available at the Ministry of Power pertaining to 22 states. The ministry's information indicates that that only 3 per cent of 500,000 consumers who used over 500 kWhr a month and only 1 per cent of 1,750,000 consumers who take between 200 kWhr and 500 kWhr are measured using smart meters (Ramesh 2017). However, in the medium to long term, an improvement in operational performance such as increased billing efficiency through feeder metering and feeder audit leading to the higher collection will be crucial for keeping the DISCOMs' finances healthy. Tighter monitoring of action plans, the appointment of nodal officers and state-level monitoring committee are also equally important for achieving the desired results (*Economic Times* 31 July 2017). Experts believe that privatization move is not going to be easy. Issues such as the review of faulty PPAs, huge investment in regulatory assets, shutting down of thermal plants in the state sector and underutilization of manufacturing capacity in India will need to be resolved to realize the full potential of this scheme. Also, the privatization of nuclear power should proceed only after ensuring its long-term safety in a transparent manner after taking the public into confidence. The government will need to be careful in importing nuclear power plant from firms which have been declared bankrupt or likely to be in the near future (Sirhindi 2017).

The privatization move is showing positive results in regards to AT&C losses. This is seen in the better results in billing and collection in the areas where private players have been incorporated. This is evident in the data of Feedback Energy Distribution Company (FEDCO), which has the distribution networks in four districts of Odisha, covering 545,000 consumers, which has made significant progress by reducing the AT&C losses by 23 per cent in the years between 2013 and 2017. It is worthwhile to mention the example of FEDCO, as a distribution franchisee of the Odisha DISCOM, CESU, is exceptional in that 90 per cent of the consumers in its area are rural. Other examples are Torrent Power which operates in Bhiwandi in Maharashtra and reduced AT&C losses by 25 per cent. Their success is attributed to the installation of smart meters that curbs corruption, irregularities and human interference in meter reading. This has been a big problem in government-owned DISCOMs (Ramesh 2017). Through these reforms, the government wants to encourage competition to push the price down. Still, consumers are not ready for this and there has been opposition to the government's privatization move. They also point out that in the Indian situation competition cannot bring the cost down. Some are of the view that the scheme will affect the public sector and can damage the power sector industry in the country.

But consumer resistance in more recent years has lessened. This is mainly because of the positive awareness in the community that has been created by the energy sector stakeholders, private companies, government and the more sensible citizens. One of the main hurdles in the privatization of the electricity has been political (Ramesh 2017). This to a great extent has been overcome by the Electricity Act 2014 which amended the Electricity Act 2003. The Act's distribution franchisee scheme has extended privatization to power distribution, bifurcated the distribution of power into the carriage (infrastructure) and sale of power. According to this concept, the government-owned DISCOM retains the assets, but the operation is outsourced to a private company. The private companies compete over the selling of electricity to consumers to do business (Sirhindi 2017). The companies get benefits on the basis of the performance and efficacy and this has lessened the political problem. Above all, how the government proceeds with these structural changes, as observed by the experts, will be the key to the success of this project.

Liberating Coal Mines

One of the more significant policy measures under the Modi government towards increasing the stakes of private stakeholders in the energy sector reforms has been in the coal sector. In what could be considered as a milestone decision in India's energy sector reforms, on 20 February 2018, the Modi government's Cabinet Committee on Economic Affairs made a decision allowing the participation of the private sector in commercial coal mining under the Coal Mines (Special Provisions) Act, 2015 (*Times of India* 22 February 2018). The Coal Mines (Special Provisions) Bill of 2015 was passed by the parliament under the Modi led-NDA government, 15 years after a similar bill was introduced under the Atal Bihari Vajpayee-led NDA government but was rejected in the upper house. In 1973, the coal sector was nationalized by the then Indira Gandhi government on the pretext of the lack of investment, innovation and interest from the private sector (Gupta 2018). The Modi government's announcement of the opening up of the coal sector ended the Coal Mines (Nationalisation) Act of 1973 which had nationalized the coal sector for almost 45 years.

This can be seen as one of the most ambitious and commendable reforms in the coal sector by ending over the past five decades moratorium on the entry of private participation. This move provides the avenues for unconnected individual mining entities to enter the coal sector. This is significant as when the Narasimha Rao government embarked on a liberalization process through a series of economic reforms in the 1990s, most of the sectors were opened up for private participation, but coal was missing from the realm of this liberalization process.

The past inaction in the coal sector reforms adversely affected economic progress. Coal dominates India's energy mix including for electricity generation. Despite the fact that around 70 per cent of India's electricity is generated in coal-fired thermal power plants, the energy policy ignored the reforms in the coal sector. Coal mining was restricted to just the public sector entities and was unable to meet demand. As a result, despite huge coal reserves estimated at about 308.80 billion tonnes and the potential for domestic coal production, India had to rely on imports which accounted for 22 per cent of the total coal demand in 2017 (*Times of India* 22 February 2018).

The private companies which generally specialized as power generators could not develop mining expertise. This hampered the development of mining skills and expertise among the private companies which could have significantly enhanced the indigenous coal production. The benefits of the opening of the commercial coal mining to the private players go beyond the coal sector. The move is expected to have positive impacts across a wide range of sectors, particularly when it comes to business stability and sophistication, investment in environment-friendly technologies for mining, job prospects, labour rights and transparency in coal block allocation.

This move has ended the state monopoly over coal mining in India. Ending more than four decades monopoly of state-controlled Coal India Ltd, the Modi government's move to open the coal sector to commercial mining by private entities is aimed at boosting domestic mining firms and global giants such as BHP Billiton, Rio Tinto and Vale to exploit the huge potential of India's coal reserves for maximum productivity by employing the state-of-the-art technology (Gupta 2018).

The move is expected to facilitate the inflow of extra investment and the latest technology in the coal sector. To attain the objectives of the plan to privatize the coal mining, the government has come up with an auction methodology with a dual purpose—increasing domestic coal production and enhancing the efficiency of India's energy economy. This could be seen in alignment with a series of recommendations envisaged by the NEP which focuses on energy efficiency.

One of the future steps in this direction could be the establishment of a coal regulator to oversee the privatization of the coal sector. In December 2013, a bill to create a regulatory authority in coal was introduced under the UPA government in the lower house but could not be passed as the house was dissolved. The government can revise this bill according to future needs (*Times of India* 22 February 2018). An independent regulator should result in transparency, accountability and speedy implementation of the privatization policy as the government holds a direct commercial stake in both coal mining and its most important consumer, thermal power companies.

Privatizing Renewable Energy

Today energy security policy cannot ignore environmental concerns. Privatization does not mean only the opening of coal mines, and oil and gas exploration to the private players, but is also related to the renewable energy sector. Given the 2015 Paris Climate Summit commitment which was ratified by more than 140 nations, India has taken steps to a green transition. According to the Paris Climate Accord commitment, the developed countries are committed to the financing of the green projects and the move to renewable energy projects in the developing countries.

Unfortunately, developed countries are not on track to provide the $100 billion of collective annual assistance promised to the developing world for climate change mitigation and adaptation purposes. Current estimates show that approximately $51 billion a year is flowing to developing economies, with $22.8 billion coming in the form of bilateral assistance and $16.6 billion stemming from multilateral organizations.[4] Additionally, according to experts, the actual amount of annual funding needed to achieve the goals set forth in the Paris Agreement is closer to $300 billion, further highlighting the need for climate action financing in the developed world. While bilateral

---

[4] MNRE–CWET with the support of GIZ has worked towards setting up of solar radiation measuring stations in 51 locations in different states of India (Mathur, Pandey, and Ray 2017).

funding, multilateral organizational funding and domestic government spending will continue to play an important role in funding climate action projects in the developing world, the flows need to be complemented by a substantial increase in private capital in order to meet the goals set forth in the Paris Agreement (Climate Policy Initiative 2016).

In this context, India's move towards a low carbon emission economy is important. Modi has not only committed to the fulfilling of Paris Climate Summit target but has gone beyond. The Government of India has taken further major initiatives.

DISCOMs are a good example in this regard. The revamping of further electricity distributions system will eventually lead to the inflow of private capital in the renewable energy sector. Provided the views of private investors and businesses are accepted, the free market will determine the electricity rates (Prasad 2016).

Studies show that the participation of private players has been hampered by a combination of factors, and it is reflected at the policy formulation and implementation level as well.

Since the beginning of the National Solar Mission in 2009, India's progress towards solar energy goals has faced many challenges which have impeded the progress. The first hurdles for the private players are in acquiring the land and getting mandatory clearance from the government. After the land is acquired under a lease or sale by a state government, the additional challenges come in the form of local claims on the land, putting solar projects in areas of high solar irradiance, close to the power grid, and keeping in mind the availability of sufficient resources, infrastructure, the grid proximity, poor grid infrastructure, transmission problems and chronic power shortages.

These challenges get further multiplied by community involvement, concern to protect the habitat and conflicting land claims (Mukherjee 2014). As stated, India's transition towards a low-carbon economy and clean energy has been hampered both at the policy formulation and implementation levels. The urgent need to address mounting energy security often leads to haphazard policy formulation. A study undertaken on India's transition towards green energy by the New

Delhi-based Observer Research Foundation suggests that a number of hurdles remain to hamper the private capital investment in the renewable energy sector. This is mainly in the form of regulatory logjams, political risks, off-taker risks, technology risks, construction risks and the unavailability of domestic debt finance. The study suggests that these challenges can be overcome by policy initiatives aimed at resolving these regulatory, political and technology issues, strengthening the judiciary to deal with construction risks and private sector involvement to better manage off-taker risk (Mathur, Pandey and Ray 2017).

The KPMG assessment points out that the Indian government will need to take a direct head-on approach, tackling stressed power assets to give the much-needed boost to privatization. The Economic Survey 2016–2017 suggestion of the setting up of a Public Sector Asset Rehabilitation Agency (PARA) if realized could be an important step in this direction. To encourage the privatization in the renewable energy sectors, particularly wind and hydropower, would need precise government support which could be provided in the form of generation-based incentives or viability gap finance. The government will need to make provision in its budget to encourage the renewable energy space. The government's previous steps in this regard were generation-based incentives for wind power which would need to be continued. The steps such as the setting up of an open coal market driven by commercial principles would be a much-needed step in the direction of privatization (KPMG 2017).

The recent report by the KPMG titled *Energy and Natural Resources: Union Budget 2017–18. A Post-budget Sectoral Point of View* provides some tax-specific measures to encourage the private participation in the energy sector. It lists steps such as the exemption from payment of Minimum Alternative Tax, non-applicability of Section 14A of the Act for holding companies investing in special purpose vehicle for power projects and no specific deduction, that is, capital-linked deduction for the power sector as in the case of other infrastructure sectors in the power sector (KPMG 2017).

In regards to oil and gas, some of the tax-relief steps could be the extension of exemption under section 10(48A) to Indian companies that have been engaged in the business of storage of crude oil/natural gas in India. To facilitate the import of LNG, the storage facility at port locations should be included in the definition of 'Industrial infrastructure' in section 80-IA. The ambiguity on the definition of 'mineral oil' under section 80-IB(9) of the Income Tax Act, 1961, should be removed and should also include natural gas and coal bed methane under the definition retrospectively, irrespective of NELP rounds. Tax holiday period for E&P undertaking should be increased from 7 years to 10 years out of 15 years. Again extend the period of Incentive (sunset clause) under section 80-IB(9) acquisition of new blocks in India. Allotment of infrastructure status to the E&P facilities to enable access to funds and availability of other benefits which have been granted to infrastructure projects, such as tax holiday and weighted deduction could be introduced (KPMG 2017).

Some of the necessary steps to achieve the solar energy target of 2022 should take into account not just the cost of the solar plant but also the transmission cost due to underutilization, system stability and operational concerns. The policy pertaining to large solar plants needs to be reviewed and the alternative of decentralized solar power plants, rooftop solar plants and solar agriculture pumps need to be seriously considered and encouraged (Sirhindi 2017).

Reducing the Burden on Hydrocarbons: Technology, Conservation and Energy Efficiency

India has 17 per cent of the world population, but its share in the world gas, oil and coal reserves are only 0.6 per cent, 0.4 per cent and 7 per cent, respectively. These hydrocarbon sources of energy constitute a major component of India's energy mix. Given the focus on renewables, the share of hydrocarbons will decrease overall in India's energy mix, but the absolute growth of these sources will continue to increase given the mounting demand for its modernizing economy and lifestyle.

A low per capita level and a large disparity between urban and rural areas characterize energy consumption in India. In 2015–2016,

India's per capita energy and electricity consumption at 670 kgoe and at 1075 KWh/year, respectively, constitute just one-third of the world average. In addition, about 25 per cent of India's population today is without access to electricity and 40 per cent without access to clean cooking fuel. In the year 2014, the share of electricity in its energy demand was only 17 per cent compared with 23 per cent in the member-countries of the OECD. This low share means that a large proportion of energy consumption comes from solid and liquid fuels, exacerbating the air quality at the demand centres. Because electricity has the virtue of delinking emission from the point of consumption, for many uses, it is a preferred form of energy (NITI Aayog, Government of India 2017c).

Thus, India has relied on imports even at a relatively low level of energy consumption. After negligence for a long time, India has begun to focus on the demand-side interventions to reduce energy consumption and enhance energy efficiency. According to NITI Aayog recommendations, there are at least two demand-side interventions that can help cut energy usage: behavioural change that results in reducing the demand for energy-based services and the introduction of alternatives that maintain the level of services but reduce the energy required for its provision. The former is called energy conservation and the latter greater energy efficiency. An example of energy conservation is the shift to fan from air-conditioning, which cuts the need for energy by lowering the level of service received. Similarly, an example of improved energy efficiency is the shift to LED bulbs from the regular bulb, which maintains the service but cuts energy consumption. Often conservation and efficiency effects come jointly (NITI Aayog, Government of India 2017c). For example, when houses are designed to allow a better flow of air and the use of air-conditioning is foregone, there is a partial decline in service (comfort level) indicating both conservation and efficiency. In January 2015, the Government of India launched the Domestic Efficient Lighting Programme under which it provided over 600,000 by November 2015 and 1,632,000 LED bulbs by March 2016 to families through power distribution companies at an instalment of ₹10 per month for each bulb (*Economic Times* 26 March 2016; Ministry of Power, Government of India 2015).

The recently adopted Nationally Determined Contributions (NDCs) as a signatory to the UNFCCC emphasize the importance of demand-side factors. In its submission, India gave particular importance to behavioural change leading to energy conservation—something that has received insufficient attention in the developed countries. The NEP aims to internalize this shift in India's energy policy (NITI Aayog, Government of India 2017c).

In order to reduce the burden on hydrocarbon sources, the Indian government's initiatives are remarkable. Often energy is wasted because of a lack of efficiency and the misuse of resources. The Government of India has taken important measures in this regard especially to reduce energy demand by conserving energy and the efficient use of energy. This is mainly more to reduce the dependence on oil for which India is heavily dependent on imports from foreign sources.

Towards an Efficient Automotive Sector

In this direction, the steps taken in the automobile sector is significant. India's automobile sector is massive and is one of the fastest growing in the world, and there is no sign that it is going to slow down in the near future. The Indian automotive industry produced 21.48 million vehicles in FY 2013–2014. The automotive sector accounts for 22 per cent of India's total manufacturing GDP. By the end of 2015, it was projected to become the world's fourth largest automotive industry by volume (Australian Trade Commission 2015). India is also a substantial exporter of vehicles, and exports are expected to grow solidly in the near term. Initiatives of the Indian government and its major automobile makers are expected to make India a world leader in the two-wheel and four-wheel markets by 2020. The government's initiatives include moves to allow 100 per cent foreign direct investment (FDI) in the automobile industry, credit facilities for farmers to buy tractors, cuts to excise duty and the promotion of eco-friendly cars—compressed natural gas (CNG)-based vehicles, hybrid vehicles and electric vehicles.

India is probably the most competitive country in the world for the automotive industry. The Indo-Japanese automobile maker Maruti

Suzuki expects sales of passenger cars to reach four million units by 2020, up from 1.8 million units in 2013–2014. The Automobile Mission Plan 2006–2016 foresees India, with an output reaching US$145 billion, accounting for more than 10 per cent of the GDP and providing additional employment to 25 million people by 2016, to emerge as the destination of choice in the world for design and manufacture of automobiles and auto components (Australian Trade Commission 2015).

This Automotive Mission Plan 2006–2016 was the collective vision of the Government of India and the automotive industry: How large the automotive industry will be by 2016 and how much it will contribute to the broader Indian economy. The plan envisioned the output of India's car industry reaching US$145 billion—more than 10 per cent of India's GDP—in 2016 and employing an extra 25 million people, as the world's destination of choice for the design and manufacture of automobiles and components (SIAM 2016). The Second AMP 2016–2026, sees India among the top three automotive industries in the world. According to AMP 2026, the Indian automotive industry will grow by between 3.5 and 4 times its current value to between US$260 billion and 300 billion; it will contribute 12 per cent of India's GDP and create 65 million more jobs. This plan makes the automotive industry a central feature of the Modi government's 'Make in India' campaign by increasing exports to 35–40 per cent of the overall output (*Economic Times* 2015).

The new fuel efficiency norms for vehicle manufacturers in India (Overdrive 2015) from 2016–2017 and the Indian government's focus on technology development under the NEMMP 2020 (Ministry of Heavy Industries & Public Enterprises, Government of India 2015) have created an imperative for Indian automotive companies to pursue improvements in technology through international collaboration.

The new fuel norms efficiency move is good news for the ever fuel-conscious Indian vehicle owner. The government has sent notices setting minimum fuel efficiency norms for passenger vehicles. The norms which are the brainchild of the Bureau of Energy Efficiency (BEE) stipulate that passenger vehicles must have a mileage increase of 14 per cent by 2016–2017 and an increase of 38 per cent by 2021–2022. The ministry of road transport and highways will be the agency responsible for implementing these norms and will levy harsh penalties for non-compliance. This benchmark will rank India among the likes of the US, Germany, Japan and China which also enforce the same regulations.

The Nano is currently India's most fuel-efficient petrol car. The new norms named 'Corporate Average Fuel Consumption' standard will be applicable on passenger and imported vehicles running on all types of fuel—petrol, diesel and gas. The standards will stipulate the consumption of fuel on the total vehicles manufactured and not on the number of models. This rule, if implemented, is expected to save India about 20 mt of fuel over 10 years (Overdrive 2015).

Another factor in the calculation of fuel consumption will be the vehicle's weight, and based on this factor, eight different categories have been proposed. Currently, passenger vehicles in India average 16 km/litre, but according to the new norms, the average fuel efficiency by 2016–2017 should be 18.2 km/litre and 22 km/litre by 2021–2022. The BEE has also proposed a star based labelling system, in which models will be ranked between 1 and 5 stars depending on their fuel efficiency. This system, if implemented, will be similar to the star rating system seen in power saving household appliances. Reacting to this development, SIAM (Society of Indian Automobile Manufacturers) claims that it will be difficult for manufacturers to comply with these norms and they would necessitate huge investment.

Industry experts have unanimously stated that complying with these norms is very difficult and will require a trade-off with vehicle. The safety aspect is a topic which is already gathering storm, because of the recent failure of top-selling hatchbacks in a safety test conducted by NCAP (Overdrive 2015).

In 2013, the Manmohan Singh UPA II government launched the NEMMP 2020. It aimed to achieve national fuel security by promoting hybrid and electric vehicles. There is an ambitious target to achieve 6–7 million sales of hybrid and electric vehicles by 2020. The Government of India aims to provide fiscal and monetary incentives to kick-start this nascent technology. With the support from the government, the cumulative sale of these vehicles is expected to reach 15–16 million by 2020. It is expected to save 9,500 million litres of crude oil equivalent to ₹62,000 crore savings.

To give a further boost to this initiative, the Modi government launched FAME under NEMMP 2020 in the Union Budget for 2015–2016 on 1 April 2015 with the objective to support the hybrid/electric vehicles market development and manufacturing ecosystem. The scheme envisioned the adoption of both and market creation of both hybrid and electric technologies vehicles in the country. The scheme is to allow hybrid and electric vehicles to become the first choice for the purchasers so that these vehicles can replace the conventional vehicles and thus reduce liquid fuel consumption in the country from the automobile sector. It is envisaged that early market creation through demand incentive, in-house technology development and domestic production will help the industry reach self-sufficient economy of scale in the long run by around the year 2020 (Ministry of Heavy Industries & Public Enterprises, Government of India 2015).

As per the data available, around 42,000 electric vehicles were sold in 2012–2013 and nearly 20,000 hybrid and electric vehicles were sold in 2013–2014. In the year 2012–2013, most of the electric vehicles sold were electric low-speed scooters. However, after the introduction of the scheme, there has been a growth in the hybrid and electric vehicles and this is likely to gain momentum for all the vehicles segments including 2 W, 3 W, 4 W, light commercial vehicles and buses.

Through this scheme, 145,618 vehicles were sold, which resulted in 35,441 litres of fuel saving per day, as of 1st June 2017. Under the FAME, around 57 per cent of the total governmental incentive was spent in the year 2015–2016 of which 95 per cent cars sold were diesel mild hybrid. Despite these records, about a 10 per cent reduction in fuel consumption was witnessed for diesel mild hybrid resulting into a slow progress on pre-existing technology. This sluggish development in embracing electric motor vehicles could be attributed to reasons such as difficulties in the broader acceptability of electric vehicles and consumer perception. Added to this are technical problems associated with electrical motor vehicles such as efficiency of batteries, driving range, the speed of electric vehicles, charging time, inadequate battery charging stations, battery recycling, slow technology development and the high cost of batteries (Mishra 2017a). Though car companies in India want to make big push towards the use of electric vehicles over the next two decades, the indigenous manufacturing of electrical vehicles remains sluggish despite government incentives. However, this drawback and the lack of enthusiasm for electric vehicles among private consumer and the government for a 100 per cent electric vehicle use by 2030 create ample scope for the use of electric vehicles in public transport (*Economic Times* 26 March 2016; *Times of India* 2007).

These plans by the government include financial help to buyers to encourage the sale of these hybrid and electric vehicles. Under the FAME, the Department of Heavy Industry came out with the details of the plan including financial help for various hybrid and electric vehicles (incentive per vehicle-technology segment). According to the scheme, the vehicle manufacturer will sell the hybrid and electric vehicle at a reduced price to the buyer at the level of the eligible predetermined incentive amount, and the same will be reimbursed to the manufacturer by the government (Ministry of Heavy Industries & Public Enterprises, Government of India 2015).

To date there is a 10 per cent increase in vehicle registration in India annually for the past decade, but the share of electric vehicle sale to total vehicle sale remains at a very low level of 1 per cent annually. A joint study undertaken by the NITI Aayog and the US-based Rocky Mountain Institute found that India could save 64 per cent of expected road-based mobility-related energy demand and 37 per cent of carbon emissions in 2030 by pursuing a shared, electric, connected mobility future. This would result in the reduction of 156 Mtoe in diesel and petrol consumption for that year and a net saving of around $60 billion in 2030 at the present oil prices. However, currently the prospects for electrical vehicles are bright as there is a low personal auto vehicles ownership in India and the use of public transport and sharing between non-motorized transport and public transport is higher when compared with countries such as the United States and China. This gives ample scope for the introduction of electric vehicles to potential motor vehicle owners in India (Mishra 2017a).

At the energy efficiency front, the Modi government's policy is at the macro level and an all-encompassing approach as reflected in the focus on efficient system of public and freight transportation systems. PM Modi's address at the inaugural session of PETROTECH-2016 exhibition clearly highlighted the need for an energy efficient system at the macro level. While speaking on energy efficiency, he said,

> India's commercial transport sector has become skewed. An increasing proportion of goods is transported by road. To increase energy efficiency, my government has given the highest ever priority to the railways. We have stepped up public capital investment in railways by more than one hundred per cent between 2014–15 and 2016–17. We are completing dedicated freight corridors. We are constructing a high-speed rail corridor between Mumbai and Ahmedabad which will be more energy efficient than air travel. We have given a big thrust to waterways, both inland and coastal. Our Sagarmala project will connect the whole of India's long coastline. We have also opened up new inland shipping routes on large rivers. These steps will improve energy efficiency. The long awaited legislation on a national Goods and Services Tax has been passed. By removing physical barriers at state boundaries, GST will accelerate long haul transport and further increase efficiency (PM India 2016).

In the draft on energy security by the NITI Aayog, a future projection of India's energy consumption in 2040 under alternative scenarios with respect to efforts towards achieving greater energy efficiency is given. The projection shows a positive trend for India's efforts towards energy efficiency.

Table 3.1 shows the estimates under a range of two sets of assumptions: A baseline effort and a significantly more ambitious effort towards achieving energy efficiency and conservation. The baseline scenario (BAU) generates the higher demand bound and the ambitious scenario is represented by the lower bound. The table reflects that in the ambitious scenario, energy consumption ends up being 17 per

Table 3.1 India's Energy Demand: Sector-wise

Sectors		2022		2040	
TWh	2012	BAU	Ambitious	BAU	Ambitious
Buildings	238	568	525	1,769	1,460
Industry	2,367	4,010	3,600	8,764	7,266
Transport	929	1,736	1,628	3,828	3,243
Pumps and tractors	237	423	388	728	592
Telecom	83	131	124	207	164
Cooking	1,072	829	684	524	467
Total	4,926	7,697	6,949	1,5820	3,192
% reduction in energy demand in 2040			17%		

Source: NITI Aayog, Government of India (2017c, 11).

cent below that in the baseline case, illustrating the power of energy conservation and efficiency. In this scenario, the share of electricity in the final energy demand at 26 per cent is also significantly higher than the 23 per cent share in the baseline case suggesting an environmentally cleaner outcome at the point of consumption. In per capita terms, annual energy consumption rises from 670 kgoe in 2015–2016 to 1055–1184 kgoe in 2040. Correspondingly, per capita annual electricity consumption increases from 1075 KWh in 2015–2016 to 2911–2924 kWh in 2040. At these levels, India's economy would still be much more efficient than many developed countries while meeting satisfactory levels of energy demand of its people (NITI Aayog, Government of India 2017c, 11).

With the GDP composition across different sectors changing with growth, energy shares of different consuming sectors shift. Buildings, industry and transport sectors together are the main gainers in both scenarios. The maximum efficiency gains accrue in the transportation sector whose share in the total energy consumption in 2040 is estimated to be 23 per cent under this ambitious scenario compared with 25 per cent in the baseline scenario. The demand-side intervention focusing on the shift to electric vehicles and public transport, investment in manufacturing of efficient vehicles/EV/hybrids and higher spends on fuel efficient vehicles, by building the infrastructure supporting the Electric Vehicles such as charging stations, hydrogen filling stations for fuel cell vehicles and more CNG stations is recommended by the NITI Aayog to achieve greater efficiency in the transport sector (NITI Aayog, Government of India 2017c, 11).

The most significant change is going to be in the cooking sector. The table demonstrates that environmentally cleaner energy will replace the current biomass fuels. Energy share of cooking drops from 22 per cent in 2012 to just 3.3 per cent in 2040 in the baseline case and 3.5 per cent in the ambitious scenario. Not only form the health point of view but also at the energy efficiency and conservation front this is a transformative change. Demand-side intervention such as improvement in efficiency of biomass cook stoves and the shift towards an environment-friendly fuel for cooking including electric cook stoves are positive trends in this direction leading to the reduced energy share in the cooking sector. The trend in the table shows the serious effort put by the government in achieving energy efficiency and conservation through the supply- and demand-side intervention (NITI Aayog, Government of India 2017c, 11).

Enhancing the Domestic Production: Oil and Gas E&P

As discussed earlier, the Indian government has taken a concerted approach to deal with its energy demand in the domestic context. Ranging from oil and natural gas exploration, to making new nuclear power plants and shifting to renewable energy sources, the Indian government has taken various measures to address energy security.

The heavy dependence on the import of hydrocarbons, particularly oil and gas, has resulted in large outflow of foreign exchange. This is highest in terms of GDP share when compared with G20 nations. India has been takings steps to explore its reserves and increase production. Nevertheless, it has not achieved much success. Some of the major oil E&P moves so far in recent years have not been impressive.

Stagnant domestic crude oil production has diverted a large share of foreign exchange to the importation of oil.

NELP

In India, the public sector oil companies mainly ONGC and OIL have been leading the petroleum and gas E&P activities since before India began to liberalize its economy in the early 1990s. The first major move in reforming the hydrocarbons E&P space by the Government of India was the launch of the NELP in 1997. This new contractual and fiscal model for the award of hydrocarbon exploration rights became effective in 1999 and remained valid for all contracts entered into by the government between 1997 and 2016. Under the NELP, the Directorate General of Hydrocarbons (DGH) became the main regulatory agency to ensure a fair opportunity for both the public and private sector companies in the E&P of hydrocarbons. Since the introduction of the NELP, the rights for E&P of hydrocarbons are contested and given only through a competitive bidding system. The NELP opened the hydrocarbon exploration sector to private and foreign investment with 100 per cent FDI and provides an opportunity for all public and private, and domestic and foreign companies to compete on an equal footing to win the petroleum exploration licences.[5]

The NELP aimed at attracting significant risk capital from Indian and foreign companies, state-of-the-art technologies, new geological concepts and best management practices for the exploration of oil and gas resources in India to address the growing demands for oil and gas. Following are the main features of the NELP:[6]

- 100 per cent FDI
- No mandatory state participation or any carried interest of the government

[5] New Exploration and Licensing Policy: http://www.arthapedia.in/index.php?title=New_Exploration_and_Licensing_Policy_(NELP)
[6] Ibid.

- Blocks to be awarded through an open international competitive bidding system
- ONGC and OIL to compete for obtaining the petroleum exploration licences
- Freedom for the contractors for the marketing of crude oil and gas in the domestic market
- Royalty at the rate of 12.5 per cent on land areas and 10 per cent for offshore areas
- Cess to be exempted for production from blocks offered under NELP
- Companies to be exempted from payments of import duty on goods imported for petroleum operations

The NELP data shows that under the NELP at the end of nine rounds of bidding completed in 2012, 360 exploration blocks were offered, and for 254 blocks, PSCs were concluded. Currently, 166 blocks are active and 88 have been relinquished. The number of companies grew to 117 of which 11 companies were Public Sector Undertakings, 58 Indian private entities and 48 were foreign companies. This was an improvement when compared with the pre-NELP period which had a total of 35 E&P companies (5 PSUs, 15 India-based private sector companies and 15 foreign companies) working in India in nomination blocks, producing fields and pre-NELP blocks, either as operators or non-operators (Director General of Hydrocarbons 2015). Overall, NELP was successful in achieving its main objectives, especially in attracting private and foreign companies in hydrocarbon E&P activities. It brought healthy competition and public participation and introduced the state-of-the-art technology and efficiency in the functioning and administration of India's hydrocarbon E&P.

However, despite its achievements and progress made in E&P for almost 18 years, the NELP suffered drawbacks and its policies became outdated in the present context. One of the biggest shortcomings under the NELP was the provision of separate policies and licences for different hydrocarbons. Separate policy regimes for conventional oil and gas, coal bed methane, shale oil and gas and gas hydrates were additional impediments. Different fiscal terms were also in force for allocation of acreages for exploration of different hydrocarbons. In practice, this resulted in the overlapping of resources between different contracts. The NELP policies did not cover the unconventional hydrocarbons such as shale gas, coal bed methane, gas hydrates, oil sands, shale oil, oil shales and tight oil as they were not discovered when NELP was launched. As a result, the fragmented policy framework resulted in inefficiencies in exploiting natural resources. For example, while exploring for one type of hydrocarbon, if a different one were found, it required separate licensing, adding to cost. The pricing of gas under the NELP witnessed many changes resulting in extensive litigation. On the issues of royalties, it could not differentiate between shallow water fields (where costs and risks were lower) and deep/ultra-deep water fields, where risks and costs were much higher.[7]

The PSCs under NELP were based on the principle of 'profit sharing'. If a contractor discovered oil or gas, the government would get its share of the profit from the contractor's endeavour, as per the percentage decided in contracts. This method of endorsement of activities and cost provided the government with extra advantage and left scope for manipulation which affected the smooth implementation of policies. As a result, many projects were delayed and held up in litigation because of differences between the government and private contractors on proportioning costs.

Another area where the NELP was silent was that E&P was confined to blocks which had been put on tender by the government. There were situations where the private entities may themselves have had information or interest regarding other areas where they were likely to pursue exploration. Those opportunities remain untapped, as they were not brought in the bidding arena.

Despite all efforts for alternative energy and renewables, India continues to be dependent on hydrocarbons. In addition, oil production has witnessed stagnation and gas production has declined in

[7] Ibid.

recent years in India. As a result, India continues to be dependent on imports for its hydrocarbon needs. The imports of oil and gas constitute a big percentage of India's total imports. At present, India imports over three-quarters of its domestic requirement of crude oil and approximately a third of its domestic requirements for gas. This puts pressure on foreign exchange reserves. This affects the availability and affordability aspects of India's energy security. Hence, the desired increase in domestic production is significant from both the economic and energy security aspects.

To give a resolute and concentrated policy push to enhance domestic production, the Modi government took policy decisions by revising the existing policy on E&P and launched a series of policy measures to boost the domestic production of hydrocarbons on 10 March 2016. The Union Cabinet and the Cabinet Committee on Economic Affairs in its meeting on 10 March 2016 took the following decisions:

- Introduction of HELP
- Focus on marketing and ensuring pricing autonomy for new gas production from deep water, ultra-deep water and high-pressure–high-temperature areas.
- Policy for grant of extension to the PSCs for small, medium-sized and discovered fields
- Cancellation of the Ratna offshore field award from ESSAR Oil Limited and assigning it to the original licensee, ONGC.

HELP

The first major initiative was the launch of the HELP. HELP replaced the extant policy regime for E&P of oil and gas—NELP, which had been in existence for 18 years.[8] HELP is a significant jolt to bring reform in the existing policies for E&P activities and a further step to boost the domestic production of oil and gas. HELP is a groundbreaking futuristic step aimed at introducing a uniform licensing

---

[8] Ibid.

system encompassing all hydrocarbons such as oil, gas and coal bed methane under a single licensing framework.

This is a momentous step, as India needed an intensive and focused approach to encourage the indigenous production, given scenarios of oil and gas dominating India's total imports, stagnation in both domestic oil and gas production. HELP has the following key features (Government of India, Cabinet 2016):

- A uniform licensing system which will cover all hydrocarbon sources under a single licence and policy framework.
- Contracts to be based on 'biddable revenue sharing'. Bidders will be required to quote revenue share in their bids and this will be a key parameter for selecting the winning bid. Quote to be at two levels of revenue called 'lower revenue point' and 'higher revenue point'. Revenue share for intermediate points to be calculated by linear interpolation. The bidder giving the highest net present value of revenue share to the government, as per transparent methodology, to get the maximum marks under this parameter.
- An Open Acreage Licensing Policy (OALP) to be implemented whereby a bidder may apply to the government seeking exploration of any block not already covered by exploration. The government examines the Expression of Interest and justification. If it is suitable for the award, government will call for competitive bids after obtaining necessary environmental and other clearances. This is to enable a faster coverage of the available geographical area. OALP allows companies to approach the government at any time and seek permission to explore any block. Under OALP, interested companies get access to the National Data Repository (NDR) maintained by the government, for bidding purposes. Under the old NELP, the companies used to wait for formal bid rounds by the government, and E&P activities were confined to only the blocks that were offered by the government for the contract (Ladislaw and Bellur 2017).
- A concessional royalty regime to be implemented for deep water and ultra-deep water areas. These areas shall not have any royalty

for the first seven years and thereafter shall have a concessional royalty of 5 per cent (in deep water areas) and 2 per cent (in ultra-deep water areas). In shallow water areas, the royalty rates shall be reduced from 10 per cent to 7.5 per cent.

- The contractor to have freedom for pricing and marketing of gas produced in the domestic market on an arms-length basis. To safeguard the government revenue, the government's share of profit is calculated based on the prevailing international crude price or actual price.

How HELP changes India's E&P policy is evident from Table 3.2 (Ladislaw and Bellur 2017).

The table shows the improvement and major changes that HELP bring in the E&P activities. The HELP replaces the PSC with a revenue sharing contract. According to this arrangement, the government will get a share of revenues instead of a share of profits, from production. In addition, the biddable work programme commitments will be the main consideration for the award of contracts in the new policy prescription. This particular change to the fiscal regime is to check the reappearance of previous disputes, litigation and adjudication matters pertaining to cost recovery preceding the sharing of profits. This also exposes lacuna in India's regulatory and administrative capability (Sen 2016).

The new policy regime marks a generational shift and modernization of the oil and gas exploration policy. It is expected to stimulate new exploration activity for oil, gas and other hydrocarbons and eventually reduce import dependence. It is also expected to create substantial job opportunities in the petroleum sector. The introduction of the concept of revenue sharing is a major step in the direction of 'minimum government maximum governance', as it will not be necessary for the government to verify the costs incurred by the contractor. Marketing and pricing freedom has many benefits that will not only lead to the simplifications of the process but will also end government discretion. It will lessen disputes and litigation, reduce the scope for corruption, minimize administrative delays and, thus, stimulate growth.

Table 3.2 HELP and PRE-HELP: A Comparison India's Hydrocarbon Exploration and Licensing Policy

Policy category	HELP	Pre-HELP
Types of hydrocarbon	Covers all conventional and unconventional oil and gas	NELP covered only conventional oil and gas; Coal Bed Methane Policy covered coal bed methane
License	A single license for exploration and extraction of all types of oil and gas	Separate license required for conventional oil and gas, coal bed methane, shale oil and gas, and gas hydrates
Revenue model	Revenue-sharing model under which revenue will be shared with the government in the ratio submitted by bidders	Production/profit-sharing model under which government received a share in the profits
Coverage	Open acreage policy under which exploration companies can apply to explore any block not under exploration	Exploration was restricted to blocks opened for bidding by the government
Oil and gas pricing	Companies have the freedom to sell their production domestically without government intervention	Crude oil price was based on import parity; gas price was fixed by the government
Royalty	Concessional royalty for deep water (5 percent) and ultra-deep water (2 percent) areas, which are difficult to explore, and reduction of royalty in shallow waters (from 10 percent to 7.5 percent)	12.5 percent for the onshore areas and 10 percent for offshore areas; 10 percent for coal bed methane

Marketing and Pricing Freedom for New Gas Production

In India, the most of the untapped oil and gas is in regions containing deep water/ultra-deep water or high pressure/high temperature (HPHT). To make things clear, on 18 October 2014, the Cabinet Committee on Economic Affairs, Government of India, approved a mechanism for pricing of domestically produced natural gas. The policy was a much-needed reform over the extant NELP provision as the new policy initiative acknowledged the need for incentivizing gas production from deep water, ultra-deep water and HPHT areas because of higher costs and higher risks involved in the exploitation of gas from such areas. In addition, in principle, approval was also given for a premium on the gas price for the gas to be produced from new discoveries from such areas.

The launching of this policy was timed to coincide with the substantial decline in the global oil and gas prices which were at the lowest level for over a decade. This was to attract potential investors in the E&P sector. Many discoveries of gas in deep water/ultra-deep water, HPTP areas could not be developed, as they were not considered economical for production. In the scenarios of growing demand, declining production and increases in imports of oil and gas, the government decided to provide marketing and pricing freedom to the gas to be produced from the new discoveries as well as existing discoveries which were yet to commence production.

After a serious review of India's strategic, economic and environmental goals and the interests of both producing and consuming sectors, the government launched a new policy containing the following main features (Government of India, Cabinet 2016):

- For all the discoveries in deep water/ultra-deep water/HPHT areas which are yet to commence commercial production as on 1 January 2016 and for all future discoveries in such areas, the producers will be allowed marketing freedom including pricing freedom.
- To protect user industries from any market imperfections, this freedom would be subject to a ceiling price based on the landed

price of alternative fuels. To the extent that domestic gas can be produced and sold at a price below import parity price, it will not only benefit the overall economy by boosting employment and GDP and reducing imports but also benefit the user industry by lowering the average price.

- The ceiling will be based on publicly available prices of substitute fuels and the method of calculation shall be communicated transparently.
- The ceiling price will be calculated as lowest of the (a) landed price of imported fuel oil (b) weighted average import landed price of substitute fuels (namely coal, fuel oil and naphtha) (c) landed price of imported LNG.
- The Ministry of Petroleum and Natural Gas will notify the periodic revision of gas price ceiling under these guidelines.
- All gas fields currently under production will continue to be governed by the pricing regime which is currently applicable to them.

These measures are an improvement over the NELP which was narrow in scope in the production sharing and marketing relationship with the government. Under the NELP, companies needed to acquire a separate licence if they discovered unconventional oil or gas that was not covered under the initial permit. The new policy is aimed to improve the economic feasibility of some of the discoveries already made in such areas and to lead to the monetization of future discoveries as well. This would also lead to the creation of jobs in the developmental phase of these discoveries. According to the ONGC assessment, in the process of the development of discoveries in the block KG-DWN-98/2 around 3,850, direct skilled workers would need to be employed. On top of that, around 20,000 persons would be employed during the construction phase.

Policy for Grant of Extension to the PSCs

Prior to the reform measures, 28 small and medium-sized fields discovered by national oil companies (ONGC and OIL) were awarded to private joint ventures through PSC between the years 1994 and 1998 for periods varying from 18 to 25 years. Out of 28 PSCs, two fields in which the duration of the PSC expired in 2013 were extended up to 2018, and the outstanding PSCs needed to be extended from 2018 to remain valid. Also, it was not possible to produce hydrocarbons from the new recoverable reserves within the remaining period of the contract for several of these fields. A uniform and transparent policy was to allow contractors to take investment decisions for exploitation of the remaining reserves. This is imperative given the need for faster decision-making and timely planning by contractors for increased hydrocarbon production. The Modi government came out with the following process and guidelines for extension of contracts for small and medium-sized discovered fields (Government of India, Cabinet 2016):

- The contractor is required to submit the application for extension of the contract at least 2 years in advance of the expiry date, but not more than 6 years in advance. The DGH will make a recommendation within 6 months of submission of application by the contractor. The government will take a decision on the request for the extension within 3 months of receipt of the proposal from DGH.
- The government share of Profit Petroleum during the extended period of contract shall be 10 per cent higher than the share as calculated using the normal PSC provisions in any year during the extended period. For example, if the current profit share is 10 per cent or 20 per cent, it shall become 20 per cent or 30 per cent, respectively.
- During the extended period of the contract, the royalty and cess shall be payable at prevailing rates and not at concessional rates stipulated in the contracts.
- The extension of these PSCs would be considered for 10 years for both oil and gas fields or economic life of the field, whichever is earlier.

The extension of PSC has twofold benefits. The first is the increase in the production of hydrocarbons beyond the present PSC, and the second is the possible employment generation in the energy sector. The PSC extension would attract additional investments in the fields leading to the creation of employment opportunities both in field operations and in service industry-related sectors. The extended term will further lead to the extension of the existing employment levels. This policy is also expected to see the much-needed transparency and fair play for the private companies in the E&P space with least government intervention but maximum governance. The policy is likely to encourage the speedy exploitation of the indigenous resources and facilitate additional investment in the E&P in the broader goal of energy security.

Fixing and Tapping Ratna Field Reserves

The Ratna offshore field, located south-west of Mumbai, was discovered in 1971 by ONGC. In 1996, the oilfield was tendered out and tentatively awarded to ESSAR Oil Ltd. However, this contract was never concluded, and the reasons attributed were mainly administrative hurdles, legal ambiguities, indecisiveness and doubts that were raised at various stages after the contract was signed. For over 20 years, Ratna field remained unproductive and without exploitation since its initial tendering. The Government of India took the decision to assign it to ONGC on the nomination basis. The move is considered as significant in removing the impediments to this long pending and proven oil reserve. This would not only facilitate the production and create new job opportunities but also address India's overall energy security efforts (Government of India, Cabinet 2016).

These measures by the Indian government are significant steps in the direction of the E&P of oil and gas. At the outset, the HELP heralds a major shift from an era of government control to government support for upstream E&P in India. The provisions of OALP clear the hurdles in E&P progress by providing companies both the data and the guidance for the exploration of the fields of their preference. India's extant policies on E&P have been lamented because of strict government control and bureaucratic red tape which hampered the prospects of oil and gas production (Ladislaw and Bellur 2017). Their removal has been a long-standing goal of the Ministry of Petroleum and Natural Gas. It now offers the private companies the opportunity and responsibility in the E&P activities. This was a much-needed reform measure as OALP was persistently postponed due to the lack of NDR without which such policies were incomplete (Sen 2016).

The new policy measures under Modi government in all likelihood is going to boost the inflow of additional capital from private and foreign entities and businesses and enhance the productivity of indigenous oil and gas, an important stride in India's energy security quest.

Summing up, the Indian energy sector is still evolving and going through a fundamental change. An assessment was taken by the KPMG on India's energy security which highlights challenges and opportunities and indicates the positive trends and achievements over the last four years under Modi government's prioritized energy security policy. Given the Indian government's impetus to regulate the energy space to achieve the desired goals, the energy sector comes with a set of challenges and opportunities. The focus on renewable energy has brought new players into the energy market. They are young and dynamic entrepreneurs and investors who are ready to tap the opportunities. On the other hand, investment in conventional power is at the risk of impairment as coal and power purchase markets remain grid-locked. Some of the main issues requiring attention that were highlighted in the KPMG assessment report are as follows (KPMG 2017):

- Achievements made in the power sector may not be long-lasting unless the financial and operational health of DISCOMs is addressed for the long term.
- The large conventional power capacity is stressed. While institutional capital is available for a financial resolution of these assets, policy support from the government has yet to pick up pace leading to little action on the ground.
- Large-scale renewable energy integration would need a higher hydro and gas share in power markets. Base load stations may also require recycling capabilities in the medium to long term.
- India's renewable energy programme is one of the largest in the world and is attracting international investors. Payment delays by DISCOMs, particularly to wind power producers, have been a dampener over the past few months for such investors.

- The 2G ethanol programme would need large-scale funding and policy support. A concerted effort would be necessary to create a multiplier effect by making the early investments successful.
- The year-on-year declining production of domestic natural gas continues to impact the business and profitability of many consumer sectors such as power and fertilizers.
- The shortage of gas transmission infrastructure is restricting the flow of gas to the demand centres, thus resulting in low utilization of the existing gas infrastructure such as LNG terminals and in higher consumption of polluting fuels such as diesel.

Notwithstanding these challenges, the Indian government's measures to fix the energy sector are commendable. One of the major policy actions is increasing domestic oil and gas production and reducing import dependence of hydrocarbons. The government-stated goal is to trim down the import dependence by 10 per cent by 2022 to be achieved during a period of increasing oil consumption (PM India 2016).

In this context, the HELP is a significant step welcomed by the energy sector. The government's initiative to hold oil and gas field auctions has shown a positive result. In the auction for the small fields conducted on 21 November 2016, 42 firms placed 134 bids for 34 contract areas out of 46 total areas on offer. This implies that bids were received for nearly 73 per cent of contract areas (*Financial Express* 2016). The new policy regime under HELP is very progressive and can go a long way to encourage domestic E&P activities, and the idea towards a 'gas-based economy' is a positive trend.

The Government of India is supporting the production of second generation (2G) ethanol from agricultural residues. This would provide additional sources of income to farmers and is aimed at reducing carbon emission amidst growing environmental concerns and supports the Ethanol Blended Petrol (EBP) programme for attaining 10 per cent ethanol blending in Petrol. The proposed EBP programme will have the plant with a capacity of 100 kilolitres of ethanol per day (Ministry of Petroleum and Natural Gas, Government of India 2016b; *Tribune* 2017).

In the power sector, the reforms have more outcomes that are positive. This is reflected in a power surplus; all the villages in India are now connected through the power grid, and the use of non-fossil fuel is growing. The energy efficiency has shown a remarkable thrust through many initiatives including the successful implementation of the LED scheme (KPMG 2017).

In addition to energy generation, progress has been made on energy transmission and distribution on both the policy formulation and implementation levels. The transmission capacity addition via public–private partnerships has shown a positive result. Above all, the UDAY is proving to be a landmark initiative which is resulting in the financial and operational turnaround of DISCOMs. The UDAY is making a significant change by uplifting the financial health of DISCOMs which has been a major long-pending issue for the energy sector.

On the concerns of climate change, the Indian government has made a strong commitment to decarbonize the Indian energy sector. It has committed to a 33–35 per cent reduction in emissions intensity of its GDP by 2030 from the 2005 level. It has also decided to attain 40 per cent cumulatively installed power generation capacity from non-fossil fuel based energy resources by 2030. Sustainable, affordable power supplies for all have been the vision of the Modi government. To achieve this goal, the government has imposed a cess on coal to fund a National Clean Environment Fund. Earlier a 175 GW renewable energy target was set for the year 2022. However, the progress made in this direction in the last four years has made R. K. Singh, Ministry of New and Renewable Energy, state that India is on the path of exceeding its 2022 renewable power goal and could exceed it by 52 GW (*Hindu* 2018).

A hydropower revival policy is in the works which inter alia is likely to include the classification of all hydropower projects as renewable energy. In parallel, to ensure a smooth integration of renewable energy into the grid, the government is also working towards the 'green corridors' initiative.

One of the daunting tasks of India has been to provide electricity to all the households in India. In April 2017, the Modi government accomplished the task of connecting all the villages to the power grid (D'Cunha 2018). Now, the next step of the government in this direction is the power connectivity programme to household levels and to electrify more than 40 million un-electrified households. The Indian government has given a considerable push to its electrification programme. On 26 September 2017, PM Modi launched a $2.5 billion project to electrify all households by the end of the year 2018. Almost one-fourth number (300 million) in India have yet to be hooked up to the electricity grid (*Reuters* 26 September 2017).

Looking at the government's commitment to addressing looming energy crises by reforming the existing policy and introducing innovative policy, it is not an exaggeration to state that remarkable progress has been made in pursuit of India's energy security and that this trend is continuing. However, the need is for the coordination of various stakeholders from implementing and regulating agencies affected by the energy policy. India's federal system poses challenges to the Indian government on how to reconcile the centre–state views, interests and relations on the differing energy policy initiatives and challenges. Though previous governments have addressed this issue, an all-encompassing strategy would need to deal with energy policy approach and implementation including centre and state, public and private energy stakeholders, focusing on the macro and micro level of energy policy, targeting specific energy sources ranging from hydrocarbons to renewable sources. This strategic planning with its visionary outlook could lead to the coordination of overall energy policy in the future. The biggest challenge remains the implementation and achievement of the goals prioritized in the NEP. The apt conditions and an improved energy structure, efficient bureaucracy and agencies involved in the regulation and implementation of energy policy, both at the national and state level, will be the key to achievement of these goals. The energy sector will remain a key sector of government policy as it derives a big chunk of its revenue from this sector.

Overall, India's energy security policy continues to face the challenge of matching up with its fast economic growth. In the domestic context, the move has been towards more self-reliance by optimizing the indigenous potential and reducing high import dependence, strengthening the supply of energy by increasing the participation of private players. Increasing investment, market-driven pricing and the independent regulatory set-up across the energy chain, plus the move towards renewable energy sources, have been the mainstay of energy policy at the domestic level.

However, the analysis by energy watchers suggests that despite government actions to open up and liberalize E&P opportunities in India, it is likely that the new policy will not have a major impact on India's oil and gas supply and demand balance. The assessment undertaken at the Center for Strategic and International Studies points to some important facts about India's current status of oil and gas reserves, demand and import dependence. Currently, India's demand production stands at 876, 000 barrels per day (bpd) of oil and 1,030 billion cubic feet (bcf) of natural gas. This is inadequate to meet India's current oil and gas needs. It has led to heavy import dependence to address the demand and supply gap which is about 80 per cent of its crude oil needs and nearly 40 per cent of its gas. The production in the domestic oilfields is on a declining trend and the current oilfields have reached their peak production. India's aim is to reduce import dependence by 10 per cent but the new domestic production can only contribute a small part of India's total oil and gas demand, and inadequate resources constrain the potential for further domestic oil and gas production. A two-pronged strategy to manage import dependence is emerging in the Indian govern-ment's strategy—increasing the production of hydrocarbons and the build-up of strategic petroleum reserves and investment in overseas oil and gas fields. The investment in the upstream is full of intense competition as given the oil prices at lower levels and companies' inward-looking approach on spending, the latest trend of shaky oil price recovery mean the government reform measures may not be able to stimulate capital inflow in the upstream as the demand curve is rising quickly (Ladislaw and Bellur 2017). Simply, the measures taken in regulating the price and encouraging the E&P of gas may not be enough to meet the mounting demand (Sen 2015). Even the move to electric vehicle, efficiency, and the shift to alternative and renewable sources of energy have limitations. In the case of oil and gas, the first two elements of four A's—accessibility and availability—remain paramount, especially in the case of oil and gas, but increasingly the affordability aspect is making India go beyond its domestic reserves. PM Modi and Petroleum and Natural Gas Minister Dharmendra Pradhan both have clearly reiterated that the government's aim is to make energy available to the people at an affordable price.[9] In addition, any oil and gas exporting country looking to tap the Indian energy market will need to consider the affordability aspects if it wants to do long-term business with India. This is affected by market requirements, innovation and technology. These dynamics make India's foreign destination of acquisition and exploration of hydrocarbons very competitive and necessarily involves diplomatic, commercial and geopolitical considerations.

In addition to the domestic approach discussed in this chapter, India has been moving to diversify its sources of oil and gas. The Indian government over the past two decades has taken steps to diversify the sources of its hydrocarbons imports and the exploration of new oil and gas reserves. This will require India to ensure the security of free navigation and security of sea routes through which energy imports flow.

For a long time, India has been dependent on the Gulf nations for its oil and gas needs. This overdependence on the Gulf nations increases India's vulnerability to disruptions and price fluctuations which in turn affects the economic development and day-to-day life of the consumers. Growing global energy demand has led to severe competition amongst the major economies in securing their future energy needs from the finite resources available. This affects accessibility, availability and affordability, which have been the fundamental elements of India's energy security policy framework of oil and gas to the Indian consumers. These energy dynamics have pushed India to have a concerted approach to energy security in the foreign policy context. This aspect of India's energy policy is dealt with in the next chapter.

[9] See the answers in interview on the questions of India's energy security, *Hindustan Times* (9 July 2018).

CHAPTER 4

Energy Security as a Foreign Policy Priority

India's Diversification of Energy Acquisition Sources and Exploration Abroad

Energy security is no more confined only to the domestic sphere. Fossil fuels dominate the energy mix of many countries. The majority of nations do not have enough domestic reserves of oil and gas, and these are unevenly distributed. Many countries in the world dependent on oil and gas as resources. Given the importance of these sources of energy in the economic, consumer and military sphere, the quest for these resources has become a top priority for any country that wants to secure itself on the energy front. Besides, the fast-growing economies, growing population, modern lifestyle-generated unstoppable consumer demand and the lack of adequate domestic energy reserves have pushed energy security in the international realm. The origin of energy security getting such prominence in the foreign policy strategy goes back to the momentous strategic decision by Winston Churchill, then the First Lord of Admiralty, before the beginning of the First World War. To outpace the German fleets, Britain decided to give up coal and shift to oil to power its navy. This implied that for energy security, Britain had to rely on Persian oil for its uninterrupted supply of oil for its fleet. During the Second World War, again the control of oil became strategically important for expansionist and ambitious powers. The supply of oil became a strategic necessity for Japanese leaders for the conquest of Indonesia, then known as the Dutch East Indies, a colony of the Netherlands, and for Germans to annex much of the USSR. Throughout the 1950s and 1960s, energy security was stable for most of the developed industrialized nations.

However, it was during the events of the 1970s in the Middle East that energy security again became a serious foreign policy issue for many of the developed nations. The energy supply to the developed countries was disrupted due to the Oil Embargo of 1973–1974. During the 1973 Arab-Israeli War, Arab members of the OPEC imposed an embargo against the United States in retaliation to the US decision to re-supply the Israeli military and to gain leverage in the post-war peace negotiations. This internationalization of energy space was further visible during the same period when efforts were made by the international community to regulate the energy space. This took the form of creation of a few international bodies, especially IEA.

After the end of the Cold War, energy security began to emerge as a foreign policy issue for both the energy-rich and energy-poor countries. But it was the developments at the beginning of the 21st century that refigured energy security in the foreign policy realm. It could be seen mainly in two broad contexts: security threat environment in the post-9/11 world and the new global energy market dynamics. In the post-9/11 scenario, the threat of terrorism, political instability in oil-exporting nations, growing resource nationalism, concern over the scramble for energy and resources and, above all, fear over the future availability of adequate energy and geopolitical rivalry pushed the energy security to the forefront of foreign policy realm. In the market context, a combination of dynamics affecting the energy world, particularly increased oil prices, uncertainty over continued supply of oil, decline in energy capital inflow to the energy sector, constrained oil market, diversification in the forms of energy, move to alternate source of energy, new supply destination and an increase in demand for oil, fuelled by the emerging economies, brought the issue of energy security to the forefront of many nations' foreign policy strategy. The global recession and financial crisis of 2008 further pushed the energy security at the forefront of international relations as those affected looked to improve the economic performance and create jobs. For many nations, the energy sector is not a significant revenue generator, but the overall role that it plays in the economic performance and development is crucial. It became paramount for most of the nations to have a stable and reasonably adequate energy supply to put the economy back on track post recession (A. Sharma 2007; Yergin 2006).

In addition, some of the noticeable developments have been in technology and discovery of energy sources in countries other than the Gulf Cooperation Council (GCC) region, the main supplier of petroleum products for decades. Instead, it was in countries such as the United States and Russia where the new forms and reserves of hydrocarbon sources of energy were discovered. The evolution of new application leading to a more efficient and effective technology with broad applications has further pushed the energy boom. Today, oil companies are doing things that they have been doing for decades more efficiently, more effectively and with much wider applications. This especially is very much reflected in the case of the discovery of shale gas. This is bringing a favourable change for the nations with rich energy reserves, but the countries with limited energy reserves have not much to cheer. These dynamics further pushed energy security at the forefront of foreign policy for both energy exporting and energy importing nations. The discovery of the new form of petroleum products and the move towards renewable energy are adding a new dimension to the energy dealings in the international context.

Overall, the dynamics of energy space in the international context is mainly visible in the vulnerability of supply due to unstable political condition in the energy-rich region of the Gulf nations. Moreover, the unstable petroleum price has pushed it further to the foreign policy front. Energy security has become the fundamental of a nations' economic and political performance, and a major component of determining a nation's power status. The world is witnessing the rise of new economies and great powers that are fast moving towards industrialization and development. The sustainability of the emerging powers' growth trajectory is very much dependent on energy security. This is very well echoed in the traditional definitions of great power given by Kenneth Waltz in which resource endowment is one of the parameters for including military strength, economic capability, size of population and territory, and political stability and competence (Waltz 1993). Energy is a major resource endowment for a modern nation.

India simply cannot afford to not act on energy security. India's pursuit of achieving its developmental goal and the sustainability of its economic growth is dependent on an uninterrupted supply of energy. India's energy demand is growing given the scenario of projects that it is not going to stop. Despite all the efforts to diversify the sources of energy such as renewable and alternative sources of energy, hydrocarbons, mainly oil and gas, continue to be significant in India's energy basket. Hydrocarbons constitute the major share of India's energy basket. Despite the recent push to domestic E&P activities for oil and gas, India needs to look abroad for these sources. The inadequate domestic reserves of oil and gas continue to push India to look for foreign sources.

To explore this, this chapter examines India's attempt to explore and diversify the new sources of hydrocarbons abroad. The first section of this chapter deals with India's NEP and its objectives in the foreign policy context. The second part is about India's energy exploration abroad and its efforts to diversify its energy import sources in which India's energy exploration in Latin America and the Caribbean (LAC), Africa, the Caspian basin, and Russia has been discussed in detail. The third section scans India's energy-driven foreign policy and the geopolitics relating to the proposed gas pipeline projects. Before conclusions, the chapter also examines India's new energy sources such as Israel and its energy security effort in neighbouring nations in South Asia, especially Bhutan and Sri Lanka, with a focus on hydropower collaboration.

In India's case, the energy exploration abroad is not only for oil and gas, and enriched uranium for its nuclear reactors but also for the acquisition of innovative technology and its application to India's domestic energy E&P and the move towards renewable and alternate sources of energy. Despite its huge coal and thorium reserves, India has inadequate oil and gas reserves, enriched uranium and clean coal, and lacks the technology to fully exploit these sources.

Though Indian government has given a massive push for boosting the E&P of oil and gas, any increased domestic production alone will not be sufficient to meet the projected needs for either oil or gas. As a result, India has given a substantial focus on ensuring energy security efforts for these resources abroad. It has become a foreign policy priority and cutting energy deals have been a frequent foreign policy achievement that the subsequent governments have highlighted and focused upon. Other dynamics are the overreliance for these resources, especially oil and gas in one region. India's two-thirds of oil imports come from one single region, that is, the GCC countries. Over the past two decades, sensing the mounting increase in energy demand, India has been taking steps in the direction of diversification of import destination sources for these resources. In fact, India is following the footsteps of other major oil-importing economies and making great efforts to obtain supplies from sources outside the Gulf. The quest for these resources has pushed India's pursuit of energy security on its external policy agenda reflected in bilateral agreements, intense diplomacy and geopolitics.

The Indian government has urged leading public sector energy companies, such as the ONGC, to secure energy resources overseas by participating directly in the global energy market, and investment by Indian companies in overseas oilfields is projected to reach $3 billion within a few years. ONGC has bought equity stakes in oilfields in Iraq, Sudan, Libya, Angola, Burma, Russia (Sakhalin), Vietnam, Iran and Syria. Other Indian PSUs have become involved not only in acquiring exploration and exploitation rights but also in establishing sales outlets for Indian petroleum products and in offering a variety of technical services (A. Sharma 2007).[1]

---

[1] Oil and Natural Gas Corporation (ONGC): http://www.ongcindia.com/.

NEP and Foreign Policy Context

This is very much reflected in India's NEP, which views energy security in terms of assured supply. Until domestic sources, particularly the share of renewable energy, increases by a large percentage of India's total energy mix, the import dependence on energy supply is going to continue to mount. Given high import dependence for commercial primary energy supplies, it is imperative for India to pursue profound overseas commitment across the various stakeholders in the energy sector. The increased import dependence offers challenge and opportunity as well. India's energy infrastructure needs capital inflow from both the domestic and foreign investors and it offers the Indian companies opportunities to explore abroad to secure energy availability. Given the inevitability of a steep rise in the import dependence over the next two decades, Indian companies get hold of a large share of the overseas energy ventures. The justification for an effective overseas energy strategy is proposed for a number of reasons:

- Accessing latest technology
- Overland energy supplies through pipelines and transmission lines
- Leveraging our large buying position to influence energy markets
- Playing a lead position in international energy organizations
- Climate policy diplomacy to protect India's energy strategy
- Collaborating in large international consortia-based research

Due to a variety of overseas energy interests, the number of stakeholders and the nature of engagement are also broad. India has developmental, commercial and strategic roles to play in the international arena, all of which call for a comprehensive strategy. The launch of the ISA by India at COP21 is one example of the emerging role of India as a world leader in promoting clean energy solutions while assisting the emerging nations to meet their energy needs. Ensuring steady, enduring and lucrative business contracts for the supply of oil and gas and generating electricity from the renewable sources in the Northern Himalayan region are features of India's comprehensive energy strategy in the external context. Given India's future energy demands that are expected to constitute about 40 per cent of the additional worldwide demand by 2040, India has no option but play a proactive role in influencing the global energy regime. This is reflected in the subsequent Indian government's ministries dealing with the energy sector and the MEA' effort to pursue energy security goals by actively participating in the global energy debates, taking a normative leadership role on renewable energy, participating in international organizations dealing with energy, sending space mission for energy exploration and by elevating energy security as one of the top foreign policy priorities.

Over a period of nearly two decades, India's energy policy structure has evolved. Now, this is reflected in an emerging all-encompassing energy sector policy in the foreign policy context too. India is now actively participating in the international energy spectrum to shape the norms and policies of international energy organizations. Traditionally, India's oil and gas policy could be seen in the context of availability and accessibility aspects. But in recent years, given the Indian government's commitments to provide energy to all, the affordability aspect has begun to figure more prominently. India has begun to raise its concern strongly on determining the oil prices to make it affordable. India has emerged as an important lucrative market for oil and gas exporting countries. India, as the world's third largest consumer of energy and as the fastest growing oil and gas-consuming nation, expects to continue to grow up in the ladder. The foreign policy context and affordability aspect are clearly visible in views of the Government of India and the energy sector stakeholders. Dharmendra Pradhan, India's Petroleum and Natural Gas Minister, clearly stated,

> Transportation fuel is an international commodity; it will fluctuate with geopolitics. India has succeeded in influencing it. PM Modi consistently has said on global platforms that India is a big consumer, and those who want to do business here must take into consideration our affordability. This has had a positive impact on producer countries. We have wanted for a long time that Asian premium should be revisited (*Hindustan Times* 9 July 2018).

This is further reflected in the fact that despite India being the third largest energy-consuming nation, it has the lowest per capita energy

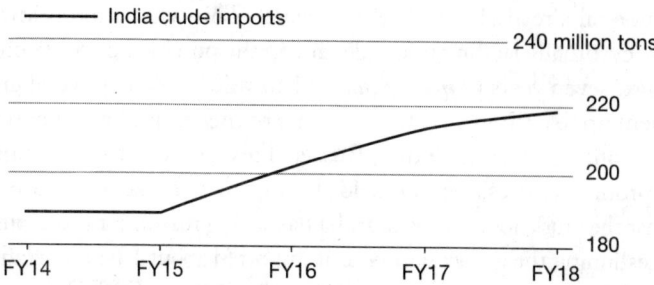

Figure 4.1 *Increasing Dependence on Import*

Source: Petroleum Planning and Analysis Cell, Ministry of Petroleum and Natural Gas.

Note: Imports surged as consumption of oil products rose.

consumption among emerging economies which implies that India's energy consumption will only increase further. This puts the affordability question on the forefront when dealing with the international market of oil and gas. As indicated, this has been tactfully pursued by PM Modi on many international energy platforms.

As highlighted in Figure 4.1, India's import dependence has increased in recent years due to an increase in the consumption of oil products.

Growing demand has made India the fastest growing crude oil consumer. The uncontrolled price rise in mid-2018 compelled the stakeholders to raise their concerns. India, the fastest growing crude oil consumer, warned that it might reduce the purchase from crude cartel if the price was not controlled. Sanjeev Singh, the chair of the Indian Oil Corporation, the country's biggest refiner, warned that India might reduce purchases from the crude cartel if the price is not controlled and may switch to alternatives such as electric vehicles and gas as more cost-effective alternatives, replacing 1 million barrels of the country's daily oil use by 2025. This could have long-term implications as India's oil consumption is expected to grow to 10 million barrels a day by 2040, which will make it the leading growing consumer worldwide. This is based on the conjecture that oil prices will be at $83 a barrel by 2025 and $113 by 2040. However, the current price of crude oil already is nearly $80 a barrel. This will be too costly for India, eventually leading to the decline in the purchase of crude oil for the next seven years (*Times of India* 11 July 2018). The price of crude oil went up to a three-year high because of the reduced production by OPEC and its allies, including Russia. This has been troubling India and prompted President Donald Trump to tweet a series of rants against the crude lobby as well. India has strong reason for involvement in questioning the growing price as it imported about 1.6 billion barrels (220.43 mt) of oil in the year 2017 largely from the OPEC countries (*Times of India* 11 July 2018).

Though India's oil refineries companies such as Indian Oil, the biggest fuel retailer in India, have been trying to compensate for this by shifting to alternatives by expanding into natural gas, renewables and electricity to power vehicles, this cannot solve the looming demand for oil. Realistically, it is not possible for alternatives to substitute for 1 million barrels a day. The dominance of oil in all likelihood is going to continue for the near future, at least until India achieves self-sufficiency in natural gas production. Thus, India has given considerable focus to its oil and gas exploration activities abroad.

So far, the performance of India's overseas oil and gas exploration activities has produced a mixed outcome. NITI Aayog in its draft on India's energy security has rightly pointed out that there has been ambiguity across the board on the goal of India's overseas strategy, and success is usually measured merely on the parameter of acquisition of overseas assets. Indian companies have been engaging internationally for over two decades, but no more than 5 per cent of India's domestic oil requirements are being produced by the Indian companies' overseas assets. However, in the more recent years, Indian companies have secured oil and gas assets that are projected to further exceed that 5 per cent. Another positive trend is reflected in the Indian energy ventures of renewable energy in the Himalayan region neighbours, especially in developing hydropower (NITI Aayog, Government of India 2017a, 76–79).

The non-performance and unyielding results of Indian companies in energy E&P is attributable to a number of reasons. At the outset, it reflects the lack of a visionary and integrated energy strategy in the foreign context that could encompass a combination of domestic and international factors affecting the energy space. This should be divided into two aspects.

The first is India's internal drawbacks, especially at the policy formulation and implementation levels.

- At the domestic policy formulation level, the major drawbacks were the lack of clarity on the goals of overseas engagement in energy exploration and a tardy approach in decision-making leading to sluggishness in acquisition tenders. As noted in the previous chapter, one of the major policy lacuna in India's energy space has been the lack of coordination between various agencies and ministries dealing with energy matters. This was reflected in India's overseas energy exploration too.
- And the second is at the external level, the unstable and uncertain geopolitics in the Middle East impeded Indian companies' overseas energy engagements. The Gulf region in the recent decade has witnessed political turmoil due to a number of reasons including the growing religious fundamentalism, terrorism, authoritarian regimes, democratic protest, Arab spring and political upheaval, overthrowing of the authoritarian regime and above all the great powers' stake in the region. This obviously hampered the Indian overseas mission's progress in E&P activities.
- This was further compounded by unproductive participation in international organizations, underutilization, and the lack of exploitation of the resources, capability and potential of the Indian overseas mission.

India's overseas engagement for energy security has been mainly focused on addressing the mounting demand for oil and gas. India aims to realize this through international collaboration vis-à-vis engagements with oil and gas firms in countries which have got these resources in abundance. Reducing imports and diversification of sources thereby facilitating the interests of Indian energy businesses to attract capital inflow in the energy sector and the constructive engagement with those international organizations dealing with the energy sector are all part of this masterplan.

NITI Aayog points out that India's energy strategy overseas has an excellent opportunity in this interconnected world, reflected in the growing commerce in energy. NITI Aayog envisions that this offers an opportunity for India to diversify its energy imports by securing long-term supplies across all energy sources—oil, gas, electricity, nuclear fuels and coal—from all regions, including India's immediate neighbourhood. Bringing energy supplies from Indian overseas assets will harmonize well with the greater interplay between market forces in the domestic energy market. Following are the scenario and strategy projected by the NITI Aayog (2017a, 76–79) in which India can pursue its energy exploration abroad:

- The over-supplied energy markets offer an opportunity for India to leverage its large buying position for several energy sources. This will be effectively used to acquire assets and seek beneficial energy supply contracts, which will provide India price and volume suppleness.
- There is a scope for India to develop a strong relationship through energy commerce in its neighbourhood, including balancing its electricity grid by interconnecting with the South Asian grid, along with obtaining renewable energy from long-distance destinations such as Central Asia and its immediate vicinities such as Sri Lanka and Bhutan.
- Overland oil and gas supplies from Central and West Asia will be secured by pursuing existing and new opportunities. The Turkmenistan–Afghanistan–Pakistan–India (TAPI) and Iran–Pakistan–India (IPI) gas pipeline projects must be pursued earnestly, keeping in mind the evolving geopolitical situation in the region.
- The No Objection Certificates (NOCs) to be provided greater freedom to take commercial decisions without the requirement of multiple clearances, consistent with the devolution of financial powers. Necessary diplomatic assistance to be provided to assist Indian energy firms.

- An effective diplomatic initiative to be pursued in its international engagement to protect the interests of Indian companies. Indian companies can also benefit from a commercial orientation where required, without losing time to lengthy government-to-government deals. This will template accountability and accelerate the decision-making process.
- Unlike many large countries that have an omnibus energy ministry, the absence of a unified face to overseas counterparts is a challenge for India. A separate coordination mechanism will be created in MEA to effectively promote India's energy strategy internationally. Moreover, there is a need to realize the benefits of the presence of the Indian technical diaspora. Indian missions could seek their assistance for technology and workforce access.
- When engaging internationally, energy security should take primacy over other considerations, especially as India's energy imports are expected to rise alarmingly in the medium term. Therefore, India will source energy to promote energy security based on sound commercial decisions.
- India's large energy programme will require high fossil fuel dependence even as India ramps up renewable energy supplies over the medium term. The energy supply and affordability agenda require greater involvement of energy experts in engagement with the international climate change community. The line ministries will play an active role in overseas engagements.
- Energy technology and policy are evolving rapidly across the world. Adoption of the aforementioned strategies offers opportunities to both suppliers and consumers. Indian ministries dealing with the energy and private sectors need to engage effectively with the international community to access these strategies through joint programmes.

The MEA has been pushing the energy security agenda on the international stage. This was visible in many bilateral and multilateral meetings and dealings including the Third Round Table of ASEAN–India Network of Think-Tanks in which the emphasis was on a comprehensive strategic partnership in all sectors, the Indian External Affairs Minister Sushma Swaraj emphasized on building energy security into the capacity building agenda of the ASEAN–India strategic partnership (MEA, Government of India 2014a). Again in January 2016, in her valedictory address at the 4th India–Africa Hydrocarbons Conference concluded in New Delhi, the Minister of External Affairs, Sushma Swaraj called for cooperation and moving towards 'energy justice' from energy poverty (Ministry of Petroleum and Natural Gas, Government of India 2016a). This has been one of the main items in the Indian External Affairs Minister's recent international dealings. In the latest significant event, the Indian External Affairs Minister Sushma Swaraj in the recent IEF Ministerial Meeting, attended by 92 countries held in New Delhi in June 2018, gave a valedictory speech at the 16th edition of IEF. Swaraj, emphasizing the international context and importance of energy as the key to the economic growth of the country, said, 'Oil and gas are strategic commodities.... We are focusing on renewable energy. We are also focusing on incremental growth in the area of energy.' Highlighting the significance of energy security amidst the environmentally acceptable concerns, she pointed towards India's joint efforts with France of founding the ISA in 2016 as a landmark step towards renewable energy by promoting solar energy at the global level. Her speech reflected the centrality of the international context of energy security,

> The ISA is now a reality today due to efforts by India and France. Energy is a key element of diplomatic engagement and outreach. Geopolitics is a complex and an evolving subject. But the geopolitics of oil is much more complex (*Economic Times* 12 April 2018).

At the same event, PM Narendra Modi observed in his inaugural address the increasing demand for energy worldwide and the move towards renewables and for less polluting energy sources to address the global climate change concerns to meet the goal set at the Paris Climate Summit. PM Modi has been one of the leading voices at the international stage in promoting norms for renewables to address the climate change concerns.

India's Energy Exploration Abroad: Diversifying the Energy Import Sources

As per the IEA, India's oil consumption is projected to increase by 6 million bpd to about 10 million bpd by 2040. Since increased domestic production alone will not be sufficient to meet the projected needs for either oil or gas, India is also increasing its efforts abroad. At the same time, since two-thirds of India's oil imports come from GCC countries, India is following in the footsteps of other major oil-importing economies and making great efforts to obtain supplies from sources outside the Gulf. The Indian government has urged leading public sector energy companies, such as the ONGC, to secure energy resources overseas by participating directly in the global energy market, and investment by Indian companies in overseas oilfields is projected to reach $3 billion within few years. ONGC has bought equity stakes in oilfields in Iraq, Sudan, Libya, Angola, Myanmar, Russia (Sakhalin), Vietnam, Iran and Syria. Other Indian PSUs have become involved, not only in acquiring exploration and exploitation rights but also in establishing sales outlets for Indian petroleum products and in offering a variety of technical services (Sharma 2007).[2]

In the gas market, the Gas Authority of India Limited (GAIL) has started to invest heavily in equity stakes in LNG plants in Oman and Iran, and GAIL is building the port facilities and pipelines needed to handle large-scale imports. GAIL is also pursuing plans for direct pipelines from neighbouring Bangladesh, Myanmar, Iran and even Pakistan.[3]

Over the past 10–15 years, Indian companies have made inroads in the exploration of oil and gas. The data shows that Indian oil companies are eventually making progress in foreign assets and ventures across the globe. According to the ONGC official website, it has stakes in 41 oil and gas projects in 20 countries, namely Azerbaijan (two

[2] Oil and Natural Gas Corporation (ONGC): http://www.ongcindia.com/.
[3] See Gas Authority of India Limited: http://gail.nic.in/homepage/homenew.htm.

projects), Bangladesh (two Projects), Brazil (two projects), Colombia (seven projects), Iran (one project), Iraq (one project), Israel (one project), Kazakhstan (one project), Libya (one project), Mozambique (one Project), Myanmar (six projects), Namibia (one project), New Zealand (one Project), Russia (three projects), South Sudan (two projects), Sudan (two projects), Syria (two projects), UAE (one project), Venezuela (two projects) and Vietnam (two projects). In all these projects, ONGC Videsh Ltd (OVL) adopts a balanced portfolio approach and maintains a combination of producing, exploring, discovering and utilizing pipeline assets. Currently, OVL has oil and gas production from 15 assets; 4 assets are where hydrocarbons have been discovered and are at various stages of development, 18 assets are under various stages of exploration and 4 are pipeline projects. The OVL ('ONGC Videsh Operating 41...') has nine overseas active assets in Canada, the US, Venezuela, Libya, Russia (two assets in Vankor and Taas), Iran, Nigeria and Gabon. OVL ventured into midstream, had successfully completed a 741 km long product pipeline project in Sudan in 2005, and is a partner in the Baku–Tbilisi–Ceyhan (BTC) pipeline in the Mediterranean.

OVL production has been witnessing an increasing trend as shown in Figure 4.2.

Another energy giant, OIL also has joint ventures and alliances (as of July 2014) in various countries, namely Libya, Gabon, Nigeria, Yemen, Bangladesh, Myanmar, Russia and Sudan. Bharat Petroresources Limited (BPRL), a wholly owned subsidiary company of Bharat Petroleum Corporation Limited (BPCL), has 17 blocks worldwide, across six countries—7 blocks in India, 6 blocks in Brazil and 1 each in Mozambique, Australia, East Timor and Indonesia. These blocks are in various stages of exploration, evaluation and pre-development. Further, BPRL has also signed definitive agreements to acquire stakes of companies in Russia that have oil and gas assets producing in their portfolio, the company added (*Times of India* 28 December 2018).

The GCC's ample oil and gas reserves are of great significance for India's energy requirements. The India–GCC relations have been

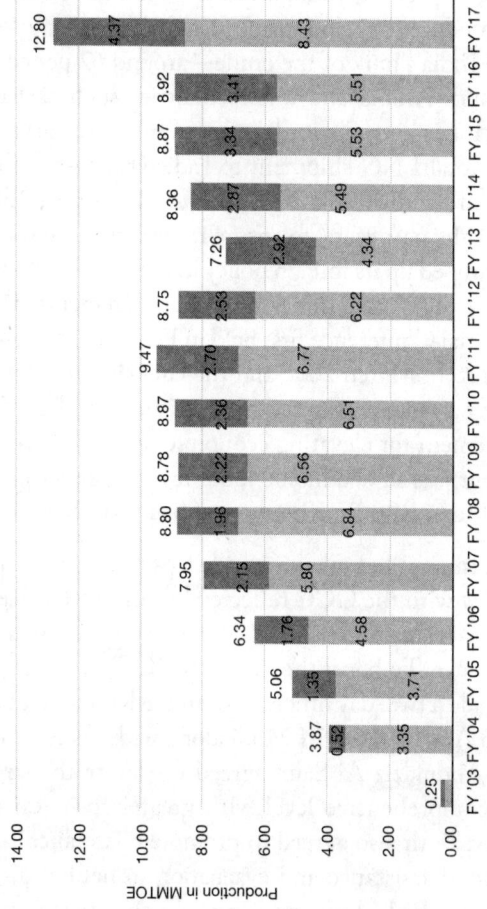

Figure 4.2 OVL Production Performance

Source: ONGC Videsh, 'Production'. Available at: http://www.ongcvidesh.com/performance/production/

significant in trade, investment, energy, manpower, with 6.5 million Indian expatriate in the GCC. But it is the energy which is the dominant component of India–GCC relations. India continues to rely heavily on the GCC nations for its oil and gas needs. India's import dependency for its crude oil needs is about 82 per cent. According to latest IEA reports, at present, India imports around 3.78 million of crude oil bpd and produces 897,300 barrels. Saudi Arabia is the leading import source and India's bulk of the crude—around 62 per cent—is imported from Saudi Arabia and other countries such as Kuwait, Iraq and Iran. Qatar has been the leading gas supplier, accounting for 61 per cent of the total LNG shipment to India (Embassy of India, Riyadh, Kingdom of Saudi Arabia 2007; *Oilman Magazine*, 'India to Diversify...'). Over the past years, to ensure its energy interests in the region, India has ramped up its foreign policy in the GCC. The India–GCC relations have progressed under various arrangements. But the India–GCC Industrial Conference first held in February 2004, second in March 2006, third in March 2007 and the latest in 2015 has been significant in giving a shape to the ties between India and the GCC. A framework agreement for elevating economic ties the India–GCC free trade agreement was signed in 2004, but it is still under negotiation (Embassy of India, Riyadh, Kingdom of Saudi Arabia 2017).

The Modi government has taken several steps to enhance the traditional engagement with the GCC reflected in summit-level leaders and high-level official visits, and bilateral conference and exchanges on trade and energy issues. To boost the ties with GCC, the Indian PM Narendra Modi made a two-day official visit to the Kingdom of Saudi Arabia from 2 to 3 April 2016. PM Modi along with his counterpart King Salman bin Abdulaziz Al Saud agreed to elevate the strategic partnership to a comprehensive level with greater focus on trade, security and energy. Both also agreed to promote bilateral collaboration for humanitarian assistance and evacuation in natural disasters and conflict situations. PM Modi and King Salman agreed to further strengthen cooperation in combating terrorism, both at the bilateral level and within the multilateral system of the UN. But significant was their agreement to elevate the cooperation to strengthen maritime security in the Gulf and the Indian Ocean regions, crucial for security of energy sea routes.[4] It has the energy geopolitics dimension as both nations want to secure the sea lines of communication (SLOC) as per the international law, which stands in contrast to the recent Chinese posture in the Indian Ocean. Also, India is treading cautiously, balancing its strategic interests with Israel and Iran but at the same time maintaining its ties with Arabian nations, all have growing energy ties with India. But, India's energy exploration abroad is diversifying and is no more confined to the GCC. India's diplomacy in the non-GCC countries with rich energy reserves has intensified. In recent years, PM Modi and his ministers have made visits to the new energy-rich nations to develop a comprehensive trade relationship in which the energy component has been a dominant component.

Exploration of LAC

The LAC region has emerged as one of the most important oil-producing regions of the world. The LAC is estimated to be home to some of the world's largest oil reserves. The known oil reserves in the LAC are estimated to be 20 per cent of the global underground oil reserves of nearly 1.7 trillion barrels. In 2009, confirmed discoveries increased by 20 per cent worldwide, while the increase in LAC was 40 per cent (Ortiz 2015). The countries with leading oil reserves in the region are Venezuela that tops the world, followed by Brazil, Ecuador and Colombia. Brazil, Mexico and Venezuela accounted for about 75 per cent of the region's total production in 2014 and ranked 9th, 10th and 12th as oil producers at the world level, respectively (Carpenter 2015). Venezuela now tops the oil-producing countries in the LAC chart followed by Brazil (*World Atlas* 2017). Given the region's richness in oil reserves, Indian diplomacy has seen a growing engagement with Latin American countries over the past decade. This engagement with Latin America is visible through trade and investment. The LAC offers immense opportunities for India, especially in energy, pharmaceuticals and the agribusiness sectors.

---

[4] For the details about the India–Saudi Arabia joint statement, see Modi (2016).

For almost five decades since India's independence in 1947 from British rule, the contacts between India and the LAC were almost negligible. In the foreign policy context, India continued its British colonial legacy by being a member of the British Commonwealth but pursued an independent foreign policy as the leading and founding nation of the Non-Aligned Movement (NAM). On both these international groupings, Latin American nations had a limited presence. Despite the commonality of being developing nations, India and the LAC could not engage on a regular basis even on the NAM platform. In addition, the perception about each-other was not conducive for a constructive engagement. Indians viewed Latin American nations as politically wobbly and inundated with coups, earthquakes and inflation. On the other hand, Latin Americans were surprised by India's widespread poverty and internal conflicts. Unlike other regions, India had the least historical and cultural contacts, and long distance, language and the comparatively high costs of trading further impeded any meaningful India–Latin America trade ties.

However, the changed post-Cold War international scenarios in the early 1990s, and the converging of political and economic interests prompted India's engagement towards the LAC nations. In the political context, India's international engagement with the LAC nations has witnessed this convergence under similar condition of democratic political systems in the majority of LAC countries, South–South cooperation, and on the agenda's in IBSA and BRICS in recent times. India's proactive foreign policy towards the LAC is driven by economic and energy interests.

As recently as 1991–1992, Indo-LAC trade was less than $500 million. Indian exports increased elevenfold over the next 10 years. By 2004, Indo-LAC trade had reached $3 billion (Heine and Viswanathan 2011). In 2007, the Indo-LAC bilateral trade was at $12.2 billion. By April 2008, the bilateral trade reached $13.2 billion that was mainly driven by India's import of oil and raw materials (India's Ministry of Commerce 2009). India with its $50 billion of bilateral trade with the LAC is still less than China with a bilateral volume with the LAC at $264 billion. However, India has tapped the opportunity in recent times and is now fast emerging as an important player in the region. As per the Inter-American Development Bank, India accounted for 50.1 per cent of the LAC's energy exports to Asia in 2013, compared with 44.8 per cent for China. Also, India's total oil imports from the LAC increased from 4.5 per cent in 2003 to 20 per cent in 2014, but China's share from the LAC was 10 per cent of its total imports in the same year (Seshasayee 'India's Rising Presence...'). Though India's bilateral trade and investment with the LAC in comparison with China are becoming more diverse, the energy sector remains the dominating sector.

With a decade of a low profile but steadily growing investment in the LAC, India is engaging this region comprehensively. Indian companies are expanding in the region. Many new initiatives to step up political engagement and increase trade and investment in LAC have been taken. Extensive interactions with the region could be seen in the exchange of visits of ministers and government officials with almost all countries in the Latin American, Central American and Caribbean region in recent times.

Though Indian IT companies have begun to show their presence as the latest entrants in the LAC, India's private oil companies have been making their inroads in the LAC steadily over the past decade. In addition to the state-run oil hydrocarbon hunters, India's private companies have also made their inroads in energy exploration in the LAC. RIL in Colombia, Venezuela, Argentina, Brazil, Mexico and Ecuador; Jindal Steel and Power Ltd in Peru and Bolivia; the acquisition of EnCana Brasil Petroleo Limitada (EBPL) by BPCL; and Essar's agreement with Venezuelan oil firms are the noticeable Indian private entities in the LAC (Moreira, 'India's Expanding Role...').

Venezuela is the top destination in India's hunt for energy in the LAC region. Venezuela is not only the largest oil producer in Latin America but also among the largest oil producers in the world, with the country producing as much as 135.2 mt of oil each year. Venezuela is also an OPEC member. It was also one of the founding members of OPEC (*World Atlas* 2017). Venezuela is estimated to have the world's largest oil reserves with 3,000 billion barrels.[5] Since 2005, Venezuela's proved oil reserves have seen an exponential rise. From the estimated reserves of 80 billion barrels in 2005, Venezuela oil reserves increased to 297 billion barrels in a decade overtaking Saudi Arabia to become the world's leading nation in crude oil reserves. However, on the production front, it has not been the same. Despite the sharp rise in proved reserves in the same period, Venezuela's actual oil production between 2005 and 2015 declined by more than 20 per cent. During the same time, US oil production rose by more than 80 per cent (Rapier 2016).[6] The reasons attributed are mainly the siphoning of money by the government that could have been utilized to develop the oil reserves and the fact that many foreign companies have backed off as they found investing in Venezuela oil reserves unprofitable, and the low oil price further impeded the investors in Venezuelan oilfields resulting in low oil production.

Clearly, Venezuela is in the need of investors and refiners for its oil reserves. Given the unstoppable demand for oil in the coming decades, India has a long-term interest in Venezuela's oil reserves. India has been eying this country for its oil needs for almost more than a decade. In March 2005, Venezuela, then ranked as the fifth largest oil producer in the world, with the largest gas deposits in Latin America, entered into an energy cooperation agreement with India for the joint production of oil and supply of petroleum. As part of the agreement, OVL had a 49 per cent stake in an oilfield discovered onshore at San Cristobal and two offshore gas fields. Since then the companies from both nations have been cooperating in exploring and developing hydrocarbons in Venezuela (A. Sharma 2007).

The visit of Minister of State for External Affairs, V. K. Singh, to Venezuela was significant. Venezuela is India's largest trading partner in the LAC and India's third largest supplier of crude energy. One of the priorities for India is to increase its exports to Venezuela where

---

[5] *BP Statistical Review of World Energy*, 2016: https://www.bp.com/content/dam/bp/pdf/energy-economics/statistical-review-2016/bp-statistical-review-of-world-energy-2016-full-report.pdf

[6] Ibid.

there is a strong demand for Indian goods. India has some significant investments in Venezuela. OVL has invested in two projects in San Cristobal and Carabobo-1, respectively (Siddiqui 2015).

Over the past few years, India and Venezuela have concluded around a dozen energy deals. The development of the Orinoco Belt, oil E&P, cooperation in the hydrocarbon sector and development in the gas and petrochemical sectors indicate noticeable progress on this front. India imports 20 per cent of its oil needs from Latin American countries, in which Venezuela dominates (Siddiqui 2015).

The growing relationship is not just one-sided. It is both ways as evident in the growing interest of Venezuela in India. Enhancing crude oil supply to India significantly and the ongoing negotiation with Reliance Industries, its major customer, for a joint venture in the OPEC country's energy industry, the visit of Eulogio Del Pino, President of Petroleos de Venezuela SA, to New Delhi in 2015 as a follow-up to the visit of Minister of State for External Affairs V. K. Singh to resolve the bilateral issues and opening the road for future investments are noticeable initiatives on the part of Venezuela. Venezuela produces heavy, viscous oil, which is usually priced lower than lighter grades. The refinery companies make much higher refining margins than other processing plants that require superior grades. Venezuela also claims that oil is competitively priced. In addition, Venezuela is stable, reliable and secure in comparison to the Middle East competitors where the political situation is much more volatile and the supply is questioned because of terrorism (Siddiqui 2015).

In 2014, Foreign Office Consultations (FOC) have been held with Colombia, Bolivia, Nicaragua, Mexico, Chile, Ecuador, Argentina, Guatemala, Uruguay and El Salvador, and a Joint Commission Meeting (JCM) was held with Mexico. In 2015, FOC have been held with Costa Rica, Honduras and Barbados and a JCM was held with Suriname.

The Modi government's senior ministers have made their visit to a number of countries in the LAC region, namely Colombia, Venezuela, Mexico, Ecuador, Guatemala, Jamaica and the Dominican Republic. There have been incoming visits by foreign ministers of

Mexico, Guatemala and Suriname. The President of Guyana and the Vice President of Cuba have also visited India in 2015. Though the focus was to increase India's outreach in the LAC nations in a range of sectors including agriculture, pharmaceuticals, IT and the automotive industry, Indian foreign economic relations have been mainly driven by the need for new sources of energy. Within a year of the Modi government's arrival, Petroleum and Natural Gas Minister Dharmendra Pradhan, accompanied by a high-level business delegation, followed by Minister of State for External Affairs, V. K. Singh, visited the LAC nations. Both reflect the foreign policy context of India's energy security approach.

Yet another important country in the region for India for oil exploration is Ecuador.

Ecuador is also an OPEC nation with estimated crude oil reserves of 8.273 billion barrels. In the year 2017, Ecuador Crude Oil Production was 531.32 thousand bpd (Y Charts 2018). Besides inauguration of an IT centre set up with the help of India, there have been talks on the expansion of the trade basket, as Ecuador offers a privileged location, extended production diversity, great potential in agribusiness, fishery, aquaculture, forestry, mining, tourism and services and also has favourable legislation on foreign investment, offers preferential commercial access to different markets and the US dollar as its official currency.

In the process of engaging the LAC, another such significant development has been an official delegation led by State External Affairs Minister V. K. Singh and his meeting on 28 May 2015, with the representatives of member-states of the Central American Integration System (SICA) in Guatemala City. India had an institutional mechanism for dialogue with the SICA. Two ministerial meetings have been held one in 2004 and again in 2008, both in India. The meeting in Guatemala was very significant as the SICA–India ministerial-level dialogue was held in a Central American country for the first time and after a gap of seven years.

The member-states of the SICA highlighted the importance of financial cooperation received from India through lines of credit

(LOC) and welcomed India's offer to increase this further to US$240 million. The region is already utilizing LOC worth approximately US$133 million. India is making further inroads in the region by providing a number of scholarships—between 100 and 200 under the Indian Technical and Economic Cooperation Programme (ITEC) and extending the scholarships exclusively to the Secretariat of the SICA in the area of capacity building under the ITEC Programme, by proposing the establishment of a Regional Barefoot Vocational Training Center for the SICA member-countries in Guatemala, which will train women in rural areas as solar engineers. Obviously, these initiatives by India need to be seen in the broader context of India's aim of getting access to the oil reserves in the region. This is beneficial for both, as India will get oil and the SICA region will get much a needed financial inflow to its various oil reserves that have yet to reach full potential.

Indian oil and gas hunters' expansion in the LAC can be attributed to various factors including the market dynamics such as currency devaluations and lower asset pricing in many of the Latin American nations, and India's need for the diversification for new sources of hydrocarbons. India's track record as a leader of the 'Global South', LAC nations' technical and economic limitation in exploiting its oil reserves, the synergies between India and the LAC, and India's capital and its energy needs are all facilitating the India–LAC ties. It will be not be an exaggeration that given the pace of Indian investment and trade in the past decade, India's oil exploration in the LAC will have a positive impact on its pursuit of energy security. India is taking a long-term approach for this emerging strategic market and energy sources abroad.

India's Energy Exploration in Africa

Over the past 10–15 years, Africa has emerged as an important region for India in its drive for energy security. However, for a long time, Africa remained at the periphery of Indian foreign policy. The account of India's current ties with Africa has been mainly framed in the historical and sociocultural–ethnic links nourished by century-old trade connections and a prosperous Indian diaspora. On the geopolitics front, India–Africa collaboration was confined to their solidarity against colonial rule and apartheid during the post-Independence era and sharing common political ground on South–South cooperation and the NAM platform. On top of that, there existed complete apathy on geopolitics and economic ties as Indian foreign policy was confined to advancing its strategic interests in the US–Soviet competition and maintaining its security interests in South Asian geopolitics.

But in the post-Cold War era and especially over the past decade and a half, the growing economies of Africa and its energy resources began to attract the attention of India. As a result, despite past negligence, India and Africa have begun to move towards a multifaceted development partnership based on equality, friendship, mutual benefit and solidarity which reflects South–South cooperation. The India–Africa relationship is today moving towards a comprehensive partnership visible in a range of issues including human resource development through scholarships, training, capacity building; financial assistance through grants and concessional credit to implement various public-interest projects including for education, healthcare and infrastructure; clean modern energy sources and climate change adaptation and mitigation; suitable employment opportunities through development of all sectors of the economy including agriculture, manufacturing and services, value addition and connectivity, blue ocean economy, and disaster management and disaster risk reduction (Third India–Africa Forum Summit 2015).

The official level interactions and growing connection between business communities, and the summit-level interaction between Indian and African leaders at four India–Africa Forum Summits (IAFS)—beginning in 2008, the second in 2011 and third in 2015—have greatly facilitated India–Africa linkages.

However, it is African energy resources, mainly oil and gas reserves, and India's urgent search for new hydrocarbon sources that has prompted India to engage Africa comprehensively and proactively. Africa is richly endowed with fossil-based and renewable energy sources. Africa remains a key net energy exporter accounting for 8.9 per cent of global gas exports and 10.2 per cent of global oil exports (BP

'Regional Insight—Africa.'). Africa is the continent which accounts for 14.5 per cent of all the proven accessible oil reserves and 13.2 per cent of the accessible gas reserves. Africa has limited capacity in industries such as fertilizer and petrochemicals, where India has an advantage. But it is the energy sector where India comes as a good package for Africa (Prasad 2015).

Notwithstanding recent discoveries and the long history of oil and gas development in certain parts of Africa, the continent remains largely underexplored. According to the United States Geological Survey, nearly 300 billion barrels of oil equivalent of oil, gas and natural gas liquids are yet to be tapped. Recent discoveries highlight the potential of using modern technologies; for example, exploration success rates of over 75 per cent have been achieved onshore Kenya and offshore Mozambique (Dhir 2015). Africa's share of global oil production dropped again slightly in 2016 moving to 9.1 per cent of global output. Africa has proven natural gas reserves of 502 trillion cubic feet (Tcf) with 90 per cent of the continent's annual natural gas production of 6.5 Tcf coming from Nigeria, Libya, Algeria and Egypt. A total of 57 per cent of Africa's export earnings are derived from hydrocarbons (*An African Energy...* 2018). Asia is the leading importer of African oil and gas. China and India have emerged as the largest importers of African oil since 2013, together accounting for a third of Africa's crude oil exports in 2016. In contrast, the US imports of African crude have slumped since 2010, falling to 10.6 per cent of the total in 2016, reflecting the growth of US shale oil and gas production (OilVoice Press 2017).

Both India and Africa have convergence points: India needs hydrocarbons, mainly oil and gas, and Africa has the potential to fill this strategic need of India. As the world's third largest oil importer, India has been engaging the African region with energy security as the main goal. African nations are aware of India's status as the main stimulator of energy demand growth for the near future and ready to engage. This is also bolstered by the fact that China, the leading importer of oil from the region and the second largest oil consumer, has been witnessing slower economic growth. African oil and the gas market in North America have closed due to the oversupply of shale gas. These scenarios have pushed Africa to search for a long-term reliable market for its crude oil products to ensure the sustainability of its energy exports. The fall in oil price, growing supply surpluses and a noticeable economic downturn in China has made Africa a favourable destination to do business in the energy sector both for import of hydrocarbons and Indian investment in Africa's oil infrastructure. Eventually, bolstered by the fall in the crude oil price, favourable market conditions have encouraged India to be more ambitious in its overseas oil exploration mission. The Indian government has stepped up its overseas oil E&P mission and encouraged its oil firms to exploit the market condition to increase the supply of oil to India.

India's oil E&P activity in Africa has gained impetus in recent years. Indian Oil Corporation, the country's largest refiner, doubled imports from Nigeria at 60,000 bpd for 2016–2017, while Hindustan Petroleum Corporation had sought similar volumes from that African nation (Verma and Phartiyal 2017). Of India's oil imports, 20 per cent come from Nigeria, which is India's most valued source of oil as its crude is ideally matched for Indian refineries. In November 2005, India and Nigeria signed an agreement enabling India to purchase about 44 million barrels of crude oil per year on a long-term basis and to become a 51 per cent shareholder partner in the Port Harcot Veri and Kunduna refineries, producing 1.50 million barrels and 1.40 million bpd, respectively.

Sudan is another country with which India has long-standing ties and has invested $750 million. Sudan has proven reserves of 635 million barrels. China National Petroleum Corporation (CNPC), Petronas of Malaysia, Sudapest of Sudan and OVL of India are the largest shareholders in a major E&P venture in Sudan.[7] The project is divided into an upstream segment, covered under an E&P sharing agreement (EPSA), and a downstream segment, covered under a crude oil pipeline agreement (COPA). The consortium members' stakes are as follows: CNPC (40%), PETRONAS Carigali Overseas Sdn Bhd (30%), OVL (25%) and Sudan National Petroleum Corporation

---

[7] See ONGC Videsh Limited: http://www.ongcvidesh.com/

(SUDAPET; 5%).[8] India is currently chalking out a customized political–economic package for Angola. But here, there has been a challenge from the Chinese and in July 2006, India/ONGC was denied a 50 per cent equity partnership in production. India recently finalized a contract in Syria for the exploration, development and production of petroleum with a Syrian company. India has also been invited by Egypt to join a proposed pipeline project (being labelled as the Suez of oil) which would connect the Mediterranean and the Red Sea. Once built, it could benefit India by bringing Caspian oil and gas to India.

Today Indian companies' presence is visible in Africa's oil and gas sector. Indian companies have a substantial presence in Africa's energy assets. ONGC, BPCL and Oil India have a 30 per cent stake in Mozambique's Rovuma Area 1 with recoverable resources of 75 Tcf of gas. In oil-rich South Sudan, exploration is being led by the Greater Nile Petroleum Operating Company in which ONGC has a 25 per cent stake. State-run Engineers India Ltd is advising Dangote Group of Nigeria for setting up a 20 mt oil refinery, the largest consulting contract for an oil refinery in that country. Indian companies are also setting up gas compression stations and LNG terminals in Africa (Prasad 2015).

African nations are looking for the Indian investment to improve their infrastructure and oil refineries, increase production and help to develop the market for African refinery products. Equatorial Guinea has offered India equity in oil blocks. Algeria is interested in collaborating with India for exploration and developing petrochemical ventures and to increase oil exports to India. Sudan has offered three oil and gas blocks for exploration and development to OVL, the overseas arm of ONGC (Verma and Phartiyal 2017).

Africa is fast becoming a foreign policy focus area for India. After initial negligence, Modi foreign policy radar began to focus on the African continent. PM Narendra Modi envisioned an enhanced and multidimensional comprehensive India–Africa relationship at the Third IAFS which was held in New Delhi in October 2015. The

---

[8] ONGC Videsh Limited: http://www.ongcvidesh.com/op_sudan.asp.

Third IAFS, attended by around 30 African heads of governments and states, could be considered as the biggest foreign policy event in recent decades hosted by India. The event was not only big in size but also covered a comprehensive range of issues. The agendas ranged through development, trade, investment, security and, above all, energy. Both India and IAFS converge on the mentioned issues. But it is the energy sector where Indian companies both public and private have shown a growing presence in pursuit of hydrocarbon E&P in Africa.

Since 2015, India has stepped up its oil imports from Africa. The Modi government has given focus on tapping the hydrocarbon sources of Africa since it came to power and embarked on a massive developmental project for the country in which securing energy became more significant. India's oil shipments from the African continent saw an increase. This is reflected in the extension of $10 billion in credit to African nations by PM Narendra Modi who called for a broad alliance for the restructuring of the oil sector at the international level. This was echoed by India's Petroleum and Natural Gas Minister Dharmendra Pradhan at the India–Africa Hydrocarbons Conference, attended by ministers and officials of 22 African nations, 'We want Indian oil companies to take advantage of the credit line extended for five years and strike deals…we should take advantage of sliding oil prices and take an active role in the development of African nations' (Verma and Phartiyal 2017). As a part of the US$10 billion concessional LOC announced during the Third IAFS 2015 by PM Modi, both India and Africa agreed to work together and identify projects in the hydrocarbon sector which can be implemented under these LOC. To boost the relationship, PM Modi announced 50,000 scholarships for African students in the next 5 years including additional 250 fully funded scholarships for African nationals for technical and professional courses in the hydrocarbon sector in Indian institutes during the IAFS on 29 October 2015. This took the positives from the Second IAFS held in 2011 (Third India–Africa Forum Summit 2015).

With the aim of intensifying India's engagement with African nations in the hydrocarbon space, and then to take the vision of the Third IAFS further, India invited ministers and chiefs of state-owned oil and gas companies from Africa in January 2016 to explore new opportunities in hydrocarbon E&P, the export of oilfield services and the construction of fertilizer plants. Within a span of five months after the Third IAFS, New Delhi again hosted the 4th India–Africa Hydrocarbons Conference on 21–22 January 2016. A total of 21 African countries participated at various levels including ministers of petroleum, petroleum regulators, CEOs of national oil companies and experts in the field of petroleum. The delegations of nine African countries—Mauritius, Morocco, Algeria, Sudan, South Sudan, Tunisia, Senegal, Equatorial Guinea and Liberia—were headed by the respective ministers. Senior officials led the delegations from Nigeria, Ghana, South Africa, Egypt, Tanzania, Kenya, Mozambique, Uganda, Libya, Cote d'Ivoire, Gabon and Sierra Leone (Ministry of Petroleum & Natural Gas, Government of India 2016a). At this two-day 4th India–Africa Hydrocarbon Conference, India's hydrocarbon explorers ONGC, OIL, gas transporter Gail and refiners Indian Oil Corporation, BPCL and Hindustan Petroleum Corporation Ltd sought stakes in oil and gas fields, opportunities for setting up petro-chemical complexes and contracts for selling engineering, procurement and construction services to African customers (Prasad 2015). India is also expanding its energy security efforts in South Africa. In May 2018, the Indian Chamber of Commerce (ICC) hosted a pavilion of more than 45 suppliers of specialized technology and services for the energy sector at the three-day event held in Cape Town in South Africa. The ICC aims to connect with over 7,000 industry professionals across the full spectrum of the energy sector to address challenges and execute strategies (IOL 2018).

These initiatives are indicators of the current government's priority focus on energy security in India's foreign relations with the energy-rich regions. But the challenge remains for India to continue the policy initiatives in these newly targeted sources of energy. The lack of institutional capacity has impeded the maintenance of the policy initiatives and implementation. The success of any summit-level decision between the heads of the governments depends on the official and institutional capacity. In Africa, over two dozen Indian embassies were conspicuous by the absence of ambassadors/high commissioners. This is of concern as this was the case not long ago but as recent as 2015.

Africa's downstream sector is going through an indeterminate phase. The market dynamics and low production of hydrocarbons are constraining Africa's inclusive progress. India is eyeing to tap the energy potential in Africa. With several mature oilfields across Africa facing a decline in production volumes due to the lack of investment, India's hydrocarbon companies' unique set of skills, superior oil recovery techniques and expertise, and investment can be mutually beneficial. India is now acknowledged as an engineering and technology hub for the international oil and gas sector. Indian companies offer an exceptional and wide-ranging value proposition to exploit the full potential of Africa's hydrocarbon reserves. India, with the fourth largest oil-refining capacity in the world, complements Africa's requirements of investment, expertise and export markets with its oil and natural gas needs (Prasad 2015). Companies with experience in India are well-versed in working in multicultural and multi-ethnic environments, not dissimilar to those in many parts of Africa with a rich mix of ethnic, linguistic, religious and tribal diversity. Success in India requires the ability to deliver low-cost projects, often leveraging local labour and building supply chains to maximize local content. India has had a very successful record in developing strong local companies, creating local employment and ensuring a broad-based economic development. A lot of this has been organic and evolutionary and linked to capacity-building programmes. This experience of companies in India will be relevant, as host nations in Africa look to help their people participate in the economic success of resource development (Dhir 2015). Also, India provides a big market base for African hydrocarbon export as India is the fastest growing large consumer of oil and gas outside Africa. India's investment can lift Africa's upstream development and provide much-needed money for the energy sector development. In recent years, the sophistication level of Indian banking and financial institutions has been ranked high and they have expertise in oil and gas investment. This can further play a significant role in African energy projects (Dhir 2015).

Over the past few years, India's collaboration with African countries in the petroleum sector has increased. India imports around 16 per cent of its crude oil from Africa, in which Nigeria and Angola have the major share. In addition, Indian companies have interests and increased their footprint in oil and gas fields across Africa, including Sudan, South Sudan, Mozambique, Gabon and other countries (Ministry of Petroleum & Natural Gas, Government of India 2016a). Africa offers good prospects for Indian companies to exploit its untapped resources. Indian businesses have begun to take advantage of their unique skill sets to support delivery of excellent technology and expand affordable hydrocarbon projects in Africa. The extra availability of oil and gas in world markets and the decline in crude oil price have put Indian companies in an advantageous position. The Government of India has also encouraged its state-run oil companies to procure more hydrocarbon assets, and build the downstream sector for the refining of crude petroleum oil and natural gas in Africa for India's energy needs.

India looks to complement Africa in which it has expertise and benefit the people of India and Africa as expressed in the valedictory address at the 4th India–Africa Hydrocarbons Conference by Sushma Swaraj, the Minister of External Affairs, who appealed for working together to ensure energy security; she alluded to the centuries-old relationship between India and the African continent, and appealed for bringing new vigour in India–Africa relations by pursuing win-win opportunities in all areas for the common good of both regions (Ministry of Petroleum & Natural Gas, Government of India 2016a). Overall, the prospects for India in Africa are mainly in the energy sector. Africa fits well in India's long-term strategy for its energy security objectives. Indian companies have been supported by their government and have good equity and are welcomed by investors and by African nations. The 4th India–Africa Hydrocarbon Conference was a further significant step for a renewed engagement with African nations in the hydrocarbon sphere as India considers Africa a valued partner for the comprehensive development of their citizens in which Africa has become an important cog in the wheel in India's quest for energy security.

India's Hydrocarbon Exploration in the Caspian Basin

The Caspian basin has emerged as an important source for energy-hungry countries. The history of oil in the Caspian Sea goes back to the 19th century when the first offshore oil wells and machine-drilled wells were made in Bibi-Heybat Bay, near Baku, Azerbaijan. In 1873, exploration and development of oil began in some of the largest fields known to exist in the world at that time on the Absheron Peninsula near the villages of Balakhanli, Sabunchi, Ramana and Bibi Heybat. Total recoverable reserves were more than 500 mt. By 1900, Baku had more than 3,000 oil wells, 2,000 of which were producing at industrial levels, and Baku became known as the 'black gold capital', and many skilled workers and specialists migrated to the city. At that time the Caspian oil production supplied around half the world's oil needs and the US accounted for the other half. The Caspian Sea is a 700-mile-long body of water in Central Asia bordered by Azerbaijan, Iran, Kazakhstan, Russia and Turkmenistan. After the breaking of the Soviet Union, Azerbaijan, Kazakhstan, and Turkmenistan emerged as independent nations in 1991. Today, the region contains some of the world's largest oil and gas fields. The Kashagan oilfield, discovered in 2000 in Kazakhstan, was then hailed as the largest oil discovery in 50 years; it is currently the largest offshore oilfield outside the Middle East. The Caspian Sea region's oil and gas potential came into prominence in the early 1990s after the collapse of the Soviet Union, when the region was freed from communist ideological baggage and began to attract international oil and gas industry as the road for capital inflow in the energy E&P was now open. In the wake of the Soviet Union disintegration, the US hoped the Caspian region would become an alternative to the Middle East as the nations comprising the Caspian, except Iran, were not part of the OPEC, and were termed as the 'New Middle East' (Effimoff 2000; Gelb 2015; Nakhle 2017).

The Caspian basin, the world's 11th largest drainage basin occupying an area of over 3.5 million sq km is home to proven or probable reserves estimated at 48 billion barrels of oil and 8.7 trillion cubic metres of gas. The Caspian basin's natural gas sector could enable the region to increase global production by 27 per cent over the next 10 years (G. Sharma 2018; Stelletti 2017). In recent years, this region has become an attractive energy source for China and India enhanced by its geographical propinquity to these two countries. This enhances the importance of the gas-rich Caspian states of Turkmenistan and Kazakhstan as the major supply sources for the world's second and third largest energy consuming nations but has also put them and the wider region in these two nations' foreign policy radar as energy sources.

Notwithstanding the historic ties among the Central Asian states, China and India, the shipment of energy resources towards the east has yet to occur. While there has been considerable Western investment in the development and transportation of Caspian oil, large-scale investments in Caspian gas projects have not yet become economically viable. High infrastructure costs are also an obstacle for both sides of the supply–demand chain. For India, the options to import gas from the Caspian basin, particularly from fields in Turkmenistan and Iran, have been in discussion since the 1990s. Although pipelines would have to traverse a distance of at least 1,500 km from Turkmenistan and Iran to India, existing networks of pipelines within Iran can be extended to partially offset costs.

Despite the long history of the Caspian Sea region as a huge potential oil and natural gas producer, the region has not lived up to its expectation as the New Middle East and achieved its full potential. Factors such as disputes between the coastal states on the competing claims over the oil and gas reserves, regulatory indecisiveness, falling oil and gas prices, and cost overruns have prevented the Caspian from achieving its full potential. In addition, the lack of energy supply infrastructure to the distant consuming destinations, transportation costs, extreme weather conditions leading to winter freezes, regulatory doubts and growing security threats have made the Caspian basin a costly hydrocarbon region requiring heavy investment to cover escalating production costs. On top of that, sanctions against Russia and Iran have made things more difficult, especially in the context of intense interdependence between the various coastal states in the Caspian basis.

These conditions in the Caspian basin require export destination investment in this energy sector. India has been eying this region for its energy needs. India is not the only market but also is seen as the major investor in the Caspian energy sector. The Caspian basin has become one of the most attractive oil domains outside the Persian Gulf,[9] where India has taken its external engagement to get a foothold in the region.

Not only oil and gas companies but also big energy consuming nations have been showing interests in the region since it was pitched as the New Middle East. The United States was one of the first countries to enter the region with companies such as Chevron and ExxonMobil. The EU has been desperately seeking to diversify its energy supplies in order to reduce its dependence on Russian gas. As a result, the Caspian Sea area has become not only a hub of constantly developing gas E&P activities but also the foreign policy priority for these energy-hungry nations—the US, China, India, EU and Japan.

To support energy security interests in Central Asia, India stationed troops in Tajikistan, provided a $40 million aid package and undertook to refurbish an air base near the Tajik capital Dushanbe. India is also pursuing relations with Kazakhstan, Azerbaijan, Turkmenistan and Iran.[10]

India's Energy Exploration in Russia and its Geopolitics

Russia's re-emergence as a great power is attributed to its energy boom. In recent years, Russia's foreign policy strategy has been driven by its energy policy. Gas exports, investment in energy, the construction of gas pipelines and establishing the monopoly in its gas market, especially with the EU, have been dominant features of Russia's energy-driven foreign policy over the past two decades (Oxenstierna and Tynkkynen 2014, 597). Russia almost holds a monopoly over gas and oil, which in turn have enabled it to use these resources for its geopolitics interests in the EU and its neighbouring regions. But in recent years, the growing discord between Russia and its Western

---

[9] For detailed study, see Caspian Studies: http://www.caspianstudies.com
[10] Energy Security: http://www.iags.org/n0121043.htm

neighbouring European Nations, the US–Russia tension on spying issues, Iran trade sanctions and Trump's Trade War have further raised concerns in Russia over its energy security supply (Oxenstierna and Tynkkynen 2014, 43). Russia's economic resurgence in recent years has a lot to do with its booming energy exports that are heavily dependent on the EU market; it exports 78 per cent of its crude oil and 70 per cent of its gas to the EU (Oxenstierna and Tynkkynen 2014). The EU's dependency on Russia for its energy security is further enhanced because of its inability to import oil and gas from the US and other Western allies. This dependency and Russia's ability to manipulate the gas prices have given it a strategic advantage over the EU. However, this Russian strategic advantage over the EU energy supply is now being challenged as the member-states have questioned the unjust pricing of gas (Oxenstierna and Tynkkynen 2014). All this has pushed Russia to look towards its eastern and southern borders for the energy market to avoid the decline in its energy export. Russia has been eyeing China and India, the two major energy consumers in the world for its hydrocarbons export.

India and Russia share seven-decade-long diplomatic ties. During the Cold War period, both the nations shared a trusted relationship which was reflected in the strong strategic and defence partnership leading to India's armoury full of Russian defence products. This was also reflected in Russia being India's largest trading partner at one point. After the disintegration of the Soviet Union, Russia began to lose its strategic and economic importance in India's foreign policy in the 1990s. However, India and Russia have prevailed over the changing international strategic dynamics and other logistics issues in their bilateral trade. The trade relations have begun to move in a positive direction, though not comparable to India's bilateral trade with the US. The India–Russia trade volume was just over $7 billion in 2016–2017 compared to India–US trade volume of over $64 billion (Gokaran 2018). The India–Russia relationship continues to be strong and one based on mutual trust and interests even today when India's ties with the US have become strong and are moving towards a comprehensive strategic partnership in the post-Cold War. The trade between the two countries has been dominated by Russian defence sales to India, but in recent years, both the nations have been working to expand their overall partnership in other areas as well. The energy sector is emerging as another strong component of India's relationship with Russia.

The India–Russia relationship is complementary in the energy sector. The Indian Ministry of Petroleum and Natural Gas' 'Hydrocarbon Vision 2025' predicts an increase in the share of natural gas in India's future energy supply to 20 per cent by 2024–2025, up from 14 per cent in 2010–2011. Given the less pollutant character of gas among all the hydrocarbon sources, India is aiming to shift towards a gas-based economy by increasing to 15 per cent from the current 6.2 per cent in its energy mix. However, India is dependent on imports to fulfil 45 per cent of its gas demands which makes it the fourth largest importer of gas in the world. Currently, Russia tops the chart of crude oil producing countries and ranks second in the list of gas producer countries in the world. Russia has one of the world's largest reserves of natural gas and has been witnessing a boom in its production of gas that increased to 690.5 billion cubic metres (bcm) in 2018, a growth of 7.9 from the year 2017.

In recent years, energy has been the focus of India–Russia summit-level meetings, joint statements and various agreements. India and Russia have a bilateral dialogue on energy cooperation, concentrating on ways to enhance energy security, opportunities for improving the process of diversification of energy supplies and for strengthening commercial energy partnerships in already identified and prospective oil and gas projects in India, Russia and in third countries. Major joint projects include the E&P of hydrocarbons in the Bay of Bengal offshore areas and the Sakhalin-1 oil and gas field project (MEA, Government of India 2005). Russia has emerged as a key gas import source for India and has been looking for a gas export destination to overcome the fluctuation of security of supply because of the issues on the EU front. This also helps both these nations to fulfil their commitment to the Paris Climate Agreement.

Pipeline transportation between the two countries has been debated, but owing to logistical challenges, swap deals and LNG have emerged as alternatives that are more promising. Possible partners for swap deals include Japan and South Korea. Progress has been already made in this direction. In October 2017, the Indian government entered into a bilateral LNG swap deal with Japan. This is a cargo-swapping arrangement for gas exchange to lessen India's logistics and transportation costs significantly for natural gas imports. India has pacts for the supply of gas with Australia, and Japan, the largest gas importer, has a similar deal with Qatar. Therefore, under the gas exchange arrangement, India has the provision that will allow the swapping of gas under which India can get gas from Qatar instead of Japan and India's quantum from Australia be transported to Japan (*Hindu* 11 October 2017). To avoid the hurdles and delay in the gas pipeline, similar gas swap agreement is possible between Russia, India and Japan and its suppliers in West Asia. Sakhalin Island is located very close to Japanese territory and the island already hosts a number of Japanese energy companies. Most of the Sakhalin oil and gas constitute long-term contracts but the opening of the Yamal LNG plant, operated by Russia's largest independent natural gas producer Novatek, and the increasing viability of the Northern Sea Route due to climate change could serve the swap purpose (Gokaran 2018).

Over the past two decades, Indian energy companies have made their presence in Russia's oil and gas sector. In 2001, OVL, the overseas arm of ONGC, purchased a 20 per cent stake in the Sakhalin-I oil and gas field for $1.7 billion. In November 2003, the Indian government approved an additional investment of $1.1 billion by OVL in the Sakhalin-I project (*Hindu* 2005). Again, in 2009, OVL acquired the Imperial Energy Corporation and in 2016, it acquired a 26 per cent stake in CJSC Vankorneft, a Russian oil company operating in Eastern Siberia, for over $2.2 billion. A consortium of OIL, Indian Oil Corporation Limited (IOCL) and BPRL acquired another 23.9 per cent of Vankorneft for $2.02 billion, along with a 29.9 per cent stake in Taas-Yuryakh Neftogazdobycha LLC for $1.2 billion. As Russia continues to explore hydrocarbon deposits in the Arctic, Indian companies such as OVL that have experience working in the region have bright prospects as potential partners. Russia's Rosneft has also entered the Indian energy market. In 2016, it acquired a 49.13 per cent stake in Essar Oil Limited for $12.9 billion, with another 49.13

per cent split between Russian and Dutch partners, making it the largest FDI agreement in India to date. It acquired Essar Oil's Vadinar refinery under the agreement, giving it a foothold in India's domestic oil market as well (Gokaran 2018). Over the past few years, Indian oil and gas companies have invested more than $10 billion in Russian projects, whereas Russian oil and gas venture-led consortium have invested a total of around $13 billion in Indian projects.

Dharmendra Pradhan, India's Minister of Petroleum and Natural Gas, rightly sees Russia as a long-term source for India's hydrocarbon imports. India has been importing gas from Gazprom, Russia's state-owned gas company, receiving 1.7 MMT of LNG between 2009 and 2016 from Gazprom. More recently, a 20-year deal between the GAIL and Gazprom has been made. The first Russian shipment of LNG from Gazprom was delivered to India on 6 June 2018 (*Times of India* 4 June 2018). By the deal, Gazprom will supply LNG to India gradually ramping up to 2.5 MMT annually in a few years. The import of Russian LNG has added a new dimension by bringing energy security to the forefront in India–Russia relations that have been mainly dominated by defence trade for decades.

India's GAIL has been struggling to find domestic buyers, hence the staggered deliveries. To put this into perspective, imports from Qatar alone are 8.5 MMT annually. India's total annual imports amount to 18.67 bcm of gas—the equivalent of 13.81 MMT of LNG. In the latest developments, GAIL has renegotiated price and volume with Russian Gazprom on the terms of a 20-year deal to import 2.5 mt a year of LNG. Accordingly, the contracted volume has been lowered from 2.5 mt to 0.5 MT in the first year 2018–2019; increased to 0.75 MT in 2019–2020; and 1.5 MT in the third year 2020–2021. GAIL is to import the full 2.5 MT a year by the fourth year to cover the initial volume reduction over the remaining length of the contract. Under the new terms and conditions, the contract has been extended by three years and GAIL has committed to buying an additional 6 mt of LNG. India's growing position as one of the world's largest energy consuming nations has enabled it to exert influence in striking better price deals for Indian companies. For example, India was able to bargain with Exxon Mobil Corporation to lower the price for 1.5

MT a year of LNG from the Gorgon project in Australia. This helped India save more than 40,000 crores of revenue in imports (*Times of India* 4 June 2018).

Coal has been another component in India–Russia energy relations. Despite India's energy security policy of shifting from coal to renewables, the import of Russian coal will continue to be significant. Russia too has been keen to promote coal mining. In December 2017, when a Russian subsidiary of the Tata Power Company Limited (TPCL), Far Eastern Natural Resources LLC, was given a 25-year coal mining licence for the Krutogorovsky coal deposit for $4.7 million, it was offered support in terms of infrastructure, laws and taxes (Gokaran 2018).

However, India's growing energy ties with Russia are not free from geopolitics. As India looks for new hydrocarbon sources, its energy security efforts abroad is not free from the complex geopolitics among energy exporting nations. This Russian LNG shipment came to India within the weeks of India importing its first ever LNG cargo from the US under a long-term import agreement. India and the US over the past two decades have fixed their relationship and overcome the Cold War ideological logjams. The India–US relationship has come out of the US–Russia strategic competition concerns. However, Russia's deteriorating relations with the US and the West amidst its annexation of Crimea and the Ukraine conflict, and the trade war under the Trump administration have revived the US–Russia strategic rivalry. This is affecting India's effort to deepen its ties with Russia on the energy front. The Trump administration has been asking its allies and partners to cut trade with Russia and Iran substantially. India's growing energy ties with Russia and its reliance on Iran for gas imports has come under the scanner of the Trump administration. This has surfaced as an issue in India's relationship with the US and Russia, especially in the context of Trump's trade war.

During the 19th India–Russia annual summit-level meeting, PM Modi and President Putin signed eight pacts in a range of sectors including one requiring Russia to build six nuclear reactors at a new site in India and a $5 billion deal for Russian S-400 Triumf air defence system with India (*Times of India* 5 October 2018a). Speculations have been made about possible sanctions on India because of its energy and defence ties with Russia and its energy ties with Iran under Trump's trade war. But prior to this summit meeting, the issue of Russia and Iran hardly had any significant impact on the first 2+2 India–US dialogue between the US Secretary of State Mike Pompeo and then US Secretary of Defense Jim Mattis with their Indian counterpart Sushma Swaraj and Nirmala Sitaraman. The issue of impending S400 Triumf deal and the purchase of gas from Russia and Iran figured as a marginal issue. President Trump imposed sanctions on China's purchase of Russia's S 400 Triumf under the Countering America's Adversaries Through Sanctions Act (CAASTA). The greater strategic convergence between India and the US amidst the assertive military posture of China in the Indio-Pacific region, the security concerns of thwarting Islamic terrorism and the growing economic synergies rendered India's trade with Iran and Russia to the periphery at 2+2 meeting. Though buying fuels from Iran and Russia and defence purchase from Russia will keep recurring in India–US dialogue, these issues will not derail either India's energy collaboration with the two Caspian basin countries or the deepening India-US strategic partnership.

India's Energy Security and Gas Pipelines Geopolitics

The gas-based economy has been focused as an important step towards the affordability and environmentally acceptable move of the Indian government's energy security policy. Some of the solid measures taken by the Modi government to increase the share of gas in its energy basket are investments for strengthening the gas infrastructure by focusing on the building of LNG import terminals, establishing a solid City Gas Distribution networks and laying the gas pipelines.

Natural gas is considered the transition green fuel for the transportation, cooking, heating and modern life. Natural gas has lower carbon emission and is environment-friendly compared to coal and oil. Apart from the above purposes, the use of natural gas for the mobility sector addresses many concerns including the environmental concerns faced by urban cities in India. The use of natural gas releases considerably lower GHG emissions per unit of energy and the lower levels of pollutants do less harm to health. In the coming years, given the pace of urbanization in India and the plan of smart cities by the present government, gas is going to be significant in providing energy to urban life amidst global warming concerns. This goes well with India's pursuit of energy security in the context of environmental acceptability concerns.

In the present scenario, India imports gas only through LNG carriers. It is believed that transporting natural gas through pipelines is more cost-effective than LNG carriers are. For example, in 2013, China received pipeline gas imports at an average price of US$9.78 per MMBtu compared to the average price of LNG import price of US$13.8 per MMBtu. LNG is costlier because the gas must be liquefied to reduce its volume and transported using specially designed cryogenic tanks. Also, at the receiving end, specialized LNG terminals have to be built to store and re-gasify. Essentially the countries which import natural gas through pipelines enjoy a cost advantage (Kumar 2017).

Over the past decade and a half, India has been taking steps in the direction of laying a gas pipeline from its northwestern region to the Caspian basin. As a result, a number of gas pipeline projects have been in progress that technically and economically are viable but delayed because of cost, lack of technology or unstable political conditions. In fact, India's connectivity to the gas pipeline to its northwestern front including India–Caspian gas pipeline projects is full of geopolitical challenges. Most of the challenges come from security issues related to Afghanistan and Pakistan. Afghanistan continues to be a security challenge and in the Post-US Af-Pak exit, the situation remains the same. Any gas pipeline passing through Afghanistan poses a serious security issue given its chronic instability and ongoing civil war, which have resulted in the almost complete destruction of the country's infrastructure. Despite many talks between India and Pakistan and the third country that has the gas resources, the proposed pipelines traversing through Pakistan have been questioned in India because of security concerns and given the India–Pakistan nuclear arms race, the long-term conflict over the Kashmir and competition to enhance their presence in Afghanistan.

On this, Stephen Cohen, the leading expert on South Asia geopolitics, views the enduring conflict between India and Pakistan as an impediment in leveraging the economic potential that could result from the cooperation on gas pipeline projects. Economic losses are acceptable to both the nations for short-term political and strategic gains in their political and military stand-off (Hill and Spector 2001).

India–Russia Gas Pipeline

Both India and Russia have agreed to explore building the world's most expensive pipeline costing close to US$25 billion to ferry natural gas from Siberia to India. The proposed pipeline will connect the Russian gas grid to India through a 4,500–6,000 km pipeline.

The shortest route will entail bringing the pipeline through the Himalayas into Northern India, a route which poses several technical challenges. Alternatively, it has been suggested that the pipeline can come via Central Asian nations, Iran and Pakistan into western India. But this pipeline route will be costlier than the long-discussed but shorter and cheaper Iran–Pakistan–India pipeline. The third option suggested is to build a pipeline through China and Myanmar into Northeast India bypassing Bangladesh (*Economic Times* 16 October 2016).

The MoU was signed between India and Russia in the presence of PM Modi and President Putin at the India–Russia Annual Summit on the sidelines of the 8th BRICS Summit. The MoU also envisages roping in OVL, gas utility, GAIL India Ltd and Petronet LNG Ltd for a feasibility study of the India–Russia gas pipeline project. According to a preliminary cost estimate, the Russia–India pipeline with the longest route of 6,000 km may cost close to US$25 billion. The cost of transporting gas via the long-discussed IPI pipeline is less than US$1 per mmBtu and for the TAPI pipeline it is around US$2 per mmBtu. A pragmatic transportation cost of gas through the proposed India–Russia gas pipeline is estimated to be US$4 per mmBtu. This excludes the transit fee to be paid to nations through which the pipeline will pass. Accordingly, the natural gas produced in East Siberian fields is to be pumped into the Russian gas grid which would be connected to India through the cross-country pipeline network (Hill and Spector 2001). This MoU on India–Russia gas pipeline will augment India's energy security measures through an enhanced engagement with the one of the world's richest oil and gas producing nations.

TAPI Gas Pipeline

The TAPI gas pipeline has been in discussion for a long time. It is almost more than a decade since India joined the TAPI gas pipeline. But the origin of TAPI goes back to the mid-1990s when the Turkmenistan–Afghanistan–Pakistan (TAP) project was started. But its progress was stalled because of the intense civil war in Afghanistan. However, the arrival of the Karzai government renewed the prospects of the project with the backing of the Asian Development Bank (ADB) as a leading project. India joined the TAPI formally in 2006 (Ahmad 2006). However, India's security concerns in regards to the

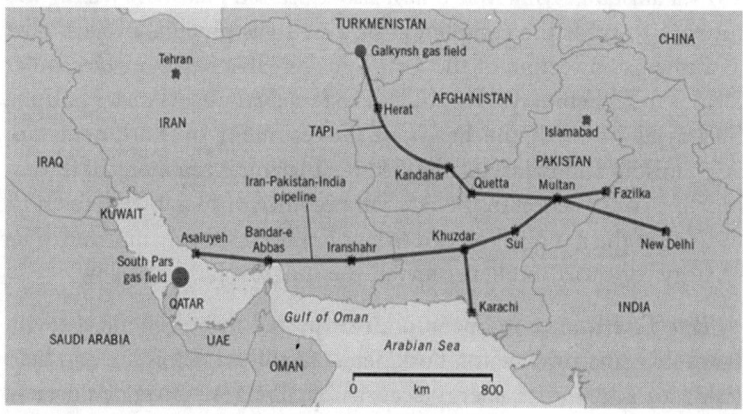

Figure 4.3 *TAPI Gas Pipeline Routes*
Source: Manish Vaid (2016)

TAPI pipeline passing through Pakistan (Figure 4.3) and deteriorating India–Pakistan relations in the wake of the Mumbai terror attack slowed down the project. Despite NATO forces' presence in the terror-affected region in Afghanistan to protect the part of the pipeline passing through terror-prone territories, security concerns because of possible terror attacks damaging the gas pipeline continued to linger. Another security concern for India was the sabotage and blackmail over the gas pipeline by Pakistan in the case of possible India–Pakistan conflict or war.

The starting point for the proposed 48-inch diameter and 1635 km long pipeline is from Daulatabad gas field in Turkmenistan which would then traverse through Herat, Kandahar, Quetta and Multan before reaching India. The TAPI once realized will supply 0.6 bcf of gas to Pakistan and 1.6 bcf to India, with an estimated cost of gas about \$2.4–\$3 per Middle Market Banking Unit (MMBU). India would get 1341.78 million cubic feet gas per day. The pipeline project could help India's move towards the gas-based economy and achieve its affordable and environmentally acceptable energy security goal. Supported by the ADB, the 1600-km US\$10 billion (AUD\$13 billion) pipeline is the largest multilateral energy project to date undertaken in South Asia.

Recently, India's interest in the pipeline began to resurface. In 2014, Pakistani PM Nawaz Sharif also expressed Pakistan's willingness to work towards the completion of TAPI gas pipeline project. The Turkmenistan section of the TAPI gas pipeline began in December 2015. On 23 February 2018, India's External Affairs Minister Sushma Swaraj along with the heads of Government of Turkmenistan, Afghanistan and Pakistan attended the inaugural ceremony of the gas pipeline construction in the Afghan province of Herat. India's effort to finalize the gas pipeline agreement has been slow mainly due to its security concerns and hence so far, the progress has been slow.

But TAPI, once in operation, is expected to be beneficial to all. It would bring revenue for Turkmenistan and Afghanistan, and help Pakistan address its energy security challenges. Given Pakistan's energy security challenges, it is too serious about the success of TAPI. Pakistan has been facing the challenge of addressing its gas deficit.

Natural gas in Pakistan is consumed for domestic and commercial purposes, as well as for electricity generation. The domestic natural gas production in Pakistan is around 4 bcf per day (bcfd) but the demand is around 6 bcfd. On top of that, the gas production capacity is expected to fall to less than 1 bcfd by 2025 as gas reserves will be exhausted, whereas demand is expected to increase to 8 bcfd. In the recent election, energy security became an important issue. Thus, the success of TAPI is significant for Pakistan's energy security as well. A fully operational TAPI would bring Afghanistan transit revenue and natural gas and pave the way for further regional connectivity and help Afghanistan to attain its true potential as a regional trade and transit hub, connecting Central Asia with South Asia and the Middle East (Rahim 2018). In the post-9/11 world, India's foreign policy has given a significant focus to enhance ties with Afghanistan. Its diplomacy has paved the way for a sound India–Afghanistan strategic partnership which is reflected in a range of sectors including growing security ties. Any attempt by Pakistan to disrupt the TAPI pipeline going to India would affect Pakistan as well given India's close ties with Afghanistan.

Two Proposed Pipelines in South Asia

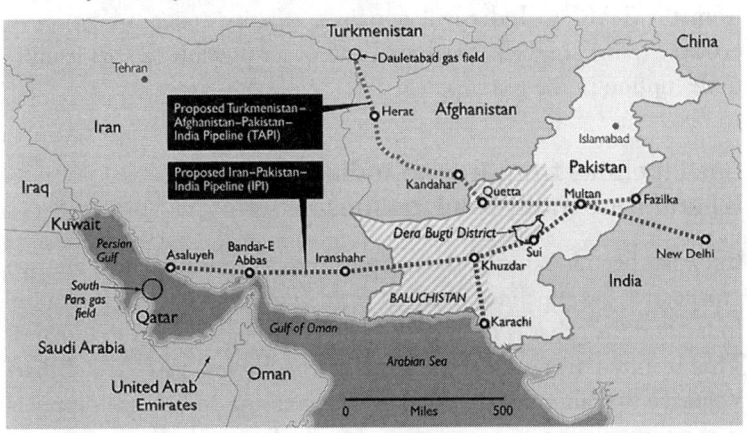

Figure 4.4 *TAPI and IPI Gas Pipelines*
Source: Curtis, Cohen and Graham (2008).

Complex geopolitics and conflicting interests of the nations have been major hurdles in achieving TAPI. Apart from TAPI nations, another geopolitical dimension is the Russian stake in the regional energy market. Russia may not be comfortable with Turkmenistan's gas being exported to South Asian countries given its own ambition of expanding its gas export market to the East. Turkmenistan gas exports to India could open a new market for Ashgabat and Russia would have to compete with Turkmenistan in the gas market in the East and again in India (Hill and Spector 2001). However, this has been of little concern so far and the diplomatic dialogues between the four nations have helped the TAPI project overcome the India–Pakistan and Pakistan–Afghanistan security concerns. Its success and speedy completion depends upon the speedy resolution of the land acquisition problem for the project in Afghanistan, Pakistan's response to the growing India–Afghanistan ties, Pakistan–Afghanistan relations, Taliban's influence in Afghanistan and, finally, India–Pakistan relations. It will be the first time a Central Asian gas pipeline will traverse to the subcontinent and if successful, it could be a game changer for India's efforts towards the diversification of foreign hydrocarbons sources. TAPI is finally progressing. However, although the gas pipeline is set to be completed in 2019, an uninterrupted gas supply from the TAPI to India will remain a security concern because of possible terrorist attack, blackmail and other possible factors leading to disruption in the gas supply.

Gas Pipelines from Iran to India: IPI and Iran–India Gas Pipelines

India has been looking at the gas pipelines from Iran. There are two prospective gas pipelines in mind from Iran. One is Iran–Pakistan–India, and the other is from Iran to India through the sea to Gujarat. This resulted from the state visit of the Indian PM Atal Bihari Vajpayee to Iran in 2001. The key to increasing India's gas supply is in the construction of one of these gas pipelines or both.

The transportation of gas from Iran to Pakistan and on to India has a sound commercial basis. Iranian gas reserves, estimated to be more than 940 Tcf, constitute 18 per cent of the world's gas reserves, second only to Russia. The reserves offshore in the Persian Gulf are geographically the nearest gas fields to the Indian subcontinent (Sourie 2005). In 1989, a project was conceived to build a pipeline from Asaluyeh to the Pakistan–India border. The distance involved would be about 1900 km long, well within the range of economical gas supply by pipeline vis-à-vis LNG. Pakistan is gas-dependent, with gas constituting 50 per cent of its energy mix; its fields are depleting, and from 2010, it is expected to need imported gas, with its demand increasing to about 400 million standard cubic metres (scm) a day by 2025. But the project fell victim to the then irreconcilable differences between India and Pakistan in the 1990s and the early part of this century (Diwanji 2000). The stalemate ended in January 2005 when India and Iran agreed to pursue the project as a straightforward purchase of Iranian gas at the Indian border, with a supplementary agreement between Iran and Pakistan covering the supply of gas to Pakistan and the transit of gas to India.

Several official, including ministerial-level, meetings among the three countries over the decade took place to address the issues related to the technical, commercial, financial and legal aspects of the gas pipeline project. To achieve a secure and high-end gas pipeline project, India, Iran and Pakistan have tried to solve the aforementioned issues linked to the gas pipeline in their quest for energy security. But the meetings have mainly focused on the pricing, structure, and security as serious matters of concern in the gas pipeline project. In the incipient years of the project, the leaders of the three countries involved repeatedly conveyed their full political support for the project and their strong interest in its successful outcome. Thus, the project was effectively removed from the domain of extraneous bilateral, regional and global political issues, and was being pursued exclusively based on commercial considerations (Muni and Pant 2005).

Since 2008, due to the fear of US sanctions and India's pending nuclear deal with the US under negotiation, the IPI was put aside. In 2015, the debate surrounding this resurfaced in the wake of possible agreement on Iran's nuclear issue with the US. After the US–Iran nuclear agreement and the consequent easing of the discord between the US and Iran, the Indian government began to work on materializing the gas pipeline projects from Iran. However, after the Trump administration severed ties with Iran and declared the nuclear deal invalid, the project again slowed. Russia has offered Gaxzpr to help build Iran–Pakistan–India gas pipes. Nevertheless, the latest sanction remains an issue. Another dimension is the India–Pakistan relations, and objections to any gas pipeline via Pakistan have been raised many times. Despite the peace efforts, the India–Pakistan relation continues to be an impasse. PM Modi has tried to reach out to Pakistan's new PM Imran Khan to resume dialogues at the official level. But the issues such as terrorism from Pakistan-based terrorist organizations and Kashmir continue to impede the progress on the proposed gas pipelines.

To avoid the security concerns evolving from Pakistan, the idea of a gas pipeline from Iran via sea lines to Gujarat has been proposed. The gas pipeline for India from its western border continues to be at risk because of its conflicting relations with Pakistan. Many energy watchers in India consider it a risk option. India is unlikely to enter into any arrangement that would involve dependency on Pakistan, as that might empower Pakistan to exert economic leverage to gain Indian concessions on Kashmir or other critical strategic issues. As a result, Iran and India have also discussed the possibilities of constructing a deepwater offshore gas pipeline to bypass Pakistan and of importing LNG from Iran's South Pars gas field by tanker. Keeping aside the Pakistan concern and the commercial viability of the Iranian gas pipelines to India, the US sanctions policy (the Iran–Libya Sanctions Act) constrains India's go ahead with Iran–India or IPI gas pipelines. So far, though, this has been a marginal issue in the greater convergence on strategic, security and economic issues between India and US.

Discussions with Iran are on for a deep-sea 868 miles pipeline via the Oman Sea and the Indian Ocean. The Iran–Oman–India pipeline from the Iranian port of Chabahar to India's Gujarat coast would transport 1098.141 million standard cubic feet of gas per day. This might compensate for the IPI project and there would be no issue of any other transit country conflict. India has also invested in the development of the Chabahar port and is funding a rail link between Chabahar and Zahedan in Iran. The completion of the rail link would connect Chabahar to the International North–South Transport Corridor (INSTC). These investments are consolidating the bilateral ties between India and Iran. This deep-sea pipeline will not only connect India to Iran's gas fields but Oman is also slated to join the pipeline at a later stage. This would give India a strong foothold to the gas trade in both Iran and Oman. This would also help India to counter China's One Belt One Road (OBOR) programme (Kumar 2017).

Recently, Russia signed an MoU with Pakistan on implementing a project to build an underwater gas pipeline from Iran to Pakistan and India. Russia is planning to sign the same contract with Iran and India. This project was put on hold in 2013 because of the US sanctions against Iran. The project was revived in November 2017 with Russia and Iran in the form of a memorandum that envisaged Russian support for gas supplies from Iran to India (*Business Standard* 2018).

India's relationship with Afghanistan has been durable and friendly. In the post-Taliban and post-9/11 era, India's foreign policy in Afghanistan was mainly to foil Pakistan and the Taliban's designs of once again rising to prominence in Afghanistan and gaining access to the energy-rich Central Asian region (Pant 2014). India has relied on soft and hard power diplomacy in Afghanistan to strengthen its position. India became the largest regional provider of aid to Afghanistan in developmental and reconstruction projects in Afghanistan. India has a comprehensive strategic partnership agreement with Afghanistan including trade, security and education. New Delhi's decade-long reconstruction effort of rebuilding Afghanistan and winning the hearts and minds of the Afghani people has not been welcomed by Pakistan. Pakistan has resisted India's growing presence in Afghanistan and the bonhomie between the two nations. This has affected the gas pipeline projects.

The geopolitics involved in India's gas pipeline is reflected in the strategic challenges that the Chabahar port project has faced. This project is expected to benefit both India and Afghanistan. The project faced difficulties when the construction of the 215 km Zaranj–Delaram road, which connects the Kabul–Herat highway to the road that leads to the Chabahar port on the Iranian border, got underway on Afghan soil. The Taliban strongly resisted the road construction and killed six Indians and 129 Afghans; approximately one human life for each 1.5 km of road (Rahim 2018). This Taliban attack needs to be seen in the larger context of geopolitics in the Af-Pak region. The Chabahar port is a gateway for India to reach Afghanistan and Central Asia. This also gives India a strategic advantage to counter the Gwadar port in Pakistan which is considered as a part of China's OBOR initiatives and seen by India as China–Pakistan all-weather friendship constraining India's strategic outreach in its North-Western region.

Myanmar–Bangladesh–India Gas Pipeline

Myanmar and Bangladesh have significant natural gas reserves and apart from India, there are other countries in the region, such as Sri Lanka and Thailand, which are major energy importers. Collaboration in the joint development of natural gas infrastructure facilities includ-ing gas field development and the establishment of natural gas pipeline networks and/or LNG facilities could result in the better economic use of energy resources, benefiting both producing and consum-ing countries. The absence of a local market and gas transportation infrastructure has prevented the optimal exploitation of India's own gas reserves in Tripura as well as the gas reserves in Myanmar, in which ONGC and GAIL have stakes. There has been a proposal for transporting gas from Myanmar to India via a pipeline through Bangladesh, a proposal from which all the parties stand to gain. In addition, Bangladesh could supplement the transit revenues from the Myanmar gas by using the same pipeline to supply its own gas to India. The pipeline with a length of around 7,000 km would link Sittwe in Myanmar's Arakan to Mizoram and Tripura in Northeast India and Chittagong in Bangladesh. The pipeline would extend to West Bengal on the Indian mainland and Assam and other northeastern states on the eastern side. The pipeline could be used by all the three countries.

Although energy ministers from India, Bangladesh and Myanmar met in January 2005 in Yangon to consider the project, there has been no progress since then as the project got stuck twice in the last decade mainly because of Bangladesh' insistence on including in the tripartite MoU references to certain India-related bilateral issues that do not directly pertain to the project (Varadarajan 2005). The Myanmar–Bangladesh–India transnational pipeline got stuck because of Bangladesh's refusal to act as a transit country for the proposed pipeline. An alternative to this route was to avoid Bangladesh and build a pipeline through Northeast India that could connect to pipelines of East India. This could not be achieved mainly because of financial constraints. India's inability to connect to Myanmar for a gas pipeline gave China a chance to exploit the gas reserves of Myanmar by laying a transnational gas pipeline to its Yunnan province (Kumar 2017).

However, the gas pipeline proposed is now being revived. The Indian government has taken fresh initiatives on this front. First, in 2015, the three-nation pipeline project resurfaced during the summit-level meeting between PM Narendra Modi and Bangladesh PM Sheikh Hasina. Then in 2017, the negotiations began under the Hydrocarbon Vision 2030 for the northeastern region which planned to connect Chittagong (in Bangladesh), Sittwe (in Myanmar) with the northeastern states. India and Bangladesh also have planned a joint LPG plant at Chittagong from where the gas will be piped to the northeastern region (R. Sharma 2017). This gas pipeline will give India an alternative gas importing source in the case of disturbances on its northwestern border. Unlike the TAPI and the IPI gas pipelines, the Myanmar–Bangladesh–India gas pipeline has no Pakistan and Taliban security threats.

Qatar–India Gas Pipeline

Qatar, having the third largest gas reserves in the world, is already an important export partner with India following the agreement signed in 1999, which provides for the delivery of 7.5 mt of LNG annually for 25 years. First shipments arrived in India in 2004.[11] A deep-sea

[11] See http://www.oilworks.com/New/i042505.html; http://www.indianembassy.gov.qa/indqat.htm.

gas pipeline from Qatar to India through Oman has also been proposed. This is going to be 2,000 km sea-deep from Qatar via Oman to Indian states of Gujarat or Maharashtra. In 2008, Qatar invested US$5 billion in India's energy sector, and in August 2009, GAIL and South Asia Gas Enterprise Pvt. Ltd signed an agreement. In his visit to Qatar in 2016, PM Modi solidified India–Qatar relations and urged the businessmen to invest in India's growing economy in a range of sectors including the energy sector.

In March 2018, Al-Kaabi, during his meeting with India's Minister of Petroleum and Natural Gas, Dharmendra Pradhan in New Delhi, reiterated Qatar's commitments to meet India's energy demands and enhance bilateral cooperation, particularly in the LNG sector. India is one of Qatar's biggest markets for LNG. Qatargas supplies multiple customers in India and is its single largest LNG supplier. In recent years, India imported more than 10 mt annually from Qatar. In 1999, RasGas (now Qatargas) and Petronet LNG Limited signed a long-term sale and purchase agreement (SPA) for 7.5 mt per year of LNG.

This was followed in December 2015 with a new SPA for the supply of an additional 1 million tpy, raising the total annual long-term commitment between the two companies to 8.5 million tpy (*Gulf Times* 2018).

Russia–China–India Gas Pipeline

OVL has even considered an ambitious pipeline route from Russia through Turkmenistan–Uzbekistan–Kazakhstan to Kashi in Western China and then along the military ceasefire line with China on the Siachen glacier in Kashmir, entering India through Ladakh in Kashmir or Himachal Pradesh and then running further down to Delhi.

India's Energy Security with Neighbouring Nations

India–Bhutan Hydropower Collaboration

India's pursuit of energy security in a carbon-controlled environment is pushing to explore the potential for renewables. As discussed in

Chapter 2, India can increase the share of hydropower in its energy mix. India has sought to explore the potential of Bhutan's hydropower capacity. India shares more than 50 years of diplomatic ties with Bhutan and is the leading developmental aid partner for the Himalayan kingdom. India's relationship with Bhutan is pursued under a range of institutional mechanisms in sectors such as trade, economy and developmental cooperation; border management and security; and water resources and hydroelectricity. Energy security is a major focus area.

India–Bhutan collaboration in the hydropower sector is mutually beneficial. India has completed three hydroelectric projects in Bhutan with a total capacity of 1,416 MW, which is operational. About three-fourth of the power generated is exported to India and the rest is used for Bhutan's domestic consumption. In 2016, the bilateral trade stood at ₹8,723 crore, with imports pegged at ₹5,528.5 crore (82% of Bhutan's total imports) and exports at ₹3,205.2 crore, including electricity (90% of Bhutan's total exports; *Times of India* 7 July 2018).

In July 2018, PM Narendra Modi and his Bhutanese counterpart Tshering Tobgay reaffirmed both nations' commitment for cooperation in the hydropower sector.

> The two Prime Ministers reviewed the bilateral economic and hydro-power co-operation, including the progress in the implementation of the on-going GoI-assisted-hydro-electric projects in Bhutan (MEA, The Government of India 6 July 2018).

Bhutan's geographical location is significant for India. Figure 4.5 shows the location of Doklam, the disputed region between China and Bhutan (*India Today* 2018).

Bhutan's geographical location is strategically important for India. Given China's effort to constrain India in South Asia, Bhutan's strategic location is crucial. The stand-off between India and China in Doklam is an example.

> Doklam's geographical position makes it a strategically important area as it is located between Tibet's Chumbi valley to the North, Bhutan's Ha valley to the East and India's Sikkim state to the West. The Doklam area carries a huge military advantage and if

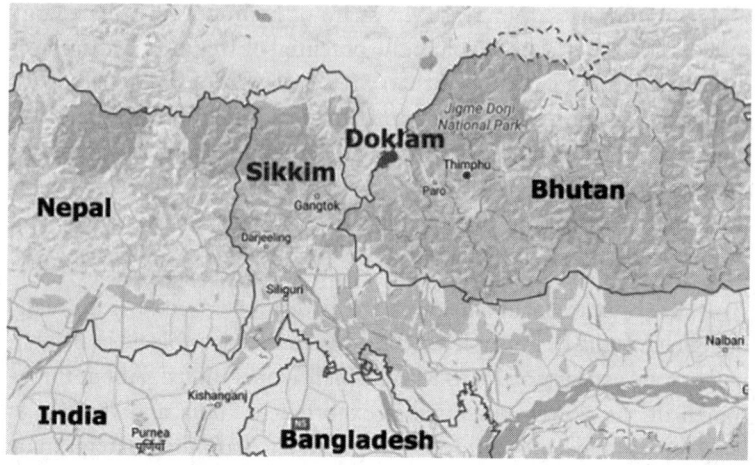

Figure 4.5 *Strategic Location of Doklam: the disputed region between China and Bhutan*

Source: Where is Doklam and why it is important for India?" India Today, 27 March 2018, https://www.indiatoday.in/education-today/gk-current-affairs/story/where-doklam-why-important-india-china-bhutan-1198730-2018-03-27

it falls into the hands of China, it will not only compromise the security of Bhutan but also of India. Also, the access to the Tri-junction area (via road from Doklam) would give China easy access to transportation of war machinery such as tanks and vehicles to the border of India. In this case, if a war breaks out between India and China, the latter will have an upper hand at conquering the Chicken's Neck of India as well as the whole of the North-Eastern region of the country (*India Today* 2018).

India is taking diplomatic and economic measures to support Bhutan. PM Modi conveyed that India deeply values its development partnership with Bhutan, and reaffirmed that as a friend and close neighbour, India will continue to partner with Bhutan in pursuit of its development goals, which are based on the priorities of its people (MEA, The Government of India 6 July 2018). Bhutan was the first country which PM Modi visited after he became PM in 2014. India's foreign policy in Bhutan is driven by both energy security and geopolitical consideration.

India–SriLanka Energy Cooperation

India is eyeing Sri Lanka for its renewable potential as well. But the potential links with Sri Lanka take a rather different form. Sri Lanka's major power source is hydroelectricity and there is not much potential for economically viable cooperation on hydro. Sri Lanka is exploring the possibilities of conserving hydro resources by importing baseload power from India and supporting India's Southern Grid during peak demand. In fact, Indo-Sri Lanka power interconnection through the land bridge is one important area where both countries can cooperate and get benefits. The Joint Study Group appointed by the governments of Sri Lanka and India (2003), while noting the potential of a regional power pool for Southern India and Sri Lanka, to be achieved by interconnecting the respective electricity grids, suggested,

> The interest of Indian companies to participate in future bids for coal-fired plants in Sri Lanka may be accentuated by the existence of a regional power pool. A regional power pool would also enable better management of peak-load demands on both sides, as well as enable faster recovery from disasters (Lama 2005).

India's state-run energy companies are strengthening their presence in Sri Lanka's push towards the renewable energy sector. Indo-Sri Lanka energy cooperation is mutually beneficial. India and Sri Lanka cooperate in wind and solar and are planning for power projects with LNG as feedstock. A 900 MW of capacity is being tendered out in phases, apart from two plants of 500 MW each, one of which India's NTPC is likely to build. India's investment in Sri Lanka's energy sector is mutually beneficial for both nations. Sri Lanka's energy sector can reach full potential and power production which would be useful for the energy security of both India and Sri Lanka.

India's Energy Exploration in Israel: A New Destination in the Middle East

The latest addition is the Indian state-run and gas explorer OVL-led consortium of OIL and Bharat Petroleum subsidiary BPRL in Israel.

The Indian consortium received a nod for one block from the Israel government in December 2017. Israel's energy ministry gave its preliminary nod on December 11 to the bid for one block submitted by the Indian consortium of OVL, IOCL, OIL and Bharat Petroleum subsidiary BPRL. This deal was the first auction of exploration licences, consisting of 24 blocks. This was the first such deal which New Delhi entered into with Jerusalem after Israel blocked foreign companies from exploration in its eastern Mediterranean waters in 2013. This could be seen in the context of a growing Indian-Israeli strategic partnership which has been in progress since the end of the Cold War and particularly during the Vajpayee government's effort to forge a strong strategic partnership with Israel. This nod for the bid of Indian consortium for gas was ahead of the Israeli PM Benjamin Netanyahu's three-day visit to India in January 2018. This could be seen in the context of India's diversification of sources of energy from the GCC nations. The same Indian consortium had discovered the Farzad gas field in Iranian waters. But Iran's refusal to award the exploring right to the Indian oil consortium has generated a widening rift over gas exploration by Indian companies (*Times of India* 28 December 2017).

Until now, oil and gas did not figure as components of India–Israel relations except India's largest private oil company leasing Eilat Ashkelon Pipeline Company's petro-products storage capacity. The collaboration in the defence industry, irrigation and water conservation has been the major component of India–Israel ties. The landmark visit of PM Modi to Israel, the first by any Indian PM, in June 2017 was a major development in India–Israel ties. The process for this began ahead of PM Modi's Israel visit when Israeli energy minister Yuval Steinitz called Petroleum and Natural Gas Minister Dharmendra Pradhan in June to invite Indian participation in Israel's exploration business, dominated by Delek Group and Isramco Negev (*Times of India* 28 December 2017).

Summing up, it is obvious that India has been taking concerted steps to overcome the overreliance on the Middle East for its hydrocarbon sources and increase the share of gas in its energy basket. PM Modi has emphasized the need for shifting towards a gas-based economy. India's dependency on import for its gas needs, the more environment-friendly and less pollutant features of gas in comparison to other hydrocarbons sources, and the low price of gas, which is likely to persist given the new sources of natural gas, have pushed the E&P of gas abroad in India's foreign policy agenda. Unless there is a drastic and dramatic discovery of gas fields in domestic exploration, India's import dependency for oil and gas will continue in the coming decades.

In recent years, India's hydrocarbon exploration has actively searched for alternative or complementary sources for natural gas. In the process, the United States and Australia have become new sources for India's imports. Both the United States and Australia have been shipping gas to India since 2016. GAIL India Ltd bought the second shipment of LNG from Cheniere Energy, Inc.'s Sabine Pass plant in Louisiana in a deal that makes it the first Asian importer of US shale gas (Chakraborty, Shiryaevskaya, and Weber 2016). In the post-Cold War era and especially over the past two decades, India's foreign relations with the US have transformed dramatically to a robust and comprehensive strategic partnership in which energy is a significant component. Over the past decade, India's foreign policy has given a consistent focus to engage Australia with the aim of tapping Australia's energy sources. (India's energy related foreign relations with the United States and Australia are dealt with in the next chapter with a focus on nuclear deal diplomacy and politics.)

An argument can be made against the costly investments in natural gas that will undermine India's remarkable progress on the renewable energy revolution. But the reality is quite the opposite—except for a low-cost, grid-scale storage technology breakthrough, any significant advancement in wind and solar energy cannot be attained without sufficient flexible power sources to address peak load power demand. Natural gas and hydropower provide by far the best source for peaking power. Given the limitations that India faces in the expansion of hydropower, gas-fired plants are essential to India's achievement of the ambitious renewables goals that it has set for itself by 2022 as committed to by PM Modi in the 2015 Paris Climate Summit (Shidore and Busby 2016).

India has clearly made a concerted effort to expand its option for gas sources and has taken a proactive foreign policy approach towards exploring for hydrocarbons, especially in the gas-rich nations. It has entered into diplomatic, economic and strategic engagements with energy-rich nations to strengthen its energy relations. India's foreign exploration for hydrocarbons has been carefully carried out and has aimed at diversification of acquisition sources to reduce the supply risk. India's energy policy in the international context is very well apt with the current scenario of political uncertainty and supply risk because of terrorism, domestic and international political disturbances. As previously stated, India's effort towards hydrocarbons explorations is under the conceptual framework of four 'As'—accessibility, availability, affordability, and acceptability amidst the global warming. However, despite its efforts in the exploration for natural gas and oil, India's energy security is still focused on search for alternative sources of energy that is affordable and acceptable environmentally. Nuclear energy has emerged as an important source of energy. India has been taking steps to increase nuclear energy in its energy basket. The next chapter deals with India's quest for energy security, its efforts to increase nuclear energy in its energy mix by concluding the nuclear agreements with countries with enriched uranium reserves or advanced civilian nuclear reactors technology and practices.

CHAPTER 5

India's Quest for Atomic Energy
Diplomacy and Nuclear Agreements

India needs to sustain its current growth rate of 7–8 per cent to alleviate poverty and meet its economic and social development goals. On top of this, India needs to provide electricity to its entire population, one-fifth of which still has insufficient access to electricity. All this require more energy. Seeing the future energy demand for its developmental process, India is exploring all possible sources of energy. Coal has been the dominant source of India's energy. However, over the past 10–15 years, India has been exploring the nuclear energy option, which has the potential to be a major source of electricity generation in India. Currently, India's nuclear power capacity is very low—less than 3 per cent of the country's total electricity generation capacity.

In the post-World War II period, the world noticed the peaceful use of nuclear power. The first time electricity was generated by any nuclear reactor was at the X-10 Graphite Reactor in Oak Ridge, Tennessee, in the United States on 3 September 1948, and it was also the first ever nuclear power plant to power a light bulb (Kanti and Sanathkumar 2016). According to the IAEA report, there are 437 nuclear power reactors in operation and 69 nuclear power reactors under construction in 31 countries around the world. Since the Fukushima triple meltdown in 2011, the worst nuclear disaster after Chernobyl in 1986, the price and demand for nuclear power went down to historic lows and turned many countries off nuclear power altogether. Though there has been a policy to move away from nuclear energy in the United States and some parts of Western Europe, the demand for nuclear energy has risen in Southeast Asia, China and India. Both India and China want to increase the nuclear energy share in their energy mix, as they want to reduce CO_2 emission. Nuclear energy is considered as environment-friendly and affordable in the long run as well.

According to the World Nuclear Association, China has 30 nuclear power stations, 24 under construction and more reactors in the planning stage. This is expected to increase nuclear power capacity by threefold to 58 GW by 2020–2021 and 150 GW by 2030. In India, nuclear power plants are the fifth largest source of electricity generation. Nuclear power plants contribute to about 2.6 per cent of its total power generation. The Indian government aims to increase its nuclear capacity to 63,000 MW by 2032 by adding nearly 30 reactors at an estimated cost of US$85 billion (McHugh 2016). India, in May 2017, announced its target of building 10 new reactors worth $14 billion, which would eventually generate 7,000 MW of electricity (BBC 2017). As of 10 October 2018, the NPCIL, responsible for the governance of India's nuclear power generation for electricity purposes,[1] operated 22 reactors with a capacity of 6,780 MW (including RAPS-1 [100

---

[1] The Department of Atomic Energy (DAE) administers the NPCIL, owned by the Government of India. NPCIL was created in September 1987 under the Companies Act 1956, 'with the objective of undertaking the design, construction, operation and maintenance of the atomic power stations for generation of electricity in pursuance of the schemes and programmes of the Government of India under the provision of the Atomic Energy Act 1962.' All nuclear power plants operated by the company are certified for ISO-14001 (Environment Management System: https://en.wikipedia.org/wiki/Environmental_management_system). NPCIL had the monopoly for constructing and operating India's commercial nuclear power plants till October 2013 when BHAVINI, another PSU of DAE, was established. NPCIL also has equity participation in BHAVINI. DAE is responsible for the implementation of the Fast Breeder Reactors programme in India. Private players

MW] under extended shutdown) and had 8 reactors with a capacity of 6,200 MW under construction. The Government of India has given administrative and financial approval for 12 more reactors with a capacity of 9,000 MW (NPCIL 2018).

India's nuclear power capacity has almost doubled from 17.7 TWh in 2006 to 35 TWh in 2017, but it is not up to its potential. Nuclear energy has the potential to contribute 25 per cent of India's energy mix by 2050. One of the main reasons for the low nuclear power capacity in India has been the international embargo that was imposed on its nuclear power programme after it conducted nuclear explosion for scientific purposes in 1974. The nuclear non-proliferation regime prevented India from obtaining nuclear reactors and enriched uranium. India needed to get rid of the nuclear embargo to enhance its nuclear power generation capacity. This needed a concerted foreign policy effort to dismantle the technological denial regime. Consequently, over the past two decades, India has consistently pursued nuclear issues in its foreign policy and concluded nuclear agreements with several countries which required intense diplomacy, lobbying and negotiations amidst its geopolitics.

In the discussed context, this chapter examines India's nuclear policy, its defiance and dalliance with the international nuclear regime and nuclear weapon states. The chapter gives an account of India's nuclear agreements with various countries starting with the US–India Civilian Nuclear Agreement, and its wider acceptance by the international community reflected in the waiver by the NSG, and safeguard agreement with the IAEA. Then the chapter deals with India's nuclear agreements in the foreign policy context with countries, namely the United States, then France, Russia, Australia, Canada and the UK. India's nuclear agreements with the these countries have been significant for a number of reasons including the influence they hold in the international nuclear regime, their legitimate status as nuclear powers, their advanced and high-tech nuclear technology, best nuclear practices and safeguards, and/or their abundant uranium reserves.

are now allowed to generate nuclear power for electricity (NPCIL, 'About Us': http://www.npcil.nic.in/content/328_1_AboutNPCIL.aspx).

India's Nuclear Energy Programme and Nuclear Non-Proliferation Regime

In the post-Independence period, India developed an excellent pool of nuclear scientists and could have soon tested a nuclear explosion. However, it chose a policy of the peaceful use of nuclear technology and made calls for global nuclear disarmament. During the Sino-India War in 1962, India suffered a humiliating defeat and its lack of defence preparedness was badly exposed. The military support that came from the United States during the war was below the state of the art-technology and not sufficient to defend against the Chinese attack. Finally, the Chinese nuclear test in 1964 prompted India to rethink the nuclear option. When the Treaty on the Non-Proliferation of Nuclear Weapons commonly known as the Non-Proliferation Treaty (NPT) was introduced in 1968 to check nuclear proliferation, India was from day 1 critical of it. India found it discriminatory in that it sought to create a nuclear order of 'nuclear haves' and 'nuclear have nots'. It allowed the 'permanent five'—the United States, USSR, UK, France and China—to continue to possess nuclear arsenals while others were condemned and denied the same. India felt that the NPT supported an unfair global nuclear order and undermined India's security concerns in the face of a nuclear China and China–Pakistan's 'all-weather friendship'. Consequently, India opposed the NPT and refused to sign it. Since then, the nuclear issue became a constant irritant in India–US relations. As the United States was at the forefront and was determined to make this nuclear regime successful, it found India the biggest hurdle to attain the goals of NPT. As India did not sign the NPT, Pakistan also refused to sign the treaty (A. Sharma 2017). Today, outside five declared nuclear powers, India, Israel, Pakistan and North Korea are the nuclear power nations.

The international rules and regulations to control weapons of mass destruction have been impressively effective, making nuclear acquisition less simple and significantly more expensive. However, a few states have still managed their way to become nuclear power nations on the ground of self-defence, prestige or strategic advantage. India, in 1974 and then again in 1998, made such a decision, testing a nuclear device and announcing its intention to become a nuclear weapons state. Since then India has been an international nuclear pariah. Its refusal to sign the NPT, the Comprehensive Nuclear-Test-Ban Treaty (CTBT) or the Fissile Material Cut-Off Treaty (FMCT) further pushed India towards this status (Indian Embassy in the United States 2008).

Though India conducted its peaceful nuclear explosion in May 1974 as a mark of opposition to the biased nuclear order based upon the NPT, India's nuclear policy has been in consonance with larger evolving non-proliferation norms and consensus. India stuck to its long-standing view of its right to the civilian use of the atomic energy and the right to nuclear technology cooperation for peaceful purposes. India faced the challenge of maintaining the balance between its values and its necessity for the nuclear energy programme. India's nuclear policy has been driven by practicability rather than by its normative predilection. The competitive tension between India's actual practice and its public policy was very much reflected in three major non-proliferation issues—nuclear safeguards, export controls and the danger of nuclear proliferation in its neighbourhood (A. Sharma 2017). Though India has been opposed to the NPT regime for its bias in the favour of P5, it has adhered itself to a strict nuclear export control regime and has supported complete nuclear disarmament from the beginning.

The NSG was brought in 1975, a year after India's peaceful nuclear explosion in 1974. The NSG consists of 48 nuclear supplier countries and seeks to control exports of nuclear materials, equipment and technology, both dual-use and that especially meant for nuclear weapons development. The NSG first came into being as an informal group of a few supplier countries in 1975 created by the United States as a direct response to India's first nuclear explosion in 1974 for the peaceful purpose. The United States halted nuclear exports to India a few years later, and worked to convince other states to do the same. In 1978, it issued a set of guidelines for nuclear transfers that did not entail the application of full-scope safeguards for nuclear exports to non-nuclear weapon states (NNWS) (Kerr 2010). This provision was incorporated into the guidelines only in 1992 after the discovery of the clandestine nuclear programme being pursued by Iraq (Sethi 2000). India's civilian nuclear energy programme was constrained by the international nuclear regime. After India's peaceful atomic explosion, stricter norms and guidelines were imposed on the export and access to nuclear technology, including the supply of enriched uranium for India's nuclear reactors for nuclear energy.

Never had India confronted the dominant discourse of the international system so directly as when it walked out of the CTBT negotiations and defied the existing international nuclear order when it conducted five underground atomic tests on 11 and 13 May 1998. India broke its self-imposed moratorium on nuclear test since 1974 for scientific purposes. The middle and major powers including Australia, Canada, Norway and Sweden strongly criticized India's atomic explosions. While China demanded full sanctions against India, it advocated for limited sanctions against Pakistan. Russia, however, was one of the major powers that did not forcibly challenge India's programme or employ trade sanctions (for detail, see Nayar and Paul 2003). The United States imposed mandatory sanctions and mobilized other nations, in particular, Japan, to cut economic assistance to India. Russia and France, sympathetic towards India, could not stop the United States creating an international framework under the UNSC Resolution 1172 in June 1998, which demanded that India sign the NPT and attempted to address the root cause of Indo-Pakistan tension—the Kashmir dispute. Also during his visit to China in June 1998, Clinton announced a new strategic partnership with China and condemned India's nuclear test. These developments were of concern to India as the UN resolution appeared to internationalize the Kashmir dispute, leading to the UN intervention that Pakistan had always wanted. The US–China convergence of interest for putting down India was again not in India's interests (for detail, also see A. Sharma 2017; Cohen 2001; Mohan 1998, 90–91).

India defended its nuclear tests on the ground of minimum deterrence to the long-term security challenges posed by its two nuclear neighbours, Pakistan and China. This was visible in its nuclear weapons programme. Despite the advancement in nuclear technology by other countries, especially its adversary nations China and Pakistan, India has not made a serious effort towards the development of an advanced and sophisticated nuclear inventory. In addition, Indian strategy is not preoccupied with nuclear balances, and there has not been any effort by higher-level authority for nuclear parity with China. This lessens the possibility of an arms race. The widely accepted belief both in official Indian and non-official strategic thinking that more atomic tests are not required shows an understanding that the higher levels of technological sophistication is not necessary for an effective deterrence. India's continuing interests in arms control and the nuclear non-proliferation regime show the depth of understanding in India about the danger and risks that nuclear conflict poses and displays in its abiding choices for strategic stability. Again, India's push towards the missile defence system has been compatible with minimum nuclear deterrence (Basrur 2006; A. Sharma 2012). In addition is its impeccable record of nuclear non-proliferation and its strict adherence to the nuclear export control regime. In the post-1998 atomic test era, India's nuclear policy and posture, its stand of no first use of nuclear weapons further strengthened the arguments for India when the nuclear exception was being made to India.

Finally, India's mounting energy demand began to push India towards nuclear agreements. This is visible in the post-1998 nuclear test phase. When Putin visited India in October 2000, agreements on defence and nuclear cooperation were made. This could be seen as the Indian government's urgency to work on the nuclear energy option for the coming decades; hence, the nuclear agreement's process began to be given a priority in its foreign policy priority and bilateral relations with the countries with high-end nuclear technology. This very nuclear energy cooperation move was not free from the NPT regime constraints. India had to work on the greater acceptability in the NPT regime, mainly the NSG and amendment to Nuclear Non-Proliferation Act of 1978 to trade internationally for its civilian nuclear energy programme. The process began after the 1998 nuclear test resulting in India's nuclear agreements with the United States and the IAEA and the NSG waiver in 2008. This ended the nuclear apartheid meted out to India for more than three decades. India's blatant defiance of the international nuclear regime and the nuclear exception made to India by the United States and the international community were unprecedented in the history of nuclear non-proliferation.

The nuclear exception to India was heavily criticized and some said that it would undermine the international non-proliferation regime. The critics contended that nuclear exception to India would encourage other aspiring nuclear nations to gain legitimacy by defying the regime. This argument made by critics was questioned on the grounds that the nuclear non-proliferation regime has been so severe in implementing the provisions of NPT that there has been no scope for countries like India to ignore the non-proliferation regime's effectiveness (Singh 2009–2010). The supporters argued that the pact would bring India under the non-proliferation regime as it would adhere to it to reap the benefits of the deal. Despite its non-signatory status, India has followed the strict export control regime of nuclear material and technology.

The NSG waiver and India's agreement with the IAEA safeguards of its civilian nuclear facilities have encouraged countries to do nuclear energy business with India. The 45-nation NSG waiver places neither any restriction on fuel supplies to India nor any curbs on its right to build strategic reserves (A. Sharma 2009).

The NSG's decision was based on commitments by India to:

- Separate civil and military nuclear facilities in a phased manner and place civil facilities under IAEA safeguards;
- Have in place an IAEA Additional Protocol on safeguards with respect to civil nuclear facilities;
- Continue India's unilateral moratorium on nuclear testing;
- Work towards conclusion of a FMCT;
- Refrain from transferring uranium enrichment and plutonium reprocessing technologies to states which do not have them and support international efforts to prevent their spread;
- Match India's export controls with those of the NSG.

The agreement between the Government of India and the IAEA for the Application of Safeguards to Civilian Nuclear Facilities was approved by the Board of Governors on 1 August 2008 and was signed in Vienna on 2 February 2009. Pursuant to paragraph 108 of the agreement, the Agreement entered into force on 11 May 2009, the date on which the agency received from India written notification that India's statutory and constitutional requirements for entry into force had been met (International Atomic Energy Agency 2009).

India has been placing its civilian nuclear facilities under safeguards pursuant to its 2009 agreement with the IAEA (IAEA document INFCIRC/754) and its plans to separate its civilian and military nuclear facilities and programmes (INFCIRC/731). India had placed all its 22 civilian nuclear reactors under the IAEA safeguards by the end of the year 2014. India had ratified Additional Protocol with the IAEA in June 2014 and the ratification entered into force on 25 July 2014 (National Interest Analysis 2014). This was important, as the nuclear reactors were not producing the power as per their capacity. There was a disparity of power and supply–demand of uranium for Indian nuclear reactors. India's placement of its reactors for the IAEA safeguards opened the road for its reactors to use imported uranium. After the NSG waiver and agreement with the IAEA, India's nuclear apartheid began to end. Today, India is integrated into the larger nuclear non-proliferation regime and is a member of multilateral organizations that deal with the agenda of the NPT, IAEA and the World Association of Nuclear Operators.

India currently has 22 operating nuclear reactors at seven plants, with about 6,220 MW of generating capacity. Six units with generating capacity of 4,350 MW are under construction, and there are plans for another 19 units over the next decade, according to the World Nuclear Association (Proctor 2018). The Indian government wants to increase the country's nuclear output. On the domestic front, Indian scientists have made impressive progress in nuclear technology. India, with its cutting-edge fast breeder technology, has the potential to become a major exporter in the coming years. India's record has been impressive on the progress on the operation of nuclear power reactors. India's indigenous PHWR Unit-1 of Kaiga Generating Station (KGS-1) made a world record in terms of longevity of operation. KGS-1 completed 895 days of continuous operation on 25 October 2018, setting the world record of long continuous operation among PHWR. With this achievement, KGS-1 stands first among PHWRs and second among all nuclear power reactors (of all technologies) in the world in terms of continuous operation. The unit has been operating since 13 May 2016. KGS-1 (220 MW) is fuelled by domestic fuel which started commercial operation on 16 November 2000. Indian nuclear power plants have demonstrated continuous operation for long periods exceeding a year 28 times so far. Three reactors, KGS-1 (895 days), RAPS-3 (777 days) and RAPS-5 (765 days) have operated continuously for more than two years. The long continuous operation of Indian atomic power reactors shows the maturity achieved in nuclear power technology (NPCIL 2018).

The Indian government's decision of allowing foreign companies to participate in India's nuclear power programme and setting up the nuclear reactors has paved the way for foreign companies to enter India's nuclear industry. Consequently, many countries with high technology in the civilian nuclear programme and nuclear reactors began to look towards India's lucrative energy market, one of the biggest in the world. In that race, apart from the United States, other leading players have been France, Russia, Canada, Japan and countries such as Australia and others with enriched uranium have also been for nuclear energy cooperation with India.

India has so far signed the following nuclear agreements that have been the result of a long and tedious process of negotiations, diplomacy and lobbying reflected in its foreign policy:[2]

- Agreement between the Government of India and the Government of France on cooperation in peaceful uses of nuclear energy (30 September 2008).
- Agreement between the Government of India and the United States on cooperation in peaceful uses of nuclear energy (10 October 2008).
- IAEA Safeguards Agreement (2 February 2009).
- Agreement between the Government of India and the Government of the Russian Federation on cooperation in the use of atomic energy for peaceful purposes (12 March 2010).

[2] Department of Atomic Energy (DAE), 'Important Agreements': http://www.dae.nic.in/?q=node/75.

- Agreement between the Government of India and the Government of Canada on cooperation in peaceful uses of nuclear energy (27 June 2010).
- Agreement between the Government of India and the Government of Argentina on cooperation in peaceful uses of nuclear energy (23 September 2010).
- Agreement between the Government of the Republic of Kazakhstan and the Government of India for cooperation in the peaceful uses of nuclear Energy (15 April 2011).
- Agreement between the Government of India and the Government of the Republic of Korea for cooperation in the peaceful uses of nuclear energy (25 July 2011).
- Agreement between the Government of India and the Government of Australia on cooperation in peaceful uses of nuclear energy (5 September 2014).
- Agreement between the Government of India and the Government of Sri Lanka on cooperation in peaceful uses of nuclear energy (16 February 2015).
- Agreement between the Government of India and the Government of the United Kingdom of Great Britain and Northern Ireland for cooperation in the peaceful uses of nuclear energy (13 November 2015).
- Agreement between the Government of India and the Government of Japan for cooperation in the peaceful uses of nuclear energy (11 November 2016).
- Agreement between the Government of India and the Government of the Socialist Republic of Vietnam for cooperation in the peaceful uses of nuclear energy (9 December 2016).
- Agreement between the Government of India and the Government of the People's Republic of Bangladesh for cooperation in the peaceful uses of nuclear energy (8 April 2017).

Except for China, India has nuclear agreements with all legitimate nuclear powers declared under the NPT: the United States, Russia, UK and France. These four countries plus China are also the permanent members of the UNSC.

India's Nuclear Deal with the United States: Ending the Nuclear Isolation

Throughout the Cold War, India's relationship with the United States was marked by missed opportunities despite sharing many commonalities in terms of political values. Even though both are continental democratic nations with a natural interpersonal affinity, and apparently connected by common values and a shared commitment to democratic pluralism, their international and bilateral issues were often seen in divergent, conflicting and incongruous ways (A. Sharma 2017). Moral anger and mutual incredulity, even at times a sense of betrayal, were the defining features of India–US relations during the Cold War period. Even on those relatively rare occasions when the two worked in tandem, bruised sensibilities and bitter recriminations soon undermined the relationship (Hathaway 2003, 7).

This can be attributed to many factors, mostly the geopolitics of the Cold War. India's ties with the Union of Soviet Socialist Republics (USSR) despite its policy of non-alignment, as well as the complex relationship between the various actors in East and South Asia, meant that India had to balance its interests between the United States, the USSR, China and Pakistan. India and the United States distanced themselves from a tranquil closeness due to their conflicting interpretations of and different strategic approaches to the Cold War. Each nation pursued different goals: India, to protect its independence by respecting national independence via non-interference, and the United States, to protect its national interests and project military power to contain communism and maintain its primacy.[3] The importance of territorial and geopolitical strategies in foreign policy during the Cold War was critical. Some prominent historians, such as Dennis Kux, Andrew J. Rotter and H. K. Brands, explored this troubled relationship and characterized the US–India relationship during the Cold War period as one of the two nations being 'estranged democracies', 'comrades at odds' or being in a state of 'cold peace', respectively (Kux 1992; Rotter 2000). The nuclear non-proliferation treaty, differing economic policies, Kashmir complications and, more significantly,

---

[3] For a detail insight, see Kux (1992).

America's economic and military build-up of Pakistan as a global bulwark against Soviet communism further created hurdles for a warm India–US relation.

However, nuclear non-proliferation and India's nuclear test became the major irritants that obstructed a sound India–US relationship for almost five decades during the latter half of the 20th century. India's nuclear dalliance with the United States goes back to the 1950s. In 1963, the United States signed a 30-year civilian nuclear cooperation pact with India. Under the pact, the United States built and provided nuclear fuel for the twin reactors at Tarapur in the state of Maharashtra and allowed Indian scientists to study at American nuclear laboratories. However, India and Pakistan's refusal to sign the NPT in 1968 alerted the White House and the US Congress, and calls for sanctions and restrictions designed to compel Indian and Pakistani conformity to the treaty began to be made. These calls became louder when in 1974 India conducted a peaceful nuclear explosion at Pokhran.

Ever since India's atomic detonation in 1974, its possession of nuclear weapons and nuclear non-proliferation became the main focus of the US policy toward India. This was reflected in the US techno-logical embargoes and economic sanctions against India (Cohen 2001, 283). The United States also threatened to stop foreign aid and reliable supplies of nuclear materials in the case of the violation of international safeguards. India questioned the US move on the grounds of United States' double standard by not imposing similar demands on China and Israel (Galbraith 1990, 72). In 1974, the Ford administration with-held fuel shipments to the Tarapur nuclear plant until it could be sure that US materials were not used in India's atomic test. But after the assurance from PM Morarji Desai of the Janata Party government, the first non-Congress PM of India, that India would refrain from further nuclear tests, the United States resumed the aid to India (Rubinoff 2001). Since the 1974 test, the United States isolated India for almost 25 years, refusing nuclear cooperation and trying to convince other countries to do the same.

Until the Soviet invasion of Afghanistan, during Jimmy Carter's presidency, the nuclear non-proliferation became the focus of American foreign policy. South Asia was the main theatre of US non-proliferation legislation that included technology denials and sanctions (Cohen 2001, 283). Congress passed the Symington and Glenn Amendments, sections 669 and 670 of the Foreign Assistance Acts of 1976 and 1977, which prohibited aid or arms sales to countries that delivered or received nuclear enrichment equipment or technology and did not accept IAEA safeguards (Rubinoff 2001). The Carter administration granted the export licences for two fuel shipments and spare parts for India's Tarapur reactor, despite objections from the Nuclear Regulatory Commission that India had not fulfilled the measures of the 1978 Nuclear Non-proliferation Act. The Reagan administration in 1982 resolved the Tarapur deadlock by bringing France to supply the fuel (Rubinoff 2001).

In the 1980s, there was some improvement in the defence sector. The Carter administration reversed its earlier policy of disapproving the use of an advanced American electronic guidance system in India's Jaguar aircraft (A. Sharma 2008). This US attempt to improve ties was visible in the nuclear sector as the United States granted two more enriched uranium fuel shipments to Tarapur. It appeared as if Carter was providing incentives to India in the wake of the Afghanistan crisis so that India did not tilt further towards the Soviet Union. This was significant because of improving Sino-US ties and the rearming of Pakistan with US weapons. As India was seeking to diversify its sources of military acquisition, scientific and technical cooperation and trade and investment sources, improvement of relations with the United States were considered important. India and the United States signed an MoU in 1984 on the transfer of technology. In exchange for alterations to India's own export-control regulations, the United States would begin allowing access to civilian and dual-use technologies as well as giving some military assistance, subject to previous restrictions imposed by US law. Under this agreement, sensitive technology transfers took place but mainly for the items that were below the level of state-of-the-art technology (for detail, see Mahapatra 1998, 63).

The Reagan administration's 'Opening to India' policy aimed to engage India. The US efforts included the visit of the US Defence Secretary C. Weinberger in 1986 and 1987 and C. Weinberger's successor Frank Carlucci in 1988. The US intention to better ties with India was made clear by these moves. India responded with the visit of Defence Minister of India to the United States accompanied by high-level civilian and armed services officials. These visits were vital towards removing mutual misperceptions and enhance mutual understanding among the policymakers in both countries.

After the creation of Bangladesh in 1971, Pakistan secretly began to work on its atomic bomb programme. China assisted Pakistan's nuclear programme development through commerce in enriched uranium and the building of reprocessing units. The United States deliberately ignored this clandestine China–Pakistan nuclear bonhomie. This should be seen in the context of Pakistan's importance in the US strategic framing. Washington's ties with Islamabad were vital as Pakistan was strategically important for the United States to fight the Soviet Union from the southern vantage point, and its role as a catalyst in the United States relationship with China, and to check the Soviet expansion. Consequently, despite the military sanctions, the United States continued with its policy to support Pakistan with arms supply by non-legal means by facilitating third-party sales of fighter jets to Pakistan, with the assistance from Iran and Jordan. The United States openly branded India as the aggressor in the India–Pakistan war during the Bangladesh crisis, and India was perceived as a Soviet ally because of the Indo-Soviet Treaty of Peace, Friendship and Cooperation on 9 August 1971, despite its open advocacy and leadership of the NAM in the tussle between the United States and the Soviet Union (National Security Archive 1972). PM Indira Gandhi signed the treaty to counter the US–China–Pakistan axis before going to war against Pakistan to stop the flood of refugees from Bangladesh—more than 8–10 million fleeing to India.

With the implementation of economic and military sanctions under the provisions of the Symington Amendment in 1977, Washington stopped economic assistance to Islamabad officially; however, the US State Department did not stop aid to Pakistan as it gave US$50 million of assistance to Pakistan yearly including significant food assistance (Kux 2001, 236). Soon after the United States terminated economic and military assistance to Pakistan based on the Glenn/Symington Amendments to the Foreign Assistance Act in 1979, to punish Islamabad for following the French reprocessing plant deal, the US administration decided to get rid of all sanctions to suit immediate US interests. In other words, the beginning of the war in Afghanistan compelled the United States to make a strategic U-turn, as Pakistani assistance had become indispensable to its strategy to check the Soviet Union (Pressler 2017, 121).

Pakistani atomic development had become a serious issue to the US legislative and executive branch. In 1985, the US Congress passed the Pressler Amendment of 1985, an amendment to the Section 620E of the Foreign Assistance Act of 1961. The Pressler Amendment required annual presidential certification that Pakistan did not possess a nuclear weapon and all US military and economic assistance to Pakistan immediately be banned if the US president found that Pakistan had attempted to illegally acquire American material for making atomic weapons. In the wake of the certainty of Pakistan's nuclear bomb nearing completion, some members of Congress tried to cut off the six-year US$4.02 billion in aid to Pakistan. To check this Robert Kasten (R-Wis.) and Daniel Inouye (D-Hawaii), lobbyists for the Pakistani embassy, were able to stop all aid to India until foreign aid to Pakistan was reinstated. The ability of Pakistan to induce Congress temporarily to cut off aid to India in 1987, when its own funding was in jeopardy for embarking on a nuclear weapons programme, was a testimony to the strength of its influence on Capitol Hill and New Delhi's lack of influence.[4] It demonstrated India's ignorance of the importance of lobbying in the US foreign policymaking, which its adversary had realized from the beginning.

After the end of the Cold War, both India and the United States, freed from the ideological barriers, began to reformulate their relations in the 1990s. This soon began to be visible in the economic and defence sector. But before the relationship could move towards a meaningful engagement, the nuclear issue again became a hurdle. The passing of the Brown Amendment on 27 January 1996 by President Bill Clinton became an issue of outrage in India and worsened the improving

---

[4] For a detailed and insightful history of the US nuclear policy towards South Asia, see Pressler (2017, 121), Mistry 2014, 169–173.

US–India relations, causing embarrassment to Washington's friends in New Delhi.[5] The BJP, then the main opposition party, renewed its call for a nuclear option. The passage of the Brown Amendment and the Indian parliament's reaction to it showed the centrality of the nuclear issue in India–US relations. In May 1998, PM Atal Bihari Vajpayee-led BJP/NDA government took the decision to conduct a nuclear weapon test in 1998.

President Clinton imposed sanctions under the Glenn Amendment, 1994, of Nuclear Proliferation Prevention Act on New Delhi on 13 May, two days after India's nuclear test. Similar sanctions were imposed on Pakistan after its atomic tests on 28 and 30 May 1998. Nevertheless, the sanctions were soon removed. The India–Pakistan Relief Act of 1998 passed in the Senate granted authority to the president to waive all sanctions except those pertaining to military assistance, dual-use exports and military sales for one year.[6] Contrary to the popular perception with the BJP in power, the India–US relationship began to move in a positive direction and the nuclear issue became an issue to be negotiated. Though heavily criticized by the Clinton administration, the nuclear tests resulted in the longest series of high-level bilateral talks in the history of the India–US relationship. US Deputy Secretary of State Strobe Talbot and Indian Foreign Minister Jaswant Singh began bilateral talks on 11 June 1998, and 11 rounds of talks with India took place. The dialogue covered broad issues

[5] Politicians from the parties on the Left criticized the Narasimha Rao government for pursuing a misguided policy of cooperation with the United States (*The Indian Express*, 25 September 1995). The setback to India–US relations was signalled by a speech that Home Minister S. B. Chavan addressed to the Rajya Sabha on 29 November 1995. According to Chavan, the selling of arms to Pakistan indicated 'evil designs' that the United States had on the subcontinent. The BJP opposition leader endorsed his unfounded assertion that the United States was interested in acquiring a 'foot hold' in Kashmir (*The Hindu*, 16 December 1995). Foreign Minister Pranab Mukherjee declared that by selling arms to Pakistan, the United States was once again forcing India to divert scarce resources to the military sector. Commerce Minister P. Chidambaram claimed that the Brown Amendment 'cast a shadow over commercial ties' because higher military spending undermined the free enterprise economy the United States desired to see established in India (*Indian Express*, 8 December 1995.

[6] Ibid.

from the questions of proliferation and nuclear policy to larger issues such as the shape of the international system, terrorism and strategic cooperation between the two states (Cohen 2001, 285). This was the first sincere mutual attempt to shape the relationship independent of Indo-Pakistani or Indo-Russian concerns. The Indian lobbying that was running parallel to the Jaswant–Talbot talks helped clarify the US apprehensions, misconceptions and misperceptions about India's nuclear posture.

In the words of the former Indian Ambassador to the United States, Lalit Mansingh, the situation required delicate handling as it was also the time when the Indian mission in the United States was having an exceptionally difficult and tough time looking for support to counter anti-India propaganda and justify its action.[7] India's nuclear defiance of the United States in 1998, and the lobbying and reconciliation after that could be considered as the most intricate, audacious and successful political exercise in the history of Indian diplomacy. The Indian mission in Washington resorted to a gigantic and conscious lobbying effort in the defence of India's atomic explosion. They lobbied before the people whose opinions mattered in US policymaking circles and could understand India's security concerns and view them objectively. Eventually, the focal point of the India–US relationship moved from non-proliferation to nuclear stability, nuclear deterrence, trade and commerce, countering terrorism, democracy promotion, the betterment of governance and energy security.

George W. Bush came to power in 2000, and had a 'big idea' of transforming the US ties with India based on the enduring foundation of common democratic values and converging geostrategic and security interests. Often mentioned during his presidential campaign, this 'big idea' was that by working together more intensely than ever before, the United States and India, two vibrant democracies, could transform the very essence of their bilateral bonds and thereby make the world 'freer, more peaceful, and more prosperous' (Blackwill 2003). In short, President Bush understood the rise of India to be one befitting world

---

[7] Interview with Lalit Mansingh, New Delhi, June 2004–June 2005.

power. Knowing that PM Vajpayee believed that the United States and India were natural allies, a strategy was chalked out in January 2001.

The Bush administration, as promised, lifted the sanctions imposed on India at the earliest possible opportunity without reference to the CTBT and set the pace for ending the nuclear embargo on India that had crippled the India–US relations (Mohan 1998, 95–96). The Bush administration, and especially the Pentagon, was firm on redefining the defence ties with India as New Delhi was now viewed a potential partner in maintaining the peace and stability in the Indian Ocean and decisive in the emerging balance of power in Asia. When the Bush administration surveyed the strategic landscape, it saw China looming large. The Pentagon saw India's support for the US missile defence systems and full cooperation with the United States after 9/11 positively. The converging security interests in the post-9/11 environment and an assertive Chinese posture in Asia put the world's two largest democracies on the path to forge a strong partnership.

Marking the first time in five years that a senior US official had come to India to discuss civil nuclear collaboration, in February 2003, Nuclear Regulatory Commission Chairman Richard Meserve toured the Tarapur Atomic Power Station and the Bhabha Atomic Research Centre. The cooperation in the defence and military sectors emerged as one of the most intense and fastest growing aspects of the India–US bilateral relationship. This paved the way for another breakthrough in the form of the NSSP which shaped the US–India relations in the coming years. In June 2005, India and the United States signed a 10-year defence agreement, which included joint production, transfer of technology and sale of high-end weapons to India. However, this would have remained incomplete without the removal of the US laws that prevented commerce in civilian nuclear energy and sensitive technology and defence cooperation. Parallel efforts regarding high-technology transfer and civil space cooperation bore fruit when PM Manmohan Singh visited Washington in July 2005 and signed an agreement to promote cooperation on civilian use of nuclear technology aimed at addressing India's energy security. Consequently, the Henry J. Hyde United States–India Peaceful Atomic Cooperation Act, 2006, was passed. On 1 October 2008, the

US Congress passed the nuclear agreement with an overwhelming majority. After the presidential assent, the law became the United States–India Nuclear Cooperation Approval and Non-proliferation Enhancement Act.

The deal facilitates the civilian nuclear energy cooperation and lifts a three-decade US moratorium on nuclear trade with India. It provides US assistance to India's civilian nuclear energy programme and expands the US–India cooperation in energy and satellite technology. The nuclear deal is considered a watershed moment in the history of US–India relations with the aim of addressing India's looming energy crisis, reducing dependency on fossil fuel and elevating US–India relations to a new level.

The Bush administration freed the relationship from the nuclear stalemate, which mainly was grounded in India's refusal to sign the Nuclear Non-proliferation Treaty and the US Nuclear Non-proliferation Act of 1978 that barred India from getting any kind of sensitive, high-technology assistance. In this context, the nuclear deal signed on 18 July 2005, known as the 'United States–India Nuclear Cooperation Approval and Non-proliferation Enhancement Act', is a landmark step (*New York Times* 2010). The nuclear agreement was part of a larger set of initiatives between the United States and India involving space, dual-use high-technology, advanced military equipment and missile defence (Tellis 2005). The India–US nuclear agreement assists India to address its energy security and diversify to an alternate and environment-friendly fuel, and ensures nuclear safety cooperation and Indian integration into the global nuclear regime to facilitate India's need of renewed access to safeguarded nuclear fuel and advanced nuclear reactors.

In fact, without the nuclear deal, the transformation of the overall US–India relations to a robust strategic and defence partnership would not have been possible. Without the nuclear roadblock being bypassed, the defence industry relationship would have remained stillborn since US laws and non-proliferation policy do not permit cooperation with a non-signatory to the NPT nuclear weapon state that has been under sanctions for three decades. The pact has led to the dismantling of the technology denial regimes of the Nuclear Non-proliferation Act of 1978 that came after India's first nuclear bomb test in 1974. The technological denial regime of the Nuclear Non-proliferation Act of 1978 had put a US ban on sale or transfer of sensitive and dual-use technology to India and constrained nuclear energy and defence collaboration.[8] The exception of India to the nuclear non-proliferation regime and the US nuclear legislation was defended by the Bush administration saying that India would be incorporated into the nuclear non-proliferation regime and that the agreement would address India's looming energy crisis. In the broader context, the nuclear deal shows the Bush administration's commitment towards helping India become a great power by acknowledging India's emerging global economic and military significance. The nuclear deal is also seen in the context of a rediscovery of mutual strategic relevance between India and the United States, resulting in a gradual paradigm shift in the balance of power in Asia that could influence the geostrategic politics at the global level in the future. During his visit to India in November 2010, President Obama gave the nod to India's inclusion in the NSG and three other multilateral export control groups. In 2011 at the NSG's plenary meeting, the United States came out with a 'Food for Thought' paper on possibilities for bringing India into the group (Horner 2012). However, India's inclusion in the NSG membership was halted because of China's opposition. This needs to be seen in the context of the India–China strategic rivalry, which is also reflected in the nuclear energy sector. (The Chinese opposition to the US–India nuclear deal and India's NSG membership is dealt with in Chapter 6.)

In July 2009, New Delhi designated two sites for United States companies to build nuclear reactors in India. However, a nuclear liability law[9] passed by the Indian parliament in August 2010 is

---

[8] United States Nuclear Regulatory Commission, *Nuclear Nonproliferation Act of 1978*: http://www.nrc.gov/reading-rm/doc-collections/nuregs/staff/sr0980/v3/sr0980v3.pdf.

[9] https://prsindia.org/index.php?name=Sections&action=bill_details&id=6&bill_id=1042&category=42&parent_category=1 (accessed on 15 March 2019).

causing a rift[10] with US nuclear suppliers. Critics of the law contend India's proposal to seek legal redress against nuclear suppliers which is a sharp deviation from the international liability regime that holds nuclear operators solely responsible in the case of an accident. India would also like the United States to relax some of its restrictions on technology transfer to India.

Though the United States has yet to make a substantial commercial deal to reap the benefits from the nuclear deal, Russia, France (Squassoni 2010) and Australia have begun to access the Indian nuclear energy market. Ultimately, the nuclear deal has helped India address its energy security concerns, facilitated US–India defence ties, enhanced India's international profile and assisted in counter-balancing China. This conclusion is fortified by the fact that India has adhered to the strict non-proliferation rules of the nuclear club despite being a non-signatory to the Nuclear Non-proliferation Treaty (A. Sharma 2013). Under the provision of the nuclear deal, India has separated its civilian and military-purpose nuclear reactors, and India's atomic reactors are under international safeguard measures and opened for the IAEA inspection. Though the US–India nuclear deal was mainly to assist India address its energy crisis by shifting to a clean energy, it recognized India's rising economic and military profile and India's strategic importance in the great power balance in Asia's unfolding strategic geometry.

The India–US Civilian Nuclear Agreement, designed for trade and commerce in the field of nuclear energy, opened doors for other countries to trade in nuclear material with India such as nuclear reactors, civilian nuclear technology and enriched uranium for reactors. Earlier, there were numerous constraints on India's nuclear energy power expansion due to the inadequate financing, technology denial regime, continued non-availability of affordable uranium and negative public perceptions about nuclear energy. In the period leading to the signing of the Indo-US nuclear deal, there were passionate debates within India on the viability of nuclear energy as a clean and efficacious

---

[10] http://www.cfr.org/publication/23305/indias_nuclear_liability_dilemma. html (accessed on 22 July 2018).

way to deal with the impending energy crisis. This eventually helped form positive public opinion towards the nuclear route to energy.

The provisions of the Indo-US nuclear deal are as follows (Council on Foreign Relations 2010):

- India agrees to allow inspectors from the International Atomic Energy Association (IAEA),[11] the United Nations' nuclear watchdog group, access to its civilian nuclear programme including its domestic reactors. By March 2006, India promised to place 14 of its 22 power reactors[12] under IAEA safeguards permanently.
- India promises that all future civilian thermal and breeder reactors[13] would be placed under IAEA safeguards permanently. However, New Delhi retains the sole right[14] to determine such reactors as civilian. India will not be constrained in any way in building future nuclear facilities, whether civilian or military, as per our national requirements.
- Military facilities—and stockpiles of nuclear fuel that India has produced up to now—will be exempt from inspections or safeguards.
- India commits to signing an Additional Protocol[15]—which allows more intrusive IAEA inspections—of its civilian facilities.
- India agrees to continue its moratorium on nuclear weapons testing.
- India commits to strengthening the security of its nuclear arsenals.
- India works towards negotiating an FMCT with the United States banning the production of fissile material for weapons purposes. India agrees to prevent the spread of enrichment and reprocessing technologies to states that do not possess them and to support international non-proliferation efforts.
- US companies will be allowed to build nuclear reactors in India and provide nuclear fuel for its civilian energy programme. (An approval by the NSG lifting the ban on India has also cleared the

[11] See https://www.iaea.org/

[12] http://www.state.gov/p/us/rm/2006/66031.htm

[13] http://www.state.gov/p/us/rm/2006/66031.htm

[14] http://www.indianembassy.org/newsite/press_release/2006/Mar/24.asp

[15] http://www.iaea.org/Publications/Documents/Infcircs/1998/infcirc540corrected.pdf

way for other countries to make nuclear fuel and technology sales to India.)

- India would be eligible to buy US dual-use nuclear technology, including materials and equipment that could be used to enrich uranium or reprocess plutonium, potentially creating the material for nuclear bombs. India would also receive imported fuel for its nuclear reactors.

Today there seems to be very little difference between Washington and New Delhi concerning nuclear non-proliferation issues. Under the Modi government, the Indian Parliament passed the Civilian Nuclear Liability Bill, which was caught in a political tussle with opposition parties strongly disapproving of certain features of the legislation. It is important for the United States and India as they have yet to reap the benefits of the nuclear deal and for the further strengthening of their deepening strategic partnership in which the civilian nuclear energy sector is a significant component.

Beyond the India–US Nuclear Agreement: The Growing India–France Nuclear Energy Cooperation

India's relationship with France has traditionally been close. In the modern period, the connection goes back to the presence of the French East India Company to the Indian soldiers fighting along with the other allied forces to defend France in the two world wars. During the Cold War period, there was not much engagement between the two nations as India pursued a non-aligned foreign policy in the world divided between the two superpowers—the US-led capitalist bloc and the Soviet Union-led communist bloc. Nevertheless, India tilted towards the Soviet Union and finally entered into a strategic and defence partnership with the Soviet Union in 1971. India's political position did not allow any meaningful interaction between it and France. However, the end of the Cold War, which led to the changed international political and economic scenario, provided the platform for both nations to engage. India began to reformulate its foreign and economic policy in the 1990s to engage the wider world including the

Western countries. It was not until 1998 when India entered into a strategic partnership with France. The partnership was based on the converging interests between France and India on political, economic and strategic issues. India's growing strategic partnership with France is now visible in a wide range of sectors, but it is the defence, space and nuclear sectors that have become most prominent.

The beginning years of India's strategic partnership with France were driven by its efforts to diversify its high-tech defence acquisition destination. Over the past two decades, France has emerged as a major defence partner of India, which is visible in the defence commerce. Scorpene submarine project for the Indian Navy and the Rafale fighter jet deal for the Indian Air Force is the core of the India–France strategic partnership. India had signed a government-to-government deal with France in 2016 to buy 36 Rafale fighter jets. The latest pacts included the reciprocal logistic support between the armed forces of the two countries and another on the protection of classified or protected information. The defence cooperation has also reflected in the regular exchange of visits at the level of services chiefs, regular joint army, navy, air force military exercises; a High Committee on Defence Cooperation (HCDC) which meets annually at the level of Defence Secretary and the French Director General of the Directorate General for International Relations and Strategy (DGRIS) (MEA, Government of India 2017b).

Besides the defence ties, it is the civilian nuclear energy cooperation that has emerged as an important aspect of India's policy towards France. Over the past decade, driven by its mounting energy demand, India has worked to build a sound relationship with France in the civilian nuclear energy sector. India's energy policy with France is in line with efforts to move to alternative and renewable sources of energy. Besides nuclear energy, India's energy engagement with France is rapidly growing in the field of solar energy reflected in the ISA amidst the concern of climate change and sustainable development. India's energy relationship with France is within the framework of four 'A's in which affordability and acceptability aspects have been dominant. The leaders of both countries are working to build a partnership and lead the world towards renewables by tapping the abundant solar energy potential in 121 nations with abundant sunlight through the platform of the ISA. This is very much in line with India's commitment to the Paris Climate Summit in 2015.

Since the strategic partnership began to unravel, New Delhi and Paris have had a regular exchange of visits at the highest level. PM Narendra Modi visited France on 2–3 June 2017 and met President Emmanuel Macron. Modi declared India's unflinching commitment to the Paris Accord and the effort to save the environment, which is inculcated in India's civilizational heritage and ancient philosophical thought. PM Modi had paid an official visit to France in April 2015 in what was his first visit to a European country, and had a meeting with then President Francois Hollande. PM Modi visited Paris on 29–30 November 2015 to attend the inaugural Leaders' Event at the COP-21 Climate Change Summit. He launched the ISA jointly with President Hollande in the presence of then UN Secretary-General Ban Ki-Moon. Along with President Hollande, US President Obama and Microsoft Chairman Bill Gates, he spoke at the event of 'Mission Innovation' for promoting renewable energy. As the Paris Agreement was reached on the conclusion of COP-21, President Hollande praised India's contribution and PM Modi's personal leadership in reaching this historic agreement.

Nevertheless, it is the growing cooperation in civilian nuclear energy that has emerged as the hallmark of the India–France relationship. Though French association with India's civil nuclear programme dates from 1949, they did not become close until after India's nuclear test in 1998. France refrained from criticizing India's nuclear test by taking an objective position on it amidst its security concerns. France along with Russia was supportive and empathetic to India's nuclear test in 1998. This perhaps could be seen in the context of business opportunities that India offered, the lack of a security threat from India and India's credentials as a democratic nation and its record on nuclear non-proliferation. By 1998, India had purchased Mirage 5000 and progress was on hand for upgradation and further defence deals. France was also in the process of making a deal to sell Scorpene submarines to India. France clearly did not want to lose the upcoming business opportunity for its defence and nuclear reactor sales to India.

French support has been crucial in India's quest for rewriting an entire international nuclear regime which subjected India to a nuclear apartheid. After India's nuclear deal with the United States, the roads were open for the transfer of technology in sensitive areas as the nuclear embargo on India was now removed, and this paved the way for France and India to cooperate in the nuclear energy field. Undoubtedly, it was the US–India nuclear deal which cleared the way for nuclear trade with India. But French support has been crucial and would remain significant not just because of French clout amidst the NSG but also for the supply of nuclear fuel and nuclear reactors (A. Sharma 2009). France became the first nation to sign a nuclear deal with India when it entered into an agreement with India in 2008 on civil nuclear energy cooperation. This was even before the United States signed a formal nuclear agreement with India after India got an exceptional waiver at the NSG despite its non-signatory status to the Nuclear Non-Proliferation Treaty (NPT). Today India's nuclear energy cooperation with France reflects both nations' willingness to develop a strong comprehensive partnership including the energy sector.

On 30 September 2008, during the visit of then PM Manmohan Singh to France, India and France signed the landmark agreement on civil nuclear cooperation on the development of nuclear energy for peaceful purposes. Following the civilian nuclear agreement, the French state-run reactor maker Areva SA signed an initial agreement in 2009 with India's government-owned NPCIL for equipment and construction at the JNPP. Subsequently, during the visit of former President Nicolas Sarkozy to India in December 2010, the General Framework Agreement and the Early Works Agreement between NPCIL and M/s Areva for the implementation of European pressurized reactor (EPR) for the JNPP were signed (MEA, Government of India 2017b).

Nuclear agreements signed during President Sarkozy's visit to India in December 2010 included the following provisions (France in India: French Embassy in New Delhi 2018):

- Commercial contract between Areva and NPCIL 'EPR Jaitapur— General Framework Agreement' for providing two EPR and supplying fuel, including the price, general terms and conditions.

- Commercial contract between AREVA and NPCIL 'EPR Jaitapur—Early Works Agreement' for preliminary works ahead of the construction of reactors.
- Cooperation agreement between the Commissariat à l'EnergieAtomiqueet aux Energies Alternatives and the DAE of the Government of India in the field of nuclear science and technology for peaceful uses of nuclear energy framework agreement aimed at covering the entire scope of cooperation on civil nuclear energy research and development.
- Agreement between the Government of India and the Government of the French Republic on the protection of confidentiality of technical data and information relating to cooperation in the peaceful uses of nuclear energy agreement on the protection of classified information and material.
- Agreement between the Government of the French Republic and the Government of India on intellectual property rights relating to the development of peaceful uses of nuclear energy agreement aimed at framing the distribution of IPR in cooperation agreements or contracts on the peaceful uses of nuclear energy.
- Arrangement between Autorité de Sûreté Nucléaire française (ASN) and the Atomic Energy Regulatory Board (AERB) of the Government of India for the exchange of technical information and cooperation in the regulation of nuclear safety and radiation protection cooperation on nuclear safety regulation.
- Cooperation agreement between Institut de Radioprotection et de Sûreté Nucléaire (IRSN) and AERB in the field of nuclear reactor safety cooperation (technical aspects as opposed to regulatory aspects covered by ASN and AERB.
- MoU between Agence Française nationale pour la gestion des déchets radioactifs (ANDRA) and BARC radioactive bilateral civilian nuclear cooperation France in India waste management. Exchange of information and experience in radioactive waste management.
- Implementation agreement between the Commissariat à l'énergie atomique et aux énergies alternatives (CEA) and Homi Bhabha National Institute (HBNI) of the DAE, on jointly supervised theses.

The Indo-French nuclear deal is mutually beneficial as it would help India to address its looming energy crises and is a move towards alternative and environment-friendly sources of energy, and opens avenues and options for nuclear trade, civilian nuclear technology and reactors. The Indo-French nuclear deal sets a precedent for companies from other countries interested in doing business in India to follow. The nuclear energy trade pact has been helping India to develop its nuclear energy capacity and equipment. The deal also brought the possibility of collaboration between India and France in mining uranium and joint investments in mines for the supply of fuel for the nuclear reactors. The deal not only opened avenues for nuclear energy cooperation but also opened France and India to elevate collaboration in defence and space sectors.

The French historic decision made with regard to civil nuclear energy cooperation with India has been justified on the ground of addressing the challenges of energy security amidst global climate change concerns. Since the signing of the civilian nuclear energy agreement, India and France have made progress. The JNPP, besides the construction of six EPR power plants, also includes other important areas of collaboration in field research, safety and security, waste management, use of nuclear energy for applications other than electricity production (for instance, desalination), as well as education/training aspects (France in India: French Embassy in New Delhi 2018).

- On the industrial front, Areva, which in 2008 sold 300 tons of nuclear fuel to the public electricity utility NPCIL, submitted a tender for two EPR reactors (2×1650 MWe capacity) for the Jaitapur site, along with their fuel supply. Eventually, the site is due to house four other similar reactors (10,000 MWe capacity).
- In the R&D field, several bilateral dealings have been signed for the exchange and interactions between the two countries, notably CEA and ANDRA from the French side and DAE, Indira Gandhi Centre for Atomic Research, BARC and HBNI from the Indian side. This is mainly aimed at the nuclear reactors' safety/security, fundamental research, radioactive waste management and non-electrical applications of nuclear energy. In addition, there is

provision for cooperation in education and training and a master's degree in nuclear energy. To advance the civilian nuclear cooperation, exchanges and links have been established between the nuclear safety authorities of the two countries and their technical support.

- On the multilateral front, India is participating in major international R&D projects: Jules Horowitz Reactor (nuclear fission) and International Thermonuclear Experimental Reactor (ITER) (thermonuclear fusion; France in India: French Embassy in New Delhi 2018).

India is concerned about the lifetime uninterrupted supply of nuclear fuel and the right to reprocess spent fuel. Thus, it is important that France transfers reprocessing technologies to India for the dedicated new facility at the JNPP. In the long run, India aims at producing locally as much as 80 per cent of the nuclear reactors by the time the last pair of EPRs is built. This would benefit India's domestic nuclear energy industry during the importation of expensive high-end nuclear technologies. Under the nuclear agreement, India is aiming at the transfer of technology from its nuclear dealer countries under the offset arrangements.

In February 2013, during his State visit to India, President François Hollande and PM Manmohan Singh expressed satisfaction about the progress on collaborative projects in R&D on the peaceful uses of nuclear energy. India and France decided to further deepen bilateral civil nuclear scientific cooperation. Recalling the MoU signed on 4 February 2009 between NPCIL and Areva for the setting up of EPR units at JNPP. President Francois Hollande and PM Manmohan Singh reviewed the progress on the first two EPR units. It was observed that NPCIL and Areva were engaged actively in techno-commercial activity and making progress. Both leaders were positive about the speedy conclusion of the negotiations. The summit-level meeting emphasized that the highest safety standards would be applied at the JNPP (France in India: French Embassy in New Delhi 2018).

During PM Modi's visit to France in April 2015, M/s L&T and M/s Areva signed an MoU to maximize localization for the manufacturing of critical and large forgings involved in EPR technology for

JNPP (including reactor pressure vessel) and M/s Areva and NPCIL signed a pre-engineering agreement. Following M/s Areva's restructuring, the French company Electricite de France (EDF) became the designated lead agency from the French side for negotiations and implementation of the JNPP. After Areva's restructuring on 22 March 2016, EDF signed an initial pact with the Indian atomic energy producer NPCIL to supply six EPR units of 1,650 MW each at Jaitapur and fuel supply for the lifetime of the reactors, which could last up to 60 years. There have been regular negotiations and consultations between NPCIL and EDF towards the conclusion of a general framework agreement on project-related factors (MEA, Government of India 2017b).

At the summit-level meeting on 10 March 2018, India and France concluded 14 pacts in a range of sectors including defence, security, nuclear energy and protection of classified information besides agreeing to enhance the counter-terrorism cooperation and enhance the strategic ties to address the emerging challenges in the Indo-Pacific region. Both the leaders reiterated their intention to start work on the JNPP by December 2018 (*Times of India* 10 March 2018). EDF was to undertake all the studies and component purchases for the first two reactors and remaining four to be given to local companies (*Livemint* 2018). The NPCIL, the owner and operator of the JNPP facility, is accountable for procurement of certifications and supervision of the construction of the reactors and plant infrastructure. In this, EDF and its associates will assist NPCIL in its role by supplying the EPR technology and leading the engineering and component procurement for the first two reactors. Eventually, NPCIL is to take some responsibilities for the remaining four units under the 'Make in India' and 'Skill India' programmes (Proctor 2018). It is almost a decade now of the talks on JNPP nuclear power plant that began in 2008. In terms of generation capacity, the JNPP project with a total capacity of 9,900 MW, once in operation, will be the world's largest nuclear power plant. India's progress on nuclear energy development has been slow also because of the reluctance on the part of international equipment makers as India's nuclear liability law requires reactor suppliers to pay for the claims for damages in the case of an accident at the nuclear site. Both India and France have welcomed the understanding between the

EDF and NPCIL in regard to the compliance of India's laws on civil liability for nuclear damages related to the JNPP.

On the progress of JNPP, Jean-Bernard Levy, the chairman, and CEO of EDF apprised,

> The industrial agreement just signed with NPCIL marks a decisive step in the development of the Jaitapur nuclear project, meaning we can now envisage with confidence the rest of this essential project for India and for EDF. We are proud to support the Indian government in its objective of achieving an energy mix that is 40 percent carbon-free in 2030. Our presence in India, already tangible in the areas of renewable energies and the smart city is a perfect illustration of our CAP 2030 strategy, which aims to develop a low-carbon mix and innovative energy services for urban and rural areas (Proctor 2018).

The steady French support for India's membership in the international organizations dealing with civilian nuclear energy has been significant. France has been firm on its support for India's entry to the NSG, the Wassenaar Arrangement and the Australia Group. France has been unwavering in its support for India's candidature for the membership of Multilateral Export Control regimes, NSG and the Missile Technology Control Regime (MTCR).

Since the signing of the strategic partnership with France in 1998, subsequent Indian governments have acknowledged the significance of French backing of India's entry in the nuclear regime and their support to India's civilian nuclear project. Recently, India's Foreign Minister Sushma Swaraj in meeting with her French counterpart Jean-Yves Le Drian, acknowledged French support, 'France's support was vital in India's accession to The MTCR in June 2016' (*Economic Times* 17 November 2017). Both support a multi-polar world order, and France has been supportive of the UN reforms and India's bid for permanent membership in the UNSC (France in India: French Embassy in New Delhi 2018). India and France are concerned about the risks to the free movement in the SLOC in the Indian Ocean and the Pacific Ocean and have decided to work together prevent the hegemonic ambition of any power with an obvious indication to China. This concern clearly indicates the strategic and economic importance of the Indo-Pacific region in the pursuit of energy security. The security of the SLOC is a major component, both for exploration and shipping of oil and gas across the Indian and Pacific Ocean. These steps are elevating the India–France relationship to one that of trust and mutual confidence. Both nations are deepening their cooperation in ensuring freedom of navigation in the Indian Ocean, which could play a significant role in peace and stability in the Indo-Pacific region.

India's Nuclear Energy Cooperation with Russia: Towards Tapping the Massive Potential

India shares a special strategic partnership with Russia. Its relationship goes back to the Cold War days. In the post-Independence period of 1947, India, after maintaining a neutral position in the initial decades of the Cold War, tilted toward the Soviet Union. Scholarly works have explained this in the context of a number of factors including India's first PM Nehru's penchant for Fabian socialism, his anti-Western colonial power tilt, the military pacts of the US–Pakistan and the USSR military assistance to India after the US denial of high-tech military weapons to India.[16]

After the breakdown of the Soviet Union in 1991, Russia retained the permanent seat of the former Soviet Union on the UNSC and control of its nuclear arsenal and establishment. On 28 January 1993, India and Russia signed the Treaty of Friendship and Cooperation for establishing a strong and long-term basis for cooperation between the two countries. India and Russia have been holding summit-level meetings annually since PM Atal Bihari Vajpayee and President Vladimir Putin launched their first annual summit meeting in 2000.

India and Russia have been witnessing a strong defence and security relationship for more than five decades. Indian armoury is full of Russian defence products, and Russia continues to be the main

[16] For detailed insight, see Mastny (2010), Bakshi (2010), Singh (2012) and Bowles (1971).

defence supplier for India's defence needs. A decade ago, Russia dominated the India armoury with an 80 per cent share. Though in recent years India's options have increased and the US arms products have made significant entry into India's armoury, the Russian defence products still continue to be more than 60 per cent. This long-standing trusted partnership and the defence ties are helping India and Russia to expand the partnership in the energy sector including the nuclear energy sector.

As a result, besides the defence industry relationship, the nuclear energy has emerged as the most noticeable sector of India–Russia relationship over the past decade. India's nuclear energy ties with Russia go back to 1988 when both signed a deal under which Russia was to build a nuclear power plant at Kudankulam in Tamil Nadu, a southern state of India.

When India conducted the nuclear test in 1998, Russia, along with France, responded objectively to India's nuclear test. Russia was sympathetic to India's security concerns and refrained from criticizing India's atomic test in 1998. This could be seen in India's long and trusted relationship with Russia. Nevertheless, economic and geopolitical interests also drove this. Russia was very aware of the fact that criticism of the Pokhran nuclear test could hamper its business presence and prospects in India's growing economy, and its lucrative defence sector, which Russia has dominated since the Cold War period. Strategically, Russia did not want to hamper its trusted strategic partnership with India developed over five decades.

When in October 2000, President Putin came to India on the first summit meeting with his counterpart Atal Bihari Vajpayee, the agreements reached on defence and nuclear were major highlights. The cooperation in the peaceful use of nuclear technology was focused on nuclear energy. Though it was a major agreement at that time for India's nuclear energy programme, it came with many challenges. Russia's commitment to the agreement was constrained by its NSG membership that strives to control exports of nuclear materials, equipment and technology, for both civilian and nuclear weapons development programmes (Sethi 2000).

Despite strong resistance from the NSG and India's non-signatory status to the Nuclear Non-Proliferation Treaty, Russia went ahead with the construction of the KKNPP in Tamil Nadu that began in 2002, first envisioned in 1988 and reiterated in 2000. After India's waiver from the NSG in 2008 and the signing of the US–India nuclear agreement in 2008, Russia became the third country to sign the civilian nuclear agreement with India. On 5 December 2008, India and Russia signed an agreement on cooperation in the construction of additional nuclear power plant units at the Kudankulam site as well as in the construction of Russian designed nuclear power plants at new sites in India. Finally, on 12 March 2010, the Government of India and the Government of the Russian Federation signed an agreement on cooperation on civilian uses of atomic energy.

After four years, the two nuclear agreements were given substantial push during the summit-level meetings between India and Russia. On 11 December 2014, President Vladimir Putin visited India to commemorate the 15th Annual India–Russia summit-level meeting. This was Putin's first visit to India aimed at giving a boost to Russia's strategic partnership with India since the Narendra Modi-led BJP government came to power in May 2014. Both leaders signed 20 agreements to improve their special partnership to a qualitatively new level. Putin's attempt to reinvigorate a long-standing friendship with India was significant amidst severe strains in Russia's relations with the United States and the West. From India's perspective, the Modi government has sought to pursue a multi-pronged foreign policy approach towards the world's major powers and Russia continues to hold a privileged partnership status.

In 2014, Russia, adversely affected by Western sanctions over the Ukraine crisis, currency difficulties and plunging oil prices, turned its attention to Asia. Russia, after signing energy deals with China to revive its economy, needed to focus on India. Russia and India were keen on increasing bilateral trade from its then current level of $10 billion; Russia–India trade is below its potential and 10 times less than the total trade volume between Russia and China. On the Indian side, the Modi government, faced with huge expectations to fix the economy, hit the development goals and created jobs by making India a hub for manufacturing, but needed to find new energy sources to power the new Indian economy.

The agreements signed were significant and comprehensive. The agreements focused on enhancing energy ties, especially nuclear energy cooperation. The main agreements signed between Indian and Russian state-controlled entities included: NPCIL and Rosatom to construct 10 nuclear reactors in India; a preliminary agreement between Rosneft and India's ESSAR to ship around 10 mt of oil to India; an MoU between Russian Zarubezhneft and Oil India; and possible coopera-tion between Gazprom and India's GAIL concerning the delivery of Russian LNG. Both nations also agreed on a joint investment fund of $1 billion for Indian infrastructure and hydroelectric projects.

However, the main highlight of the summit meeting was the strategic vision for strengthening cooperation in the peaceful use of atomic energy. Both sides acknowledged the Treaty of Friendship and Cooperation between the Russian Federation and India signed on 28 January 1993. Based on their long-standing friendly relations and the trust developed between their people and the governments, India and Russia decided to enhance mutually beneficial cooperation in the peaceful use of nuclear energy. To attain this goal, India and Russia acknowledged the significance of the two previous agreements signed between their governments, namely:

- Agreement between the Government of the Russian Federation and the Government of India on cooperation in the construction of additional nuclear power plant units at Kudankulam site as well as in the construction of Russian designed nuclear power plants at new sites in the Republic of India, signed on 5 December 2008, and

- Agreement between the Government of the Russian Federation and the Government of the Republic of India on cooperation in the uses of atomic energy for peaceful purposes signed on 12 March 2010.

The 2014 nuclear agreements for peaceful provision included the following main points (MEA, Government of India 2014b):

- Both sides acknowledged the importance of the 'Roadmap for the serial construction of the Russian designed nuclear power plants in the Republic of India' signed on 12 March 2010.
- Both acknowledged the importance of the following two MoUs between the State Atomic Energy Corporation 'Rosatom' (Russia) and the DAE, Government of India: (a) concerning broader scientific and technical cooperation in the field of peaceful uses of nuclear energy signed on 21 December 2010 and (b) on cooperation with the Global Centre for the Nuclear Energy Partnership of India signed on 20 June 2011.
- Both nations welcomed their high level of bilateral cooperation in the field of peaceful uses of atomic energy and its significant achievements. At the same time, the sides recognized the substantial potential for the broadening and strengthening of their cooperation in the nuclear power sector, research and development in nuclear power and non-power applications of atomic energy, and engineering works.
- This potential was emphasized during recent high-level political exchanges between the sides, including the meeting between the President of the Russian Federation, Vladimir Putin, and the PM of India, Narendra Modi, on the sidelines of the BRICS Summit in Fortaleza, Brazil, on 15 July 2014.
- Acknowledging the importance of on-going cooperation, and in order to provide guidance for future cooperation, State Atomic Energy Corporation 'Rosatom' (Russia), and DAE, Government of India, have prepared the present document, entitled 'Strategic Vision for Strengthening Cooperation in Peaceful Uses of Atomic Energy between the Russian Federation and the Republic of India', with a view to provide strategic guidance for strengthening their cooperation in peaceful uses of atomic energy.

On the generation of nuclear power, the 2014 nuclear agreement has the following provisions (MEA, Government of India 2014b):

- Taking India's ambitious economic growth strategy, which would require a significant enhancement of power-generating capacity, the two-sides decided to fast-track the implementation of agreed cooperation projects for nuclear power plants. Both sides decided to work towards the completion of the construction and commissioning of not less than 12 units in the next two decades, in accordance with the nuclear agreement of 2008. Towards this objective, the Indian side agreed to expeditiously identify a second site, in addition to Kudankulam, for the construction of the Russian-designed nuclear power units in India. Both sides were to combine their expertise and resources to minimize the total cost and the time of construction of the nuclear power units.

- Both sides envisaged that the issue of the construction of further Russian-designed nuclear power plants in India would be considered taking into account India's demand for power, the then available nuclear technologies including those that may be developed jointly, mutually acceptable technical and commercial terms, and the prevalent electricity tariffs.

- India and Russia acknowledged the future sustainability of their robust cooperation, which would progressively and significantly enhance the scope of orders for materials and equipment from Indian suppliers and establish joint ventures, including the transfer of technology, as mutually agreed. This was to include manufacturing of both capital equipment and spares, with special priority for spares, for Russian-designed nuclear power units in India. The Joint Working Group on Nuclear Power was assigned the responsibility to consider the proposals from both sides to this effect.

- Both decided to explore opportunities for sourcing materials, equipment and services from the Indian industry for the construction of the Russian-designed nuclear power plants in third countries.

- Acknowledging the importance of the maintenance of nuclear power plants for uninterrupted operation, both sides agreed to put maximum effort into cooperation in such areas as nuclear power plants, technical maintenance and repair, modernization and retraining of personnel.

- From a long-term perspective, India and Russia visualized their collaboration in the decommissioning of nuclear power plants.

- The agreement included the provisions on the nuclear fuel cycle and both decided to work on a priority basis on the necessary arrangements for the fabrication in India of the nuclear fuel assemblies and their elements to be used in Russian-designed units. As envisaged in the agreement of 2008, the possibility of technical cooperation in mining activities within their territories, collaboration in exploration and mining activities in third countries, developing a framework for collaboration in the field of radioactive waste management and the Joint Working Group on the Nuclear Fuel Cycle, to be set up under Section 5 of this document, will elaborate possible approaches to the cooperation in the above-mentioned areas including facilities for spent fuel management.

- Both expressed their desire for cooperation in fields of scientific and technical and radiation technologies and the creation of public awareness and educational activity by collaboration in the development of human resources in their countries as well as in third countries through advance training in all aspects of the civilian nuclear sector, and to promote understanding and create a positive public perception of nuclear energy.

- The agreement also chalked out the ways of implementation by agreeing to establish a coordination committee for cooperation in the peaceful uses of the atomic energy (hereinafter referred to as the 'Committee'), which will oversee the entire range of bilateral cooperation, including the achievement of the objectives envisioned in this document. The committee will be headed by the Secretary, DAE (India) and the Director General, Rosatom (Russia). The committee will meet once every year.

India's collaboration with Russia in the nuclear energy sector is aimed at the construction of additional Russian-designed nuclear power units in India, collaboration in research and development of innovative nuclear power plants, and the manufacturing of equipment and fuel assemblies in India. The 2014 agreement was significant as it set the pace and gave direction to further development of nuclear energy cooperation. Given the scale of collaboration in the nuclear power sector, India's efforts to increase its nuclear power generation with Russia will play a crucial role in enhancing nuclear energy in its overall energy mix. Above all, the fact that India is potentially the world's biggest energy market calculates conspicuously in Russian minds and India is not averse to the deals given its appetite for energy.

Both sides took note on the status of the progress on cooperation on atomic energy cooperation. Both affirmed their mutual desire to further strengthen and enhance cooperation in this important area of their strategic partnership. They expressed satisfaction over the progress towards putting into commercial operation Unit-1 of the KKNPP, which achieved full-rated power in July 2014. Both nations also agreed to take steps to expedite commissioning of Unit-2 of the KKNPP.

The KKNPP is expected to generate a total power output of 6,000 MW, which is only just 12 per cent less than the combined power generated by India's 22 reactors which generate 6,780 MW. The Russian-built VVER-1000 reactors at Units 1 and 2 at the KKNPP have been in operation in full capacity since 2014 and 2017, respectively. Units 3 and 4 of KKNPP, under construction since November 2017, have a target for commercial start-up in 2025 and 2026. The Reliance Infrastructure Limited grabbed the contract for engineering, procurement and construction for the project. The work for the Units 5 and 6 are in progress and Russia has agreed to provide six more 'Generation 3-plus' 1200 MW units for a second site, which will generate more energy and operate for a longer period. All reactors are supplied by Atomstroyexport (ASE), a unit of Atomenergoprom, itself a holding company dealing with civilian nuclear technology under Rosatom (Gokaran 2018).

Russia has not only been supplying fuel for the lifetime of the first two KKNPP reactors but since 2009, Russia has also been the fuel supplier to India's indigenous reactors. The TVEL Fuel Company, a part of Atomenergoprom, is the main fuel supplier to India's nuclear reactors. One of India's main goals in the civilian nuclear agreement has been the development of indigenous capability in the civilian nuclear programme. Russia's support has been significant in this goal of India, as it has been willing to help India on increasing the localization. Currently, local manufacturers make about one-fifth of the equipment. India aims to increase local manufacturing to 50 per cent in addition to the construction of fabrication plants. The prospects for the two nations to collaborate in breeding reactors are bright. Russia is the only country to have commercially operating breeding reactors (Gokaran 2018). Despite its progress in its indigenous nuclear programme, India has not been able to develop the technology to process thorium. India is eager to exploit this opportunity and collaborate on the breeding reactor to exploit its abundant reserves of thorium to convert to nuclear energy. India's thorium reserves rank second in the world after Australia.

The economic weight of US-led sanctions has been compelling Russia to the look to the east in search of market and business avenues. This was reflected in the outcome of President Putin's summit-level meeting with PM Modi that saw him ink 20 high-profile agreements in around 24 hours. India and Russia inked US$100 billion commercial dealings. The deals were heavily dominated by energy dealings with $10 billion in a number of areas including defence commerce, space, fertilizers and diamonds, but $90 billion were in the energy sector, $50 billion in crude oil and gas and $40 billion for the construction of 12 new reactors at a new site to be decided by India (R. Sharma 2014).

Russia agreed to build 12 new nuclear reactors in India at a site other than Kudankulam in the southern state of Tamil Nadu. The cost of each of these new nuclear reactors is estimated to be worth $3 billion, three times costlier than what India spent on the two Kudankulam plant units. This increase in cost is attributed to India's stringent nuclear liability law that was passed in the Indian Parliament in 2010.

The business deals signed in the nuclear energy sector during President Putin's 2014 India visit are significant, as Russia became the first country to invest in India after India's nuclear liability law was implemented. The deal is mutually beneficial, as Russia became the early beneficiary of India's lucrative energy market and it also paved the way for other countries such as the United States, France, Canada, UK and Australia to invest in India's nuclear energy sector.

India offers Russia a large and unexploited nuclear energy market and Russia is keen on entering in new areas of business in India. Russia is facing the wrath of the US-led Western sanctions and it has no option but to look for new avenues and destinations for business. Russia's long-standing trusted relationship with India is helping it to get a foothold in India and in turn India gets investment and technology in its nuclear energy sector as it looks to increase the nuclear energy share in the overall energy mix.

India's relationship with Russia in the nuclear field and Russia's support for India's nuclear test need to be seen in the context of its long-standing strategic ties with Russia that date back to the Cold War days. Russia continues to be the major defence supplier of India, despite India's diversification of defence import sources. The trust and confidence, and the fact that India continues to hold a special relationship with Russia are now helping India to have strong ties with Russia in the nuclear energy sector as well. After the demise of the Soviet Union, many countries began to reposition themselves in the new international structure dominated by the United States. India reformulated its foreign policy towards the United States; freed from the ideological baggage of the Cold War, India pursued a multi-pronged foreign policy approach and has been able to maintain good relationships with both the United States and Russia.

In the wake of 9/11, Russia and the United States converged on the issue of terrorism and Moscow supported the United States in its war on terrorism. However, the stand-off over the crisis in Syria, Russia's annexation of the Crimea and its support to the pro-separatist movement in Ukraine brought Russia into confrontation with the United States and the West. Though India's deepening strategic partnership with the United States is no longer seen through the prism of Cold War dynamics, the US–Russia logjam has created unease for India. Despite the strains, India, a partner of Russia in the BRICS group of emerging-market economies, has refused to support any US- and EU-backed sanctions towards Moscow.

Notwithstanding Russia's strategic and economic interests in India, and the latter's multi-alignment foreign policy, there is a possibility of strain due to India's growing closeness with the United States. India's nuclear energy ties with Russia in early years of this century became contingent on many factors including the NSG guidelines and the US–Russia relationship. India's nuclear energy cooperation with Russia was not only driven by its desire to access Russia's high-end civilian nuclear technology but Russia's nuclear collaboration with India would have also set an example for other countries keen on tapping India's nuclear energy market. At that time, France had already shown its interest in nuclear energy cooperation with India after it refrained from criticizing India's nuclear test and had signed the strategic partnership with India in 1998. India was held back by the NPT and the NSG guidelines, and its non-signatory status to the NPT. Though Russia and France did not want to annoy the United States, in the NSG plenary meeting held in May 2000, neither agreed to the US proposal for the improvement of the safety guidelines nor advocated for the safeguards for less strict nuclear exports. It was evident that Russia and France were looking for means and reforms in the system through which they could enter India's nuclear energy market.

The global situation is a complex one; the United States and Russia cannot afford another Cold War. The United States also has strategic challenges greater than Russia, such as the rise of China and global security threats including environmental security and terrorism. Today a nation can have good relations with multiple countries even when they disagree on certain issues. The interdependence and the strategic complexities of world politics do not allow nations to have exclusive alliances. While India is getting closer to the United States and the West, its special relationship with Russia can help it play a positive role in reducing global tensions similar to the non-aligned position that India took during the Cold War era. Furthermore, India is concerned about Russia's improving relationship with China and it fears that sanctions might push them closer; a strong Russia–China partnership would not be strategically favourable for either India or the United States. India's support of Russian sanctions would add to any unease in India–Russia ties due, in part, to a reduced Russian share in India's defence market.

The subsequent governments have lauded the significant contribution of Russia in India's progress since India got independence. During the 2014 summit meeting, it was fittingly highlighted by PM Modi:

> We have a Strategic Partnership that is incomparable in content. The steadfast support of the people of Russia for India has been there even at difficult moments in our history. It has been a pillar of strength for India's development, security and international relations. India, too, has always stood with Russia through its own challenges. The character of global politics and international relations is changing. However, the importance of this relationship and its unique place in India's foreign policy will not change. In many ways, its significance to both countries will grow further in the future (MEA, Government of India 2014b).

Again, in the 2018 annual summit meeting, President Putin and PM Modi concluded an agreement to build six Russian-designed third-generation VVER nuclear reactors on a new site in India, to enhance nuclear energy cooperation in third countries and new nuclear technologies, increase the level of participation of Indian companies in the project and considered the idea of joint construction of nuclear plants. Both sides took note of the progress on the flagship project of Kudankulam for the civilian use of atomic energy. Russia reiterated its commitment to building 12 new reactors in the next 20 years.

The joint statement stated,

> The sides noted the progress achieved in the construction of the remainder of the six power units at Kudankulam NPP as well as the efforts being made in the components manufacturing for localization. The sides welcomed consultations on the new Russian designed NPP in India, as well as on the NPP equipment, joint manufacturing of nuclear equipment, and cooperation in third countries (*Times of India* 5 October 2018b).

The joint statement also reiterated their commitment to further strengthen global non-proliferation, and Russia expressed its support for India's membership of the NSG.

Rosatom has become the world's largest nuclear reactor builder as the financial problems of the two big Western firms Westinghouse and Areva have crimped their ability to develop nuclear plants abroad. Rosatom operates 35 reactors in Russia with a combined capacity of 28 GW and it has a portfolio of 36 nuclear power plant projects in 12 countries. Westinghouse and Areva, now owned by EDF, have for years negotiated deals to build reactors in India but have made little progress, partly because Indian nuclear liability legislation gives reactor manufacturers less protection against claims for damages in the case of accidents (Miglani and De Clercq 2018).

However, the construction of new sites is not without hurdles. In recent years, the nuclear sites have become controversial and an increasing number of public protests against these sites have been witnessed in India. This is in the wake of the disastrous effects that nuclear accidents can cause to the nearby population in case of accidents or natural calamities such as earthquake or tsunami. Once in operation, the India–Russia nuclear pact could emerge as one of the biggest nuclear industry deals in recent years, and would fasten the two countries for coming decades. Any foreign company or nation entering in India to tap India's nuclear energy market will need to consider these concerns. Russia is fast becoming India's reliable supplier of LNG and now the progress on nuclear energy is further making it a new energy source destination. India's long-trusted relationship with Russia is helping its nuclear energy generation plans bear fruit.

India's Nuclear Agreement with Australia: A New Beginning in the India–Australia Relationship

Despite sharing many commonalities in political, cultural and commonwealth legacy, India and Australia had a distant and uneasy relationship. This was mainly due to the Cold War strategic and political ideological dimension which did not allow India and Australia to form a close bilateral relationship. India's choice of pursuing a foreign policy of non-alignment (later, siding with the Soviet Union) was in contrast to Australia's close ties with the United States security alliance under the ANZUS pact. The demise of the Soviet Union, however, brought an end to these ideological differences. In the post-Cold War world, India took steps to liberalize its economy and restructure its foreign policy. Over the past decade, India's economic progress, expanding military prowess and the growing concerns over the emergence of a militarily assertive China began to be noticed globally. Indian and Australian interests began to converge on many issues. As a result, the India–Australia relationship has warmed, and trade, diplomatic and people-to-people ties have begun to flourish.

However, besides the Cold War politics, India's stand on the NPT was another point of disharmony in India's relationship with Australia. This continued after the end of the Cold War despite the warming of India–Australia relations in trade, diplomatic and people-to-people ties. Despite these positive developments in bilateral ties, India and Australia were unable to solidify their improved relationship because of the nuclear issue. India's non-signatory status to the NPT and its 1998 nuclear test for military purposes prevented both nations from coming together on many shared diplomatic and political interests. India's relationship with Australia was mainly dominated by the export of Australia's energy and resources to India.

Similar to the India–US relationship, the incongruity over India's non-signatory status of NPT obstructed the formation of a sound India–Australia relationship. As a result of its non-signatory status, until recently, Australia distanced from entering a civilian nuclear agreement which would allow the sale of Australian uranium to India. The first major breakthrough came in 2006 when then Australian PM John Howard visited India. PM Howard's India visit, alongside US President George W. Bush's 2006 India visit, reversed the long-standing policy of not selling uranium to India, an NPT non-signatory nation. Three years later, however, the Kevin Rudd-led Labour government overturned the Howard government's 2006 decision to sale Uranium to India, and again the nuclear issue became an irritant blocking the progress of India–Australia relations especially in the nuclear energy sector. Adding further complexity to its stance, the Rudd-led Labour government supported the nuclear exception that was made to India in 2008 when Australia voted along with the United States in the NSG to help India solve its energy security challenges.[17]

In short, the Rudd government supported India's admittance in global nuclear trade despite its non-signatory status to NPT but was not ready to sell uranium to India for civilian nuclear energy purposes. Rudd's incompatible stance on the sale of Australian uranium was criticized by the Indian government and in Australia by the groups within his own Labour Party and the opposition, the Liberal Party. Australia was selling uranium to China (an NPT signatory nation, but one with a suspect record of nuclear proliferation to Pakistan and North Korea), but banned uranium sales to India. To many, this was hypothetical and created distrust in India over the reliability of the Australian government to stick to its commitments. However, things changed when PM Julia Gillard became PM in a leadership challenge in Labour Party. The discussions on uranium sales recommenced in 2011. Gillard's government lifted the ban on uranium sales and acknowledged the centrality of the nuclear issue to India's mounting energy needs in building closer ties with India (Sharma 2014b; Wesley 2012. For a detail insight, see Medcalf 2011a, 2011b).

Finally, in September 2014, during his visit to India, the Australia PM Tony Abbott and PM Narendra Modi concluded the long-pending civil nuclear agreement, which would allow for the export of Australian uranium to India. The nuclear deal ended the confusion and mistrust and was pitched as a new beginning in the India–Australia relationship. This summit-level meeting covered a broad range of issues—nuclear energy, trade and education, and the defence relationship. In a major step towards realizing its nuclear energy ambitions, India signed a civil NCA with Australia.

The 2008 decision of the NSG to allow nuclear supplies to India was the basis for Australia to supply uranium to it as a country not a party to the NPT. The proposed agreement requires that Australian uranium supplied to India and derived nuclear material will be subject

---

[17] A. Sharma (2014b). For a deep insight on the Australian debate on uranium sale to India, see Thakur (2013).

to safeguards under the IAEA safeguards agreements (National Interest Analysis 2014).

The negotiation of the nuclear agreement has been accompanied by the development of a bilateral dialogue on related international issues. On 12 February 2014, Australian and Indian officials held an inaugural disarmament and non-proliferation dialogue, which covered a wide range of issues including nuclear doctrine, disarmament and non-proliferation and export controls. There was an agreement to hold such talks annually, with the next meeting to be held in India. The PMs of India and Australia met in the side-lines of the G20 Summit on 15 November 2015 and formally announced the ratification of the agreement (MEA, Government of India 2016).

Australian uranium and nuclear material derived from its use (such as plutonium) are termed Australian-Obligated Nuclear Material (AONM). Australia's bilateral NCAs provide assurance that AONM is used solely for peaceful purposes and is not diverted to nuclear weapons or other military uses. At present, Australia has 23 such agreements in place, providing for the transfer of AONM to up to 41 countries, plus Taiwan. These agreements build on the IAEA's safeguards system in order to assure the peaceful and non-explosive use of AONM. They also serve Australia's nuclear non-proliferation security interests by establishing a high standard of safeguards and accountability for a significant proportion of the world's uranium for peaceful use. Australia's bilateral agreements require that AONM be subject to IAEA safeguards for the full life of the material (National Interest Analysis 2014).

The provisions of Australia's NCAs, while broadly similar among the various agreements, differ according to the requirements of each case, including the status of relevant countries under the NPT. The provisions of the proposed agreement would implement Australia's policies for the safeguarding and accountability of supplied nuclear materials for the case of India, including consistency with Australia's international obligations. The international framework, within which Australia and India have negotiated the proposed agreement, is unique to India, as is the IAEA safeguards model for India in INFCIRC/754 (National Interest Analysis 2014).

The proposed agreement includes the following key elements (National Interest Analysis 2014):

- Assurance that AONM supplied to India will be used exclusively for peaceful purposes and will not be used for any military purpose (Article VII);
- Assurance that all civilian nuclear facilities in India, and any AONM in India, will be subject to the agreement between the Government of India and the IAEA for the Application of Safeguards to Civilian Nuclear Facilities done at Vienna on 2 February 2009 and the Protocol Additional to that agreement, done at Vienna on 25 February 2009 (Article VII);
- The requirement for maintenance of safeguards on AONM when under the jurisdiction or control of India (if the IAEA decides that the application of IAEA safeguards is not possible, the parties shall consult and agree on appropriate verification measures) (Article VII);
- Assurance that adequate physical protection measures are applied to all AONM during use, storage and transport and that such measures accord with accepted international standards (Article VIII);
- The requirement for prior Australian consent before any transfer by India of AONM to a third party (Article IX) and for any enrichment to 20 per cent or more in the isotope uranium-235 (Article VI);
- The requirement that reprocessing of AONM to separate plutonium is limited to facilities dedicated to reprocessing safeguarded nuclear material under IAEA safeguards and under modalities described in the 'Arrangements and Procedures Agreed between the Government of the United States of America and the Government of India' pursuant to Article 6(iii) of their Agreement for Cooperation Concerning Peaceful Uses of Nuclear Energy, done at Washington, DC, on 30 July 2010 (Article VI);
- Rights for either party to terminate the proposed agreement on one year's notice and to cease further cooperation in the interim (Article XIV), which could be exercised if, for example, IAEA safeguards were violated or following a nuclear test explosion;

- The provision for an administrative arrangement to be established between the appropriate governmental authorities of the parties to facilitate implementation of the proposed agreement (Article III); and the provision for consultation between the parties in order to ensure the effective implementation of the proposed agreement (Article XI).

The main obligations on both parties under the nuclear agreement are to ensure that (National Interest Analysis 2014):

- Any supplied or derived nuclear material, non-nuclear material, equipment, components and technology (items subject to the agreement), as well as by-products, shall only be used for peaceful and non-explosive purposes (Article VII);
- Safeguards shall apply to items subject to the agreement in accordance with specified agreements each side has entered into with the IAEA (Article VII);
- Adequate physical protection measures are applied to items subject to the agreement consistent with international nuclear security standards (Article VIII).

Finally, in December 2016, the Australian Parliament passed the final legislation enabling the sale of uranium to India after years of discussion within Australia amidst the larger nuclear non-proliferation norms and rules about supplying uranium to India which has a strategic atomic weapons programme but has refused to sign the NPT. On the first shipment of Australian uranium to India, Julia Bishop, the Foreign Minister of Australia, reiterated the confidence in India's impeccable record of nuclear non-proliferation and said the Australian government was confident that Australian uranium would not be misused,

> We cleared all of the parliamentary requirements for the civilian nuclear supply agreement, and we see India as a country that has adhered to its non-proliferation assurances (Bennett 2017b).

As a result, India became the 'first' non-NPT signatory country to receive Australian uranium, indicating the global acceptability of its nuclear programme. In this context, the landmark deal is mutually beneficial to both nations, and will help enhance India's nuclear safety, address India's looming energy crisis and boost the trade between the two countries.

- *Ensuring India's nuclear industry safety*: After the Fukushima catastrophe in 2011, there was growing concern for the safety of the nuclear power industry. India's nuclear industry is at a unique point of transition where there is both the political will and the industrial momentum to improve its nuclear safety (Wong 2014). India's nuclear industry, however, still lacks expertise in its public engagement and risk communication. Australia can help India improve the safety of its nuclear industry by sharing expertise in this risk management and communication.

- *Towards India's energy security*: Prior to the nuclear deal with Australia, India had inked nuclear energy agreements with 11 countries and imported uranium from France, Russia and Kazakhstan. However, Australia has about one-third of the world's recoverable uranium resources and exports about 7,000 tonnes of it annually, and the sale of Australian uranium can give a significant boost to India's energy security efforts.

- *Boosting India–Australia trade ties:* India–Australia bilateral trade in the energy and mining-related sectors can be further strengthened. The potential of India–Australia relations is massive but needs to be tapped. Its two-way trade which is reflected in the Indian companies such as Adani Group, GVK Group and Tata Consultancy have made large investments in the service, energy, mining and infrastructure sectors of Australian economy. The bilateral trade at Aus\$5.1 billion in 2003 tripled in 10 years to Aus\$15.2 billion in 2013. This could increase dramatically once the mining starts. The changes brought to state and federal government laws have opened up much of the country to uranium mining. With its huge uranium reserves, Australia can make a considerable difference in India's energy security efforts while advancing its own employment target. According to a parliamentary report, the Australia–India nuclear deal 'could add up to \$1.75b to the Australian economy' and double the number employed in uranium mining to 8,000 people (Thakur and Sharma 2018).

Despite these positive benefits, the deal was criticized in Australia on the ground of concession made without India signing the NPT. The nuclear concession made was deemed unnecessary with possible dangerous results in the future. It was criticized also on the ground of entering a new era of uncertainty and would harm Australia's bilateral relations with other countries. The deal was slammed as the wrong foundation for the future of the Australia–India bilateral relationship (Rovere 2014).

Just a day before, Australia signed a landmark free trade agreement with China more than 10 years in the making. Australia–India trade stands at around $15 billion a year, just one-tenth of Australia–China trade. India and Australia are, therefore, keen to expand the volume of their bilateral trade.

The agreement got further push when PM Narendra Modi made an official visit to Australia from 16 to 18 November 2014 during the G20 summit. This was after almost three decades when an Indian PM made a visit to Australia. While 'addressing the Australian Parliament', Modi emphasized the synergy between the two countries as well as the challenges that both nations face. He emphasized the need for early completion of a deal committing Australia to sell uranium to India, under suitable safeguards, to promote cleaner energy (A. Sharma 2014c).

The removal of a long-standing nuclear irritant after the conclusion of the nuclear agreement has unshackled the potential for security cooperation between India and Australia. The India–Australia relationship has moved well beyond a mutual fondness for cricket, as the two biggest naval powers of the Indian Ocean begin to forge a comprehensive bilateral relationship underpinning their sturdy economic, strategic and security partnership. In this, the nuclear agreement has been significant. After that nuclear agreement, Modi and Abbott signed a Framework for Security Cooperation covering a number of security areas including: maritime security, counter-terrorism, increased military exchanges and joint operations, technology transfers and border security. Both agreed to hold regular meetings at the level of defence minister, annual meetings of foreign policy leaders and to convene a regular exchange of talks and meetings with all the three branches of the armed forces. India and Australia also decided to deepen their counter-terrorism cooperation (Australian Government 2014).

India and Australia converge on a range of security challenges: containing the spread of Islamist fundamentalism, defeating international terrorism, securing Afghanistan, stabilizing Pakistan as a civilian-controlled secular democracy, ensuring the freedom of sea-lanes, preventing the domination of Asia by China, and stopping nuclear proliferation to other states and terrorists. (Despite the fact that both have substantial economic interests in China.) Although, neither India nor Australia expects the other to provide military support in the event of Chinese aggression, both hope that India's strategic presence in the region and its strong relationship with other maritime democracies will lead to a new regional security architecture that can prevent Chinese domination.

The growing strategic and defence partnership of India and Australia is enhanced by India's deepening strategic partnership with the United States and the US–Australia security alliance, and the US rebalance to Asia. A significant step in this direction was collaboration among the four maritime democracies in the Asia-Pacific region under the 'Quadrilateral Initiative', an idea that emerged after the naval exercise between the United States, Japan, Australia and India during tsunami rescue operations and the Malabar naval exercise in 2007 (A. Sharma 2010). The India–Australia bilateral relationship was elevated to the status of a Strategic Partnership in 2009. It has yet to achieve full maturity, but India's importance is growing in Australia's strategic and defence policy framing. This is reflected in the statistic of the dramatic jump in the number of times India is mentioned in Australia's Defence White Papers of this century compared with those of the last century. The nuclear deal is helping to enhance both their energy and defence ties. In 2015, the two conducted first ever bilateral maritime naval exercise.

Australia and India have a shared strategic interest in stable Indo-Pacific Asia that also links them to Indonesia and South Africa around the Indian Ocean rim. While Australia–India ties are not yet as deep as the ties that bind Australia to China, Japan and Indonesia, there are also fewer potential points of friction in the future. The single most critical geopolitical backdrop underpinning the growing security cooperation between Australia, India, Japan and the United States is the China factor. But what made a closer security relationship between India and Australia possible was India's dramatic switch from Moscow to Washington as its main strategic interlocutor in the post-Cold War era (Thakur and Sharma 2018).

The nuclear agreement should further build mutual trust and confidence between both nations to work towards a close defence partnership in the unfolding geopolitics in the Indo-Pacific region. 'Modi's promise of greater cooperation on regional security contained an implied swipe at China' over disputes with its neighbours regarding islands in the South China Sea (SCS). He said that peace and stability in the Asia-Pacific region could not be taken for granted and urged India and Australia to collaborate in international forums to enhance universal respect for international law and global norms, particularly in the SCS (A. Sharma 2014c; Parliament of Australia 2014).

While signing the new uranium deal, PM Abbott remarked, 'We signed a nuclear cooperation agreement because Australia trusts India to do the right thing in this area, as it has been doing in other areas…. That is why we are happy to trust India with our uranium in the months, years and decades.'[18] This shows the growing confidence and mutual trust that Australia and India have developed in their strategic partnership. Finally, one should note that Abbott was the Modi government's first state guest and that Modi is the first Indian PM to visit Australia in the last three decades. In the future, India will host summit-level meetings between the PMs of India and Australia annually, a gesture that India has extended only to Russia and Japan. Today, Australia is no longer confined to the periphery but has risen to the

---

[18] See http://www.saisreview.org/2014/10/22/australia-and-indias-nuclear-deal-a-new-beginning-in-the-india-australia-relationship/#_edn9

top tier of India's foreign policy priorities. This assists India's nuclear energy ties with Australia, enhances nuclear energy in its energy mix and pursues energy security amidst environmental concerns.

India's Nuclear Agreement with Japan: Normative, Technological and Strategic Significance

India and Japan have a long history of a good relationship. It goes back to the cultural exchange in the ancient times to Japan's support for India's independence movement, and India' support to Japan in the post-war period reconstruction. Both are democracies and share the Asian solidarity. But despite their convergence, they could not form a sound relationship as both were separated from a meaningful relationship in a comprehensive way because of the ideological political battle of the Cold War period where India and Japan were not on the same side of the two poles led by the United States and the Soviet Union. However, both shared a compatible relationship, especially in terms of cultural exchange and economic ties which was mainly in the form of India's export of raw materials in the form of iron to Japan and Japan's export of technological goods to India. The India–Japan relationship was marred by two dynamics: First, the Cold War politics and strategic dynamics and second, India's non-signatory status to the NNPT. Even after the end of the Cold War, the nuclear issue kept the ties between the two nations weak.

India's nuclear ties with Japan are important given the fact that Japan is the only nation which has faced the wrath of the atomic bomb and also in recent years because of the Fukushima accident. The nuclear issue is important in Japanese domestic politics and any move to delete article 9 can lead to the overthrow of the government. Japan has been at the forefront of the nuclear non-proliferation regime and prevention of the spread of nuclear bomb. Its normative role in the NPT is valued highly internationally (for detail, see Rublee 2009). Obviously, given this context, India's nuclear energy programme needed to be endorsed by Japan as well. Japan is also high-end in nuclear reactor technology and uranium processing. Japan is heavily dependent on nuclear energy sources for its energy needs.

When India conducted its nuclear weapons test in 1998, Japan was at the forefront in criticizing India. Japan led the UN sanctions against India. However, after negotiations and talks, the US–India nuclear agreement, India's agreement with the IAEA and the NSG waiver to India, the development on nuclear energy relations between India and Japan began to move in a positive direction.

The civil NCA was signed during PM Narendra Modi's visit to Japan in November 2016. The nuclear agreement should further enhance their strategic partnership in the energy sector. This landmark deal between India and Japan that provides for collaboration between their industries in the civilian nuclear energy field came into force in July 2017 (*Economic Times* 17 November 2017).

On 19 May 2017, the lower house of the Japanese Parliament endorsed the nuclear agreement and the House of Councillors approved the civilian nuclear cooperation pact between the two countries. This ended the long-standing nuclear sector disagreement and opened the way for the massive realization of the potential between the two nations. This was a welcome move by the Japanese parliament in enhancing the energy cooperation in particular and for the overall India–Japan relationship. It remains significant and symbolic (Pandey 2017).

It is significant from many perspectives. The pact helps to address India's energy security amidst the CO_2 emission concerns.

The deal is symbolic and brings both the nations together on the nuclear non-proliferation goals. Given the normative role that Japan plays in the NPT regime, the nuclear pact endorsed India's faultless record and unwavering commitment to nuclear non-proliferation and the strict export control regime. Since, India's refusal to sign the CTBT and its nuclear test in 1998, the nuclear issue has been a wedge between India and Japan. Tokyo has been very critical of New Delhi's atomic explosion in 1998 and worked proactively to impose the sanctions on India under the UN provisions. The nuclear issue hindered the prospects of the commerce in the nuclear sector between the Asian giants. Japan is the only nation, which faced the wrath of atomic bombings in 1945 on its city Hiroshima and Nagasaki. But

Japan has also been the leading nation on the front of stopping the nuclear proliferation and advocating complete nuclear disarmament. Japan was also the first country with the capability to become a nuclear weapon state to ratify the CTBT. India's non-signatory status to the NPT further made Japan hesitant to enter in the nuclear cooperation with India. With the conclusion of the deal, Japan has ended its nuclear apartheid towards India.

Japan acknowledges India's nuclear energy requirement. The approval of the civilian nuclear pact opens up a pathway of civilian nuclear commerce. The two countries can now engage on the matters of civilian nuclear cooperation, ranging from reactors construction to applied research in nuclear energy and safety and even reprocessing as per the text of the agreement. The deal in operation allows India to obtain the construction of nuclear reactors in order to meet its clean energy needs (Rajya Sabha 2017).

This is important given the status of India's nuclear energy industry, which is still at a developing stage with the aim at ways to expand its commercial interaction with other countries. This deal opens the avenues for cooperation in the nuclear energy sector. It goes well with India's nuclear energy policy that is looking for expansion of the nuclear fuel in its overall energy mix. With the commencement of the civilian nuclear agreement with Japan, India concluded the civilian nuclear deal with all the main nuclear supplier countries of the world. This is noteworthy and shows the success of India's nuclear diplomacy, especially given the Japanese opposition and hesitancy in signing the nuclear agreement with India because of its non-signatory status to the NNPT and CTBT. Shyam Saran, India's Former Foreign Secretary and nuclear deal interlocutor, hailed the deal as a win for Indian nuclear diplomacy as the decision to negotiate a deal of peaceful nuclear energy was taken back in the year 2010 (Saran 2011).

Additionally, Japan as one of the oldest players known for its technological prowess can offer its expertise to India's goal to enhance the nuclear energy share and control carbon emissions. Japan manufactures about 80 per cent of critical nuclear plant components and the Japanese cooperation is important for achieving the full potential of India's nuclear agreement with all other countries including the

United States and France (Nagao 2015). Japan specializes in producing nuclear power reactor cores and is highly specialized in the steelwork that it requires (Nuclear Power Industry in Japan 2012). Furthermore, the duration of the deal is also significant. It is important to note that this will likely last for the next 40 years with an automatic extension for 10 years thereafter, as mentioned in 'Article 17 of the text of the agreement' (Ministry of Foreign Affairs, Government of Japan 2016).

The deal also allows the United States and French nuclear firms, which have alliances with Japanese companies, to do business in the nuclear sector with India. Apprehensions were raised on the prospects of India's nuclear deal with the United States after the US atomic giant Westinghouse filed for insolvency. The Indo-Japanese civilian nuclear deal now has addressed this concern as Toshiba, which has the partner ownership with Westinghouse, can export the reactors to India for impending and new nuclear projects at Kaiga (World Nuclear Association 2017). This is significant as the United States and France had concluded nuclear agreements with India before Japan but their nuclear companies were facing problems as they had partnerships with Japanese nuclear companies. India's nuclear pact with Japan allows them to venture into India's nuclear energy market. French and the US companies were facing problems in transferring some technology to India, as their export to India needed Japanese companies for some crucial nuclear technologies by supplier consortiums such as Mitsubishi, Toshiba and Hitachi (Jain 2012). Japan is one of the four supply lines for the US Westinghouse AP 1000 reactors, which being constructed at Kovvada in Andhra Pradesh, beside, Toshiba, have significant investment in Westinghouse and the reactors at the French site in Jaitapur need Japan Steel Works for all the critical reactor pressure vessel (Chaudhary 2018).

The nuclear accord is strictly restricted to civilian purpose and gives Japan the right to break the deal if India goes against its no first use of nuclear weapons and ban on its nuclear weapons explosions (BBC 2016). But the deal does not put any restrictions on India's nuclear weapons programme as the deal does not hold for signing for the CTBT and reprocessing. This is evident from the absence of CTBT in the main agreement text. Furthermore, the pact does not put any restrictions on the reprocessing of the spent fuel from the Japanese-built nuclear reactors in India as long as India adheres to the IAEA safeguards and the Additional Protocol to safeguards is followed. The agreement also does not make any mention of the cancellation of the agreement in the case of any further atomic explosions by India. Instead, Article 14 of the text of the agreement mentions the termination clause, which means the 'right to either party to terminate the agreement prior to its expiration by giving one year's written notice'. In addition, any party seeking notice is also required to provide the reason for seeking such a course of action. In addition to this, clause 2 of the same article provides for a consultation mechanism between the two parties to carefully consider relevant circumstances and address the reasons seeking the termination (MEA, Government of India 2017a; Ministry of Foreign Affairs, Government of Japan 2016). A clause in the deal states that it will be up to Japan to assess whether the situations that may result in scraping of the deal was because of India's reply to its adversary nation's action that may have posed security threat to India (*Japan Times* 2017). In contrast to the popular notion, Japan's nuclear pact is not conditional and does not put any restriction on India's security concerns and its separate nuclear weapons programme as long as it is pursued under the IAEA guidelines. Japan is a significant nation in the realm of nuclear energy and it fully acknowledges India's non-proliferation commitments.

The nuclear deal with India shows the Japanese understanding of India's security threat environment and its nuclear weapon test in 1998, especially in the context of India's two nuclear-armed neighbours Pakistan and China. This degree of understanding is further enhanced in the context of security concerns that India and Japan share on Chinese military assertiveness in the Indo-Pacific region. There has been a growing strategic partnership between the two nations on the issues of maintaining stability, free navigation, protecting the SLOC and protecting the rule-based order in the Indian and the Pacific Oceans. Both India and Japan are concerned about the aggressive Chinese posture. The nuclear agreement would further enhance their mutual trust and confidence and help promote stability in the region.

In this context, former foreign secretary and interlocutor for the Indo-US nuclear deal, Shyam Saran (2011), rightly pointed out '...People in Japan are not always familiar with India's stand on nuclear security issues....' Today, both converge on greater security issues and work towards nuclear non-proliferation. In recent years, India and Japan have collaborated on matters of nuclear non-proliferation, disarmament and security. In August 2016, India and Japan held their 5th round of discussions on nuclear non-proliferation and disarmament issues. In fact, today some years after India's entry into the nuclear non-proliferation order, it can be argued that both the countries match up on various broader nuclear security issues.

India–Canada Nuclear Agreement: Revival of Four-Decade-Old Nuclear Relationship

Both the nations are commonwealth members and share a long history of bilateral ties. After the end of the Cold War, freed from the international political constraints of the ideological battle between the two blocs, and India's move to liberalization and its reformulation of its foreign policy have elevated the India–Canada relationship. The relationship is comprehensive and backed by a strong and influential Indian diaspora presence and growing economic ties. But it is the nuclear energy relationship which has been the significant component of India–Canada relationship since India's independence.

India's nuclear relationship with Canada goes back to the 1950s. India's advocacy of complete nuclear disarmament and commitment to the civilian use of nuclear energy by PM Jawaharlal Nehru was reassuring to Canada's leaders Louis St. Laurent and Lester B. Pearson. Though there were some hesitations on the benign intention of India's nuclear programme, Canada went ahead with civilian nuclear energy cooperation with India.

Canadian leaders took India's advocacy of complete nuclear disarmament positively, and their rapport with Nehru worked in India's favour. India was able to get Canadian reactors for civilian nuclear energy purposes. India received the Canadian nuclear reactor,

Canada–India Reactor Utility Services (CIRUS) of 40 MW, in 1954 under the 'Atoms for Peace Programme' for civilian nuclear energy purposes. The Canadian policy of nuclear cooperation with India was based on the assumption that if India were not able to get the nuclear technology from Canada, it would acquire it from other countries. By allowing the transfer of nuclear technology to India, Canada was also allowed to monitor required safeguards measures to prevent proliferation, and the IAEA safeguards would ensure the peaceful use of nuclear technology by India. In addition, Canada assumed that Indian nuclear scientists did not possess the scientific knowledge to follow a nuclear weapons project based on the Canada Deuterium Uranium (CANDU) design (Singh 2009–2010). In fact, Canada played a significant role in the development of India's nuclear programme. As stated, India's first nuclear reactor CIRUS was supplied by Canada along with uranium and nuclear technology. But this nuclear supply was stopped when India used Canadian technology and plutonium from the CIRUS to carry out a peaceful atomic explosion in 1974.

India's nuclear test upset Canada, as the Canadian reactors were meant for civilian purposes only (Government of Canada 2015). Canada's strong response to India's atomic test was the result of its own failures, as the Canadian government felt India's nuclear proliferation was the product of its failed attempt to monitor compliance, assure verification and insistence on safeguards on the CANDU reactors (Singh 2009–2010). After India's atomic tests, Canada decided not to do any nuclear commerce with countries that did not sign the NPT and stressed on unconditional full safeguards on all Canadian nuclear transfers, and implement a full termination of trade for nuclear-testing states. India's nuclear test was considered by Canada as a betrayal and attempts to revitalize the relationship did not work. When India conducted its atomic explosion in 1998 for military purpose, Canada was severe in its criticism along with the United States, Japan and EU. Canada immediately recalled its high commissioner, revoked CIDA programmes, shelved commerce talks, resisted India's request for World Bank loans and questioned India's bid to the UNSC permanent membership, and advocated an anti-India stance in consequent G8 meetings (Rubinoff 2012).

However, the changed international scenario, especially in the post-9/11 Islamic terrorism security threat environment and the Chinese military build-up, and India's growing economic and military profile led to India being perceived positively. India moved closer to the US-led political West. Finally, when the US–India nuclear deal was announced on 18 July 2005, Canada took it positively. Canada began to work on a civilian nuclear deal with India in August 2008, following negotiations at the IAEA and the NSG. Canada is significant for India's nuclear energy programme given the Canadian influence and normative influence in the Nuclear Non-Proliferation regime, and its advanced nuclear technology.

Canada supported India's deal with the IAEA and the NSG in 2008. Under the support of a ministerial trip to India, coupled with other Canada–India initiatives, including a potential free trade agreement, Canadian officials began the first round of negotiations towards a new nuclear deal with India (Singh 2009–2010). Finally, on 27 June 2010, the Indian PM Manmohan Singh and Canadian PM Stephen Harper signed the India–Canada Agreement for Cooperation in the Peaceful Uses of Nuclear Energy (*Hindu* 2010).

Both governments decided to work together to strengthen the existing friendly relations by recalling the existence of past cooperation between them in the use of nuclear energy for peaceful purposes. Nuclear energy provides a safe, environment-friendly and sustainable source of energy. Canada with comprehensive capabilities in advanced nuclear technologies agreed to full civil nuclear cooperation and to promote the use of nuclear energy for peaceful purposes. This was to be done with due respect for each other's nuclear programmes and in accordance with the principles governing their respective nuclear policies and their respective international obligations. Both agreed to bilateral cooperation for the development and use of nuclear energy for peaceful purposes with a view to achieving sustainable development and strengthening energy security on a reliable, stable and predictable basis. Both affirmed their international cooperation in the peaceful uses of nuclear energy and agreed to work towards nuclear non-proliferation (Government of Canada 2014).

Some of the main areas of cooperation in the agreement are as follows (Government of Canada 2014):

- Basic and applied research regarding the peaceful uses of nuclear energy;
- Cooperation between persons in Canada and India, in conformity with regulatory requirements and including the design, construction, operating experience, maintenance and decommissioning of nuclear reactors;
- The development and use of nuclear energy applications in the fields of agriculture, health care, industry and the environment;
- The supply of uranium and other natural resources;
- Nuclear fuel and nuclear fuel cycle management and nuclear waste management;
- Nuclear safety, radiation safety and environmental protection;
- Supply of material, nuclear material, equipment and technology, as well as facilities and services;
- Technology transfer on an industrial or commercial scale.

The 2010 NCA was followed by a suitable arrangement in 2013, which states that Canadian nuclear materials can be transferred to the Indian nuclear facility, which is covered under the IAEA safeguards. After two years of prolonged negotiations (since 2013), nuclear agreement between India and Canada was achieved by PM Modi and PM Harper for the commercial supply of uranium from Canada to India in 2015 during Modi's two-day visit to Canada. With the civil nuclear deal, the Indo-Canadian partnership entered a new phase and the NCA became operational.

The NCA between India and Canada and the supporting 'appropriate arrangements' allow for the effective implementation of the NCA and ensure appropriate oversight with respect to the information required by Canada.

- These agreements provide the framework through which nuclear cooperation for peaceful purposes is conducted, in accordance with Canada's nuclear non-proliferation policy. One commercial agreement involving Saskatchewan-based Cameco will see the company supply India with over seven million pounds of uranium over the next five years. This deal was made possible due in part to the India–Canada NCA (Government of Canada 2015).

- The Canadian Nuclear Safety Commission (CNSC) is responsible for ensuring that nuclear material, equipment and technology will only go to facilities in India under the IAEA safeguards regime that includes verification, physical inspections and detailed annual reporting. Canada will receive the necessary assurances from India and the IAEA safeguards programme on the peaceful use of each Canadian export of nuclear material, equipment and technology to India. These assurances will come directly from IAEA personnel who will be physically inspecting and verifying that India is meeting its international safeguards commitment that its nuclear facilities are only used for peaceful purposes (Government of Canada 2015).

The Modi–Harper uranium supply agreement is important from the commercial point as it is worth $350 million. According to the deal, Canada's biggest uranium producer, Cameco Corporation which produces about 16 per cent of the world's uranium, will export 3,220 metric tons of uranium concentrate for Indian nuclear power reactors for the period of 2017–2022.

In November 2016, at the fourth round of the India–Canada NCA Joint Committee meeting in Ottawa Canada agreed to help modernize the NPCIL, a development that comes 42 years after it withdrew from India's civilian nuclear programme in the wake of India's 1974 atomic test. Canadian firms are expected to produce nuclear reactor components in India and investments in the civil nuclear sector under PM Narendra Modi's 'Make in India' initiative (*Economic Times* 9 November 2016).

India and Canada are exploring cooperation in technology, subcontracting and collaboration in power generation involving NPCIL. With Indian reactors built on CANDU technology, India may also seek Canadian assistance to upgrade its reactors. India's indigenous nuclear reactors are based on CANDU technology. India's nuclear pact with Canada is designed to lower coal use, enhance nuclear energy in India's energy mix, diversify its acquisition sources and pursue energy security in a carbon-controlled environment. Canada's significance is further enhanced given the fact that India is trying to reignite its bid to the NSG membership based on its impeccable record of nuclear non-proliferation. Canada is backing India's entry into the 48-member NSG club.

During the conclusion of the deal with Canada on 15 April 2015, PM Narendra Modi characterized uranium as 'not just a mineral but an article of faith (for India), and an effort to save the world from climate change'. This is clearly in consonance with the Paris climate commitments of India and its urgency in reducing CO_2 emissions and the move away from coal which is very much in line with pursuing energy security under environmental acceptability under the conceptual framework of the four 'A's.

India's Nuclear Agreement with the UK

On 12 November 2015, India signed a civilian nuclear agreement with the United Kingdom reaffirming the importance of addressing climate change and promoting 'secure, affordable and sustainable supplies of energy'. Indian PM Narendra Modi's three-day official visit to Britain sealed the deal which was hailed by PM Modi: 'The conclusion of the civil nuclear agreement is a symbol of our mutual trust and our resolve to combat climate change,' in a joint statement with then British Prime Minister David Cameron (*Business Standard* 2015).

An MoU between the Indian and the UK departments for atomic energy was signed aimed at promoting joint training and experience sharing on civil nuclear activity within the Indian Global Centre for Nuclear Energy Partnership. The deal is considered as a 'comprehensive package' of collaboration on energy and climate change with the goal of supporting economic growth and energy security. The $4.9 billion package of commercial agreements aims to encourage the research, development and eventual deployment of clean technology, renewables, gas and nuclear power. The significance of this deal was that both nations emphasized the promotion of secure, affordable and sustainable supplies of energy and the need to address global climate change concerns (*World Nuclear News* 2015).

In a joint statement, Cameron and Modi also reaffirmed their commitment to take action on climate change at the domestic level. Cameron underlined UK's commitment to reducing its GHGs emissions by at least 80 per cent by 2050, as set out in the 2008 Climate Change Act, meeting its carbon budgets in the most cost-effective manner. Modi highlighted India's commitment to reduce its emissions intensity by 33–35 per cent by 2030 compared to 2005 levels and put in place 40 per cent cumulative electric power installed capacity from non-fossil fuel-based energy resources by 2030 through nationally determined development measures and priorities (*Business Standard* 2015; *World Nuclear News* 2015).

India's nuclear pact with the UK is significant as the UK severely criticized India's nuclear test in 1974 and 1998 and strongly backed the United States, Japan and other countries to impose sanctions against India. The UK continues to be a significant high-tech nation with immense influence in international organizations with permanent membership in the UNSC. The UK and India share a long history that goes back to the colonial period, and commonwealth legacy continues to nurture their relationship after India's independence in 1947 from the British rule. During the Cold War period, their relationship was strategically divergent. But in the post-Cold War period, there has been an increasing convergence on strategic and security issues between the UK and India. The UK support to India's nuclear energy efforts also needs to be seen in the context of India's closeness to the US-led Western democratic countries to tackle the challenges of the rise of militarily assertive China in the Indo-Pacific region. The United States and UK continue to be significant defence and security allies and both are concerned about China's assertiveness in the Indo-Pacific region as both stand for rule-based order and protect the global commons in maritime affairs.

India and Britain have of late enhanced their defence and military cooperation. PM Modi reaffirmed this during his visit that was the first by any India PM for a decade. The Indian PM said,

This cooperation will grow. I am also pleased that the UK will participate in the International Fleet Review in India in February 2016. The UK will also be a strong partner in India's defence modernization plans, including our 'Make in India' mission in the defence sector (*Business Standard* 2015).

India attached great value to defence and security cooperation with Britain, including regular exercises and defence trade and collaboration. The nuclear agreement will build mutual trust between the two nations. PM Modi, while issuing a joint statement along with British Premier David Cameron, expressed,[19] 'The conclusion of the civil nuclear agreement[20] is a symbol of our mutual trust and our resolve to combat climate change.' This not only strengthens the ties between the two Commonwealth nations in the nuclear energy sector but also strengthens their partnership in geopolitics where both converge on global security challenges of maintaining stability and securing the sea-lanes of communication in the Indo-Pacific region as per the rule of international laws.

Summing up, India's approach rests particularly on expanding its nuclear energy basket by incorporating the foreign and indigenous mix of technologies to enhance its nuclear power capacity in pursuit of energy security. In this context, India's nuclear agreements acknowledge its mounting energy requirements and the role that nuclear energy can play in addressing India's present energy insecurity. These deals are the testimony of India's consistent and concerted foreign policy efforts to turn around the nuclear technological regime and end nuclear apartheid. This has involved intense lobbying, negotiations and diplomacy visible in India's foreign policy during the process of the conclusion of nuclear agreements. India has overcome the nuclear apartheid and is progressing on nuclear energy power by entering into cooperation with nuclear technologically advanced and uranium-rich countries. These deals have put India into a new domain of collaboration in the field of civilian nuclear cooperation where India's energy needs

[19] https://www.business-standard.com/search?type=news&q=david+cameron
[20] https://www.business-standard.com/search?type=news&q=nuclear+agreement

and good citizenship and adherence to the nuclear non-proliferation norms have been acknowledged. Overall, India's nuclear agreements sit very well amidst the four A's concern, especially affordability as it is cheaper in the long run and acceptable as it is environmentally less carbon-polluting than hydrocarbons fuels.

India's nuclear deals with all the signalled countries, on the one hand, recognize India's energy needs and on the other hand, acknowledge India's atomic bomb compulsion in the current geopolitics. Regulatory oversight since 1974 has improved significantly and IAEA safeguards control provide credible assurances that countries are honouring their international obligations. The nuclear agreements with India reflect the new era in India's foreign relations with major nuclear power countries and the changing realities in India's new status in the emerging geopolitics in which it figures significantly as not only an economic player but also a major strategic player.

But nuclear deals have not been without opposition, not only from nuclear ayatollahs but also there was significant opposition to India's civilian nuclear agreements from China. This is dealt with in the next chapter, which focuses on energy security in the context of India–China competitive relations and their strategic rivalry amidst unfolding geopolitics.

CHAPTER 6

India's Quest for Energy Security Abroad
India–China Energy Geopolitics and Great Game

Securing and controlling energy sources have been a major aspect of modern geopolitics. Be it coal for Great Britain during the world war or oil for the United States in the post-war period of the second half of the 20th century, the security of supply of these energy sources has been a major priority and shaped geopolitics. The interaction between the energy producing and exporting countries and energy importing and consuming countries; the quest for a foothold, possession and/or stake in the energy-rich geographical regions/nations; and the global energy system among energy-hungry nations have been significant factors in shaping the geopolitics since the 1970s. The strategic importance of energy-rich regions has pushed energy-hungry nations to exert their influence. Given the strategic importance of conventional hydrocarbon energy sources—oil, natural gas and coal—in a modern nation's prosperity, development and overall power, it has been for a long time subject of geopolitics. Lately, the alternatives such as enriched uranium for nuclear reactors and renewable energy sources have become significant energy sources amidst climate change concerns and have begun to influence geopolitics. Overall, the quest for energy security has been a vital factor in the modern-day geopolitics.

Exploring and developing conventional energy sources require huge capital investments and military set up to control. As a result, in an age of increasing scarcity, producer, transit and consumer countries are positioning themselves geopolitically to safeguard their energy security. In pursuit of their energy security overseas, nations are interconnected and interdependent. The energy geopolitics becomes intense when two strategic rival nations compete for resources in the same geographic region, especially in times of energy shortage (Criekemans 2018).

In the past, the US–Soviet energy geopolitics during the Cold War which continues to exist between the US–Russia in the latest Syrian crisis and Eurasia reflects this. But over the past two decades, the two Asian rising powers China and India, the world's second and third largest energy consuming nations, have been competing for the energy security intensely. India's pursuit of energy security has put it in a competitive relationship with China, thus aggravating its existing strategic rivalry.

The energy geopolitics traditionally has been reflected in hydrocarbon sources, mainly oil and gas. The energy geopolitics has been concentrated in the Middle East and Caspian Basin, and of late in Africa, Latin America and the Indo-Pacific region, particularly in the waters of the SCS and the Indian Ocean. These regions have about 80 per cent of the world's oil and gas, including potential reserves. Both India and China are heavily dependent on these regions for their oil and gas needs. Their need to address their pressing energy demand has pushed them towards exploring new sources of energy and import destinations. India–China geopolitics is now being shaped by both their exploration of new sources of energy other than the Middle East and their quest for alternative and clean energy such as nuclear and renewable energy. Though India–China energy quests so far have been limited to commercial competition, diplomatic tussles and lobbying, it has the potential to grow in intensity due to their adversarial relationship.

This chapter examines India's energy security abroad in the context of its complex and conflicting relations with China. The chapter aims to illustrate how India's energy security pursuit overseas is putting it in direct competition with China, thus further intensifying their great power competition and long-standing strategic rivalry. The chapter deals with these first by a brief snapshot of India–China bilateral relations and the geopolitical rivalry, and then an overview of their overlapping oil and gas import sources, their energy security geopolitics in the Middle East, Africa and Latin America, mainly for oil and gas E&P, gas pipelines, renewable energy geopolitics, and finally by discussing nuclear energy geopolitics in the context of the US–India civilian nuclear deal and India's bid to the NSG.

India–China Strategic Rivalry

China and India are pitched as emerging great powers with growing geo-economic and geopolitical significance in the Indo-Pacific region. Since the beginning of the 21st century, China and India, with their rising economic and military power, have intensely ramped up their foreign policy, which is reflected in their bilateral strategic partnerships and alliances, joining of regional forums and multilateral organizations to advance their national interests and great power aspirations.

India and China, two ancient sister civilizations with deep cultural links since ancient times, developing countries home to two-fifths of the world's population and the world's two fastest growing economies, share commonalities in the form of Asian solidarity. Notwithstanding these similarities, their relationship has largely been conflicting instead of cooperative. Despite the attempts taken by both sides to improve the relationship by increasing bilateral economic engagement and sharing concerns at multilateral forums on global issues such as climate change, tackling regional political crises and restructuring the governing structures of the international financial institutions, both remain stuck in a legacy of mistrust and suspicion.

Today, both are fierce competitors in a world of global economic rebalancing, global power shifts, environmental degradation, other transnational security threats and their hunt for energy and resources. The relationship between India and China has been conflicting and often tense. This arises out of their border dispute, economic competition, scramble for energy sources and their big power aspirations. At the outset, this is due to their unresolved border tussle dating back to the days of tension arising out of the territorial claims of China over Tibet when in 1959 Dalai Lama fled to India and set up the Tibetan government-in-exile.[1] Since the Chinese attack on India in 1962, China has laid territorial claim over Tibet, Aksai Chin and Arunachal Pradesh. In Ladakh (Aksai Chin), China is in the occupation of 38,000 sq km of Indian territory. Besides this, the Shaksgam Valley (5,180 sq km) was ceded by Pakistan to China in March 1963 under a boundary agreement that India does not recognize (Kanwal 2008).

The Sino-Pak 'Border Agreement' established the Pakistan Occupied Kashmir (POK)-Xinjiang territorial division (van Kemenade 2008). A few years back, the Chinese decision of giving stapled visas for Indians from the state of Jammu and Kashmir (J&K) created a diplomatic problem. This implies China considers J&K as disputed territory and vindicates the Pakistani standpoint. India has raised serious objection to the People's Liberation Army's (PLA) involvement in several developmental plans in POK. India considers POK within its legal dominion and brands the Chinese activity as unlawful.

Over the past decade, China has intensified its claim over Arunachal Pradesh by asserting several times that the Tawang tract of Arunachal Pradesh is part of Tibet and the merger of this area with Tibet is non-negotiable. China's increased military activities, the construction of roads and railway tracks on the border, opposition to an ADB project in Arunachal Pradesh on the grounds that the territory is 'disputed' and asking Indian leaders to refrain from undertaking even official state visits have been concerning for India.

China's often-stated official position is that the reunification of Chinese territories is a sacred duty. PLA has adopted an enhanced

---

[1] For a detail insight on India–China relations and strategic rivalry, see Malik (2011) and Ogden (2017); For intractable India–China border dispute, see Guruswamy and Daule (2011) and Tuteja (2008).

military build-up by amassing a large number of troops in Tibet, constructing the Western Express Highway and railway lines for the faster mobilization of troops and has established two major missile bases in Tibet that can aim their missiles towards India (Kanwal 2008).

China–Pakistan 'All-Weather Relationship' is another factor in the India–China conflicting relationship. China is using its more than six-decade-long nurtured relationship with Pakistan to constrain India (*China Daily* 2011; Dixit 2006). The Chinese design to use Pakistan as a front for waging asymmetric war against India goes back to the 1950s when Zhou Enlai advised Ayub Khan that Pakistan should organize itself for a protracted conflict with India instead of short-term wars. He advised Pakistan to raise a militia force to act behind enemy lines. On this line, Pakistan raised this militia in the form of 'jihadis', indoctrinating youths in the region with jihadi ideology, and started planting armed modules in India back in 1992–1993. Today, China is Pakistan's main arms supplier, selling advanced fighter aircraft, supplying missile development capability, upgrading Pakistani submarines and jointly producing the Joint Strike Fighter-17 aircraft. The Trans-Karakoram Highway, vital for commercial and strategic purposes, connects the northern areas of Pakistan to Xinjiang Province in China.[2]

The support to Pakistan is a key aspect of China's 'encirclement of India' policy and a means of preventing or delaying India's ability to challenge Beijing's regional influence (Guruswamy 2010). In return for Chinese military and nuclear assistance to Pakistan, Beijing got observer status in the South Asian Association for Regional Cooperation (SAARC) against the will of India. The export of reactors to Pakistan, ignoring NPT norms, shows China's strong support to Pakistan's nuclear programme. China justifies it on the grounds of strategic reasoning, including the US–India nuclear deal and the NSG waiver for India in 2008 (Hibbs 2010). The prospect of its two rivals uniting to form a hostile nuclear and defence partnership is concerning for India. This is not just scaremongering: 54 per cent of Chinese

---

[2] This is revealed in a book authored by a Pakistani and published in Karachi in 2000 (Katoch 2010).

arms exports already go to Pakistan (Stockholm International Peace Research Institute 2014).

Over the past decade and a half, China has been pursuing a strategy of encircling India through the so-called 'String of Pearls'—a strategy of bases and diplomatic ties with the countries along the SLOC. It is a manifestation of China's rising geopolitical influence through efforts to increase access to ports and airfields, develop special diplomatic relationships, and modernize military forces that extend from the SCS through the Strait of Malacca, across the Indian Ocean and on to the Persian Gulf (Pehrson 2006).

These 'pearls' are interwoven into a strong chain or 'string' by virtue of their strategic positioning and placement to each other (Kim 2011). China's 'pearls' consist of an upgraded airstrip on Woody Island in the Paracel Archipelago, a container shipping facility in Chittagong, Bangladesh; the construction of a deep-water port in Sittwe, Myanmar; the construction of a navy base in Gwadar, Pakistan; a pipeline through Islamabad and over the Karakoram Highway to Kashgar in Xinjiang province; construction of a port in Pasni, Pakistan; intelligence gathering facilities on islands in the Bay of Bengal near the Malacca Strait; and the Hambantota port and refuelling stations in Sri Lanka (Lin 2008), and rail–road links with India's bordering countries. The strategy uses such things as economic aid and development projects to form common ground on which to construct relationships with the nations surrounding India.

China's 'String of Pearls' strategy is expanding into the Indian Ocean, which it wants to be recognized as being a Chinese sphere of influence, managed by Chinese nuclear submarines and aircraft carriers. Its strategic designs can be seen even in the Seychelles. Not only does it have the biggest embassy in the Seychelles, but it has also built refuelling facilities in the outlying islands. China has also offered to refit and refurbish the Sri Lankan Navy. With the exodus of a large number of Indian business persons preceded by the anti-India wave in Fiji, a large number of Chinese moved to the island nation (Katoch 2010).

The OBOR initiatives launched in May 2017 by China is a development strategy which is aimed at infrastructure development and investment by linking Asia, Africa and Europe through road and sea links. The OBOR, also known as the 21st-century Maritime Silk Road, is seen as China's expansionist strategy which India finds constraining and harmful to its interests. The OBOR was supported by a majority of nations but India was conspicuous by its full boycott. India has strongly condemned the OBOR and continues to refuse to acknowledge it as China–Pakistan Economic Corridor (CPEC); a vital component of OBOR violates India's sovereignty. The CPEC links China and Pakistan through the disputed territory of POK. Their bilateral trade relationship is also fraught with the balance of payment in favour of China. The gap has been increasing despite India's repeated call for opening up of the Chinese market for Indian goods and services.

China defends its action in India's vicinity and the Indian Ocean Rim on economic grounds and states that this is to ensure a harmonious ocean. However, India views the 'String of Pearls' strategy as increasing the sea power of China threatening India's interests in the Indian Ocean, constraining India in its own backyard and dominating the Indian Ocean Region (IOR).

The enduring India–China strategic rivalry has been further intensified by India's move to counter China by growing strategic ties with the United States (for details, see Ganguly, Scobell, and Shoup 2006). India is strengthening its strategic ties with the United States, Japan, Australia and China-wary countries in the Indo-Pacific region. India has taken resolute efforts to counter the Chinese 'String of Pearls' by entering into a full-spectrum bilateral economic, political and strategic engagement with almost all the major countries and powers which are concerned about the rising Chinese military assertiveness in the Indo-Pacific region (A. Sharma 2014a).

To counter China's military preparation and encircling strategy, India has also taken significant steps. On the border front, India is beefing up its military by raising a 50,000-strong mountain strike corps along the eastern sector to boost India's offensive capability to counter China (*New India Express* 2013). In the immediate neighbourhood, India has enhanced its ties with Nepal, Bhutan, Bangladesh and Myanmar. PM Modi under the 'Neighbourhood First Policy' has sought to pursue friendly relationships with India's northeastern neighbouring countries. He has taken official visits and signed many agreements, launched development projects, pursued economic engagement and given aid to enhance the bilateral ties and resolve the issues that hampered India's relationship in the past. A good relationship with its neighbouring nations is imperative for India to balance Chinese influence in its neighbourhood. And to counter China–Pakistan dual influence in its northwestern front, India has pursued a strong and comprehensive engagement with Afghanistan. The strategy aims not only at the reconstruction of Afghanistan—India is the biggest regional aid provider—but also at helping Afghanistan security forces through training and enhancing economic ties with the India–Afghanistan preferential trade agreement.

On the eastern front, India's response to China's encircling strategy is the 'Look East Policy'. This policy evolved from isolation and neglect following India's humiliating defeat in the 1962 war with China to a more pragmatic policy since the early 1990s to engage the region economically. Though the economic aspect has yet to achieve its full potential, it is the strategic dimension which reflects India's political, diplomatic and military counterweight to the growing Chinese assertiveness in the region (Acharya 2015; A. Sharma 2014a). PM Modi has pursued this policy as the 'Act East Policy' which has cultural/historical, economic, institutional and strategic dimension. But it is the energy security which is becoming the major component of India's economically driven strategic engagement of the region. India's growing economy needs massive energy and India is not averse to the idea of exploration of oil and gas reserves in the resource-rich sea areas between the Indian Ocean and the Pacific Ocean. The Act East Policy reflects that India is ready to engage with the region proactively and ready to shun its past habit of cutting off itself from the external economic and strategic trends in the region. India's eastward engagement, in the beginning, was focused on the economic engagement of the region but now reflects a strong geopolitics-driven interest in which energy security has become a significant component.

Over the years, India's approach to its eastern region has matured into a broader strategic engagement with the ASEAN and its related frameworks such as the ASEAN Regional Forum (ARF), East Asian Summit (EAS) and ASEAN Defence Ministers Meeting (ADMM), and also with countries further east, including Japan, South Korea, Australia and the Pacific Islands (MEA, Government of India 2018). India's Act East Policy and balancing strategy in the Indo-Pacific region manifest imperatives from both India and neighbouring nations in the eastern waters. This is more visible in its growing strategic engagement with the ASEAN nations. Given no territorial disputes between India and the ASEAN nations, India's deepening strategic partnership with the United States amidst the concerns of the Chinese naval expansion in the region, and India's growing naval capability and presence in its Eastern waters make India a strategic asset to the China-wary countries in the Indo-Pacific region and ASEAN (Acharya 2015).

India's growing economic power and political and strategic move-ments over the past decade and a half, visible in the Indian and Japanese bids for a permanent UNSC seat; the formation of East Asia Summit (that includes India, Australia and New Zealand); strategic implications of 'India's Look East Policy' now being implemented under the 'Act East Policy'; and its growing engagement with coun-tries wary of Chinese influence (Mongolia, Taiwan, South Korea, Myanmar, Vietnam and the Philippines) in the Indo-Pacific region have alarmed China. India's nuclear deal with the United States and a deepening India–US strategic partnership, the revival of the Quad, consisting of maritime democracies, namely US, Japan and Australia, India's growing naval capability, and its 7–8 per cent average growth rate since the beginning of the 21st century have further alerted China and necessitated it to reassess India's strategic movement (Malik 2011; A. Sharma 2014a).

Looking at the rising economic and political profile of India and the strategic trends, some Chinese strategic analysts caution Beijing about New Delhi's ambitions. They are of the view that although there is the possibility of makeshift management of China's bilateral relationship with India, in the long run, the prospect of cooperative relationship is overshadowed by India's continued nuclear and missile programmes, territorial disputes and the absence of mutual trust and confidence. Chinese strategists consider that 'the enhancement of India's strategic position will reduce China's strategic influence to some extent, especially in the Third World, [which] thus will weaken China's strategic role, making it more complicated for China to deal with major powers' (Jiali 2002, 14; Krishnan 2010). The attacks on India have grown considerably in the state-owned Chinese media, a reflection of some nervousness in Beijing about India's growing assertiveness. This has happened even as China's military and economic embrace of Pakistan is almost complete with CPEC on one side and a potential military base in Gwadar on the other (Pant 2017). The *Global Times*, the Chinese state-run newspaper, has repeatedly projected India as a threat to China's strategic interests that must be countered. The Chinese media sees India as a country not posing an immediate threat to China, but a country that could transform it into a formidable adversary in the long run.

India's perception of China oscillates between one of the Asian solidarity and bad memories of humiliating defeat by China in the 1962 Sino-Indian border war. On the one hand, India is amazed by Chinese economic progress and desire to emulate the Chinese economic success, but on the other hand, also the more dominant view, China is seen as revanchist, aggressive and harming India's interests and march to its great power status.

Over the past 10–15 years, their complex and conflicting relationship has been further intensified as both Asian powers try to outdo each other in their scramble for resources and attempts to project power on the international stage. The need for high economic growth rates, insufficient domestic energy resources and demand for energy have compelled India and China to compete for resources in energy-rich areas. Their geopolitics is further driven by energy competition because of overlapping hydrocarbon exploration destinations.

India–China Energy Geopolitics: Scramble for Energy Sources and Overlapping Hydrocarbon Acquisition Sources

India needs an economic growth rate of around 8 per cent to meet its economic and social development goals. With 17 per cent of the world's population, India has just 0.8 per cent of the world's known oil and natural gas resources. India's crude oil import dependency is expected to increase to 90 per cent (A. Sharma 2007). (India's energy demand and supply and its energy mix have been discussed in Chapter 2.) Though India aims to move towards renewable sources, its oil dependency is not going to decline in the near future.

China, with 1.3 billion people, is the world's second largest oil consumer. Economically, it has been developing at an average growth rate of 8–10 per cent a year for the past three decades. It is the world's second largest energy consuming nation and grew by 1.3 per cent in 2016 in terms of energy consumption. The growth during 2015 and 2016 was the lowest over a two-year period since 1997–1998. Despite this, China remained the world's largest growth market for energy for a 16th consecutive year as stated by the 2016 *BP Statistical Review of World Energy* (BP 2016). The latest BP statistics show that China's demand for the gas and oil continues to be the driving factor for global consumption. The year 2017 saw a surge in Chinese gas demand, where consumption increased by over 15 per cent, accounting for around a third of the global increase in gas consumption. This was mainly because of China's move to shift to gas amidst the growing need to control carbon emissions. China's Environmental Action Plan announced in 2013 set the goal for improvements in air quality over the subsequent five years. With that five-year deadline looming, the Chinese authorities in 2016 came out with an enhanced set of measures for Beijing, Tianjin and 26 other cities in the Northeast provinces of China. These measures, which were further reinforced in the autumn of 2017, focused on the use of coal outside of the power sector. In particular, a combination of very sizeable carrots and sticks were used to encourage industrial and residential users to switch away from coal to either gas or electricity, with the vast majority opting for gas (BP 2018).

Both India and China rely heavily on imported oil and have enormous demands for energy. Both have insufficient resources of oil and gas and are looking for these resources in not only traditional regions of import such as the Middle East but have E&P ventures in Africa, the Caspian Basin and Latin America. While the Middle East, especially the Persian Gulf, continues to be the centre of the geopolitics of energy, in recent years the Caspian Basin, Africa, Latin America the SCS and the Indian Ocean have become contested regions for oil and gas exploration. These hydrocarbon destinations are becoming the new centre of energy geopolitics between India and China.

India–China Oil and Gas Exploration Destination: Overlapping and Competition

Globally, China and India are major oil consuming countries. China and India are the world's second and third largest oil consumers, respectively. Since the beginning of the 21st century, their oil consumption has rapidly increased due to economic growth. Dependency on fossil fuel in their energy mixes shows a similar trend. Coal is the primary energy source, followed by oil. Since 1965, the proportion of oil in the total non-renewable energy consumed in India has been higher than that of China. In 2015, oil accounted for 18.6 per cent of the total non-renewable energy consumed in China. This ratio was 27.9 per cent in India. China's oil consumption rose from 112.9 MMT in 1990 to 559.7 MMT in 2015, with an average annual growth rate of 15.8 per cent. During the same period, India's oil consumption rose from 57.9 MMT to 195.5 mt, with an average annual growth rate of 9.5 per cent. In 2015, oil consumption grew by 6.3 per cent and 8.1 per cent in China and India, respectively. Both countries maintained a relatively rapid growth rate of oil consumption. In 2015, China and India accounted for 12.7 per cent and 4.5 per cent of the total oil consumed worldwide (Zhang and Xing 2018).

Both have insufficient domestic oil reserves and production to meet demand. Since 2000, China's dependence on foreign oil has gradually increased. India's dependence on foreign oil has remained at or above 70 per cent. In 2015, China's oil dependence was 62 per cent;

for India, it was up to 79 per cent. The share of China and India's crude oil imports in the world rose from 9 per cent in 2000 to 27 per cent in 2015. During the same period, this trend was the opposite in Europe, United States and Japan as they witnessed a decreasing trend of import of crude oil (Zhang and Xing 2018).

A comparative study undertaken by Zhijie Zhang and Wanli Xing on overseas oil cooperation between China and India based on crude oil trade flow shows there is overlapping in both nations' oil and gas import sources. The study found that the competition over oil sources between China and India is mainly concentrated in eight countries (Saudi Arabia, Iraq, Iran, Venezuela, Kuwait, Brazil, United Arab Emirates and Angola) and that Africa, and Central and South America, will become important competitive regions for energy resources in the future (Zhang and Xing 2018). China and India import crude oil mainly from politically unstable countries and regions. To ensure the security of supply, both nations have taken steps to diversify their oil import sources.

The study by Zhang and Xing uses UN trade data from 2006 to 2015 which demonstrates that India's oil-importing destinations jumped from 26 to 44 countries and regions. Consequently, in 2015, 94.5 per cent of India's oil and 98.5 per cent of China's oil came from the same countries and regions. China and India's crude oil imports mainly depend on the Middle East, West Africa, and Central and South America; in 2006, these three regions accounted for 45.2 per cent, 24.4 per cent and 5.1 per cent of China's oil imports, respectively, compared to 72.1 per cent, 12.7 per cent and 1.5 per cent, of India's oil imports. In 2015, China's percentages were 50.7 per cent, 14.9 per cent and 12.4 per cent, respectively, whereas, for India, the percentages were 58.3 per cent, 16.4 per cent and 15.0 per cent (Quan and Liu 2012; Zhang and Xing 2018).

Table 6.1 shows the list of eight overlapping countries between China and India—Saudi Arabia, Iraq, Iran, Venezuela, Kuwait, Brazil, United Arab Emirates and Angola. The import of oil from these eight countries by China and India in their total oil imports grew steadily. The trend also shows India's entry in Latin America and Africa resulting in competition to China's energy security efforts.

Table 6.1 China and India's Crude Oil Imports to the Total Production of Major Oil-Exporting Countries

Oil Import	Saudi Arabia		Angola		Iraq		Iran		Venezuela		Kuwait		Brazil		UAE	
	Ch (%)	Ind (%)	Ch (%)	Ind (%)	Ch (%)	Ind (%)	Ch (%)	Ind (%)	Ch (%)	Ind (%)	Ch (%)	Ind (%)	Ch (%)	Ind (%)	Ch (%)	Ind (%)
2006	5	4	34	1	1	10	8	6	2	1	2	6	2	0	2	4
2007	5	5	30	2	1	11	10	9	2	1	3	8	2	0	3	7
2008	7	5	32	2	2	12	10	10	4	4	4	10	3	0	3	10
2009	9	6	37	8	6	11	11	12	3	3	6	13	4	2	3	9
2010	9	6	44	9	9	11	10	8	5	7	8	11	7	2	4	9
2011	10	6	37	9	10	17	13	6	8	6	7	12	6	3	4	9
2012	10	6	46	11	10	16	12	8	11	13	7	13	5	4	6	10
2013	10	7	46	10	15	17	13	6	11	16	6	14	5	2	6	9
2014	9	7	49	9	18	14	16	8	10	16	7	12	6	4	7	9
2015	9	7	49	9	16	16	15	6	12	17	10	8	11	3	7	9

Source: Zhang and Xing (2018).

o address the growing dependence on oil and gas and reduce the import dependency on the Middle East, India and China have diversified their hydrocarbon exploration to Africa, Latin America and Russia. In the process of exploring overseas oil supplying sources, both nations are competing for the same assets in the same countries. This overlapping has further aggravated their energy geopolitics as both are trying to outbid each other. Figure 6.1 shows this trend.

Due to insufficient domestic oil and gas resources, both nations' dependence on foreign oil continues to grow. According to the 'India Energy Security Scenario 2047' report, India's reliance on imported oil will continue to grow to approximately 90 per cent in 2030. In pursuit of new import sources of hydrocarbons, India has ramped up its foreign policy engagement with LAC and African nations (discussed in Chapter 4 in detail). India's effort to access overseas oil resources is affecting China's energy security interests and efforts abroad. This has triggered energy competition between the two nations.

The scramble for energy sources has put them in direct economic and maritime competition. The state-sponsored Chinese companies and the Chinese government together have been intensely searching for energy sources and outbidding Indian state-owned and private companies by discounts and exaggerated payment. In many instances, India finds it difficult to match overflowing Chinese bids which are deliberately intended to grind down India's ability to enter any significant economic partnership with the host nation.

Since the beginning of the 21st century, India and China have been engaged in cut-throat competition for access to some of the world's richest oil and gas deposits. Both the nations' heavy dependence on imported oil and the need to diversify the import sources have forced them to search for overseas oil and gas assets. The solution adopted by both countries has been to seek the so-called equity oil and gas, which is sourced from exploration and development contracts overseas. China's national oil companies are ahead of India's OVL in bidding and have beaten India on several occasions (Larkin 2005). Both competed for oil contract in Angola and Ecuador (Zhou 2006). China has also used its veto power in the UNSC in its bid to outmanoeuvre India

Figure 6.1 Overseas Oil Suppliers of China and India

Source: Indian Oil Company Annual Report; China Petroleum Company Annual Report; Zhang and Xing (2018).

to secure access to energy and business opportunities. China blocked the UNSC efforts to place sanctions on Ecuador by exercising its veto power to help a Chinese company get oil concessions in Ecuador. With these leads and strategies, China is ahead of India in the scramble for energy sources in foreign countries.

But India also on many occasions has been successful. For example, the ONGC won exploration rights in Egypt and Qatar in 2005. In the same year, India's ONGC and China's biggest oil producer, PetroChina Company Limited, competed to take over Canadian oil company PetroKazakhstan. Their rivalry has resulted in an increased cost of assets abroad and made some of the investments unsafe. In May 2011, India declared a generous economic aid package for African countries. India is no longer complacent in this scramble for resources. As discussed in Chapter 4, India has taken concerted efforts to engage Africa by investment in the energy sector, its own line of credit and India–Africa summits.

Over the past decade, China has pursued a concerted effort to engage African littoral states in the IOR. To counter the Chinese growing economic presence in Africa and maritime build-up, India has taken steps to engage African nations including those littoral states in the Indian Ocean. This is strategically significant as it concerns vital shipping lanes through which oil and cargo ships pass through.

The most recent example of India–China energy-driven competition for influence in Africa was during the 10th BRICS summit held in South Africa on 25–27 July 2018. PM Modi visited Uganda and Rwanda before the BRICS summit in Johannesburg to engage these two African nations. The significance of the visit could be seen in the context of energy geopolitics to enhance India's presence in the region. Rwanda is China's biggest trade partner and both have a strategic partnership which encompasses energy as well. China and Rwanda inked a strategic partnership in 2017. Modi was the first Indian PM to visit Rwanda where he signed several agreements on trade, agriculture and defence. In Uganda, Modi delivered the first-ever address for an Indian premier and announced two key LOC alongside support for

African states across sectors such as agriculture, health, education, infrastructure, defence and energy (*Economic Times* 24 July 2018).

Their efforts in Africa and Latin America to outbid each other are a testimony to the growing India–China energy competition for oil and gas sources abroad. In the future, Africa and Latin America will be intensely contested regions for oil and gas exploration. Both nations have been looking to these two regions for their overseas oil dependency. India has strongly focused on these regions by engaging Latin American and African nations comprehensively in bilateral relations to get rights and a good deal for oil and gas E&P.

India–China Competition in the Indo-Pacific Region: Indian Ocean and the SCS

The traditional India–China geopolitical rivalry has spread into the maritime dimension as it is the main route to ensure energy security. China's naval strategy is part of a well-coordinated plan to ensure energy security and neutralize India's efforts to gain any traction in business, military and political influence in various energy-rich host states. The focus of this strategy is various choke-points in the SCS, the Strait of Malacca, the Indian Ocean and the Strait of Hormuz in the Persian Gulf.

The Indian Ocean

The IOR has become crucial in energy security geopolitics. The significance lies in the role Indian Ocean states play as energy suppliers, and more importantly, it is the strategic location which makes it a vital energy route (for a detailed insight, see Rumley and Chaturvedi 2015). The Indian Ocean is the energy route to the maximum number of ships carrying oil and is the busiest maritime transportation link in the world. Almost a hundred thousand ships a year pass through these waters, carrying about half of the world's container shipments, one-third of the world's bulk cargo traffic and two-thirds of oil shipments. The fact that three-quarters of this traffic is headed for destinations beyond the region highlights the strategic significance of the Indian

Ocean beyond the shores of the littoral states (MEA, Government of India 2018). But the Indian Ocean has several choke-points which if blocked can hamper the transportation of oil. The IOR has three crucial sea lanes of transportation through which ships carrying energy pass through. Any disturbance in these can be detrimental to the security of supply and affect price stability in the global energy market.

These choke-points are narrow channels along widely used global sea routes. They are the critical part of global energy security due to the high volume of oil traded through their narrow straits. The Strait of Hormuz, originating out of the Persian Gulf, and the Strait of Malacca, connecting the Indian and the Pacific Oceans, are considered as the two major strategic choke-points. Both are in the Indian Ocean. Another important passage in the IOR is the Bab-el-Mandeb which lies between the Horn of Africa and the Middle East and is a strategic link between the Mediterranean Sea and the Indian Ocean. As the international energy market is dependent on reliable transport, the blockage of a choke-point, even temporarily, can lead to substantial increases in total energy costs. As a result, the international community has been concerned about the security of this regional SLOC. One of the examples of energy geopolitics in the Indian Ocean was a notable confrontation between the US Navy and the Iranian Navy in the Straits of Hormuz. Iran threatened to seal off the Strait of Hormuz to wreak havoc in oil markets. In response, the United States deployed its 5th Fleet in Bahrain across the Persian Gulf to prevent any such move. Again, Iran has recently threatened to respond to economic sanctions against its oil exports imposed by the United States with military action to shut down the Strait of Hormuz (Ho 2011; Lockie 2018).

The Straits of Malacca provide the main corridor between the Indian Ocean and the SCS and are a major energy route used by tankers from the Middle East, carrying oil through the straits daily. The threats to shipping in the straits include maritime terrorism, and more recently rampant piracy, especially in the Gulf of Aden and in the Indian Ocean. The Malacca Strait is conducive to pirate attacks due to the narrowness of the passage and its closeness to numerous channels and islands where attacks can be launched from (Ho 2011).

Several strategies and arrangements have been adopted by the littoral countries to ensure the smooth flow of oil ships through the Indian Ocean. These include air and sea patrolling by the littoral countries of Malaysia, Indonesia, Singapore and Thailand, and the setting up of the ReCAAP Information Sharing Centre in Singapore which gathers and analyses piracy incidents in the Asian region, and air and sea patrolling by all the major powers in the IOR including the United States. There have been many instances where big powers have jointly worked to thwart terrorists and piracy attacks.

China in recent years has increased its activities to control these sea routes in the Indian Ocean. In addition to the mentioned three major choke-points, China's strategy has been to create choke-points in several littoral states in the Indian Ocean through building ports and bases. China is using its money power and expanding its military build-up in the region. China defends its actions on economic grounds. Its bases in strategically located island states can be used as choke-points to control the shipping of oil (Ho 2011).

But the Indian Ocean is critical in India's prosperity and development. Today, 90 per cent of India's trade by volume and almost all of its oil imports come through the sea. The significance of the Indian Ocean was very well stated by the Indian External Affairs Minister's speech at the Third Indian Ocean Conference,

> Over the millennia, the Indian Ocean has shaped the destiny of India and India has always been a maritime nation. India is home to some of the oldest seaports in the world and historically has had extensive maritime links with Africa, Gulf, Mediterranean, South East Asia, and the Far East. The waters of the Indian Ocean have not only carried commerce but have borne India's culture, religion, and ideas everywhere. India's location at the very centre of the Indian Ocean has linked it with other cultures, shaped our maritime trade routes, and influenced our strategic thought (MEA, Government of India 2018).

Given the Chinese assertiveness in the Indian Ocean, over the past decade, India too has increased its foreign policy and naval build-up. PM Modi has come out with the 'Security and Growth for All in the

Region (SAGAR)'. The SAGAR has distinct but interrelated elements and underscores India's engagement in the Indian Ocean. SAGAR is aimed at enhancing capacities to safeguard land and maritime territories and interests; deepening economic and security cooperation in the littoral states; promoting collective action to deal with natural disasters and maritime threats such as piracy, terrorism and emergent non-state actors; working towards sustainable regional development through enhanced collaboration; and engaging with countries with the aim of building greater trust and promoting respect for maritime rules, norms and peaceful resolution of disputes. The SAGAR aims at the security, stability, economic revival and the sustainable harnessing of the wealth of the seas, including food, medicines and clean energy (MEA, Government of India 2018).

Under the SAGAR strategy, PM Modi has made a visit to the small Asian and African island nations in the Indian Ocean including Sri Lanka, Mauritius and Seychelles, and invited Mauritius and Seychelles to maritime security cooperation agreement. In the latest development, Indian President Ram Nath Kovind visited Mauritius and Madagascar, two key littoral states in the IOR, and India announced a new $100 million line of credit for defence procurement by Mauritius.

The other aspects of India's Indian Ocean strategy include the 'Blue Economy' initiatives, 'Sagarmala' initiative and 'Neighbourhood First Policy'. Under these initiatives, India has engaged the region economically, taken initiates to harness the ocean energy and implemented targeted programmes for re-energizing economic activity and extending port connectivity among the littoral states of the Indian Ocean and beyond. India has taken action on a range of projects to improve maritime logistics in Sri Lanka, Maldives, Mauritius and the Seychelles.

In March 2018, India took a significant step by reviving 'Milan', which is a multilateral-level naval exercise hosted by the Indian Navy, first held in 1995 when navies of five countries participated—Sri Lanka, Indonesia, Singapore and Thailand. The 2014 Milan included 17 nations. In March 2018, Milan hosted 16 countries from the Indian Ocean Rim (*Hindustan Times* 26 February 2018). The biennial exercise was hosted at the Andaman and Nicobar Islands which included 16 countries from the IOR—Australia, Malaysia, Maldives, Mauritius, Myanmar, New Zealand, Oman, Vietnam, Thailand, Tanzania, Sri Lanka, Singapore, Bangladesh, Indonesia, Kenya and Cambodia. The exercise was to encourage regional cooperation and combat unlawful activities in critical sea lanes. The naval exercise was held against the backdrop of the growing Chinese maritime influence in the region and to protect this important energy sea route from any potential threat from any single country harbouring the ambition to dominate the region. This figured in the deliberations among navy chiefs of the participating countries. Today 'Milan' has elevated itself from a regional to a prestigious international event and encompasses participation by maritime forces from the larger IOR.

India's Energy Hunt in the SCS

Over the past few years, in its hunt for oil and gas, India has increased its presence in the SCS. The SCS has become an important focus of India's maritime strategy under the Act East Policy. This is reflected in India's growing economic and strategic engagement with nations in the SCS. India's oil and gas exploration in the SCS has emerged as a conflicting issue in India–China energy geopolitics.

The SCS is strategically and economically significant. China's economic rise has enabled it to strengthen its military build-up and exert a greater strategic influence in the Indo-Pacific region. In recent years, China has taken steps toward a greater control over the SCS and the Indian Ocean by enhancing its naval capabilities, militarizing island by installing anti-aircraft and anti-missile systems, and taking ownership of port facilities in the Island nations to mark its presence in the region. To reinforce its claim over the SCS, China has erected around 250 islands. The intensity of Chinese military assertiveness in the SCS has made it a zone of competing claims. China claims the right to it by pointing to historical maps that prove its ownership of almost the entire sea. However, the Chinese claim is disputed by competing claims of countries such as Indonesia, Philippines, Malaysia, Brunei and Vietnam. The SCS, a huge stretch of waterways, is rich in oil and gas resources, and more than $5 trillion dollars' worth of international trade is shipped through it every year (A. Sharma 2017a).

India's SCS strategy could be seen in its growing strategic engagement with the United States, Japan and Vietnam (Granados 2008), and its growing relationship with the other four claimants of the SCS—Indonesia, the Philippines, Malaysia and Brunei. However, India's fast-growing strategic partnership with Vietnam has emerged as the centre of India's energy exploration in the SCS. Over the past few years, India has taken oil and gas exploration activities in collaboration with Vietnam in the SCS and repeatedly called for freedom of navigation and rule-based order. India–Vietnam trade ties go back to the early 1990s but oil and gas exploration has become an issue for China. In 2011, India expanded its relationship with Vietnam in the oil and gas exploration field. When the report of India's ONGC oil and gas exploration in collaboration with Vietnam surfaced, China reacted strongly through its state-run newspaper *Global Times*. China warned India to refrain from exploration activities and urged that China should use 'every means possible' to counter India's oil and gas exploration activities in the SCS (*Times of India* 11 January 2018). It is instructive that India entered the contested region of the SCS via Vietnam. In October 2011, former Indian PM Manmohan Singh and Vietnamese President Truong Tan Sang concluded six agreements; which covered a range of sectors including security and the pact between India's ONGC and Vietnamese state-owned oil companies for investment and exploration. India has snubbed China's objections on the grounds of free navigation principles and as part of the 1982 United Nations Convention on the Law of the Sea (UNCLOS).

China is alarmed about India's growing strategic partnership with Vietnam. India and Vietnam have enhanced their partnership in recent years in which energy exploration and defence cooperation have emerged as the two significant components. In a recent development in January 2018, Vietnam's Ambassador to India Ton Sinh Thanh welcomed the Indian investment in the SCS for oil and gas exploration. China reacted that it is not opposed to the bilateral relations between India and Vietnam in its neighbourhood, but it is firmly against any development by them on the pretext of bilateral ties that violates China's legitimate rights and interests in the SCS. Despite India's repeated assertion that ONGC oil exploration is very much commercially driven, China has objected to it as a strategy to infringe upon China's legitimate rights in the SCS. India's ONGC has been exploring oil in the SCS waters claimed by Vietnam for years.

To give a boost to its energy hunting in the SCS, in March 2018, India signed agreements with Vietnam in the areas of oil and gas, nuclear energy, and defence. In their joint statements, Indian PM Narendra Modi and Vietnamese President Tran Dai Quang vowed to expand the India–Vietnam partnership in the defence and energy sectors. The summit-level meeting also reflected their concerns regarding China's aggressive posture towards India's growing energy ties with Vietnam. PM Modi emphasized rule-based order in the SCS in the presence of President Quang and said (*Times of India* 3 March 2018),

> We will jointly work for an open, independent and prosperous Indo Pacific region where sovereignty and international laws are respected and where differences are resolved through talks...both sides are committed towards expanding the bilateral maritime cooperation and for an open, efficient and rules-based regional architecture.

Both nations are working to increase connectivity via sea route. Speaking at the 3rd Indian Ocean conference, Indian external Minister Sushma Swaraj emphasized speedy progress on building direct shipping routes between the seaports of India and Vietnam under the ASEAN–India Maritime Transport Cooperation Agreement, and said,

> It is only appropriate I mention this today because India and Vietnam are connected not only by the common waters that wash our shores but also by a shared vision for peace and prosperity. Hanoi is, therefore, a particularly appropriate setting for us to discuss developments in the Indian Ocean and the Indo-Pacific region (MEA, Government of India 2018).

Vietnam has become a vital country in India's Act East Policy. Since the emergence of the Act East Policy, PM Modi has steadily worked to enhance India's relationship with Vietnam with mutual commercial interests and strategic concerns against the backdrop of militarily assertive China. To elevate the partnership with Vietnam, in 2016, PM Modi during his Vietnam visit elevated strategic partnership to the level of comprehensive strategic partnership. Vietnam has become a pivotal nation in India's energy exploration strategy in the SCS. India–Vietnam trade relations have increased to more than $7 billion with an aim to take it to $15 billion by 2020. India and Vietnam have conducted joint naval exercises in the SCS, and India is ready to help Vietnam in enhancing its military capability including defence commerce and transfer of defence technology to Vietnam.

India–China Cooperation in Their Energy Security Quest

China and India are the two major oil and gas importing nations in the world despite all the efforts taken towards the renewable and alternative energy sources. For their hydrocarbon needs, both rely on the import of oil and gas from the Middle East, Africa and Latin America. These regions are facing many problems—intense geopolitics, unstable political condition, terrorism, lack of good governance, corruption, insufficient funding and underperformance or underdeveloped energy infrastructure. This poses risks to the security of supply. Both India and China are concerned about the stability in oil and gas prices, security and constancy of supply. Their competitions in these regions for the same assets have not been beneficial to both. The India–China energy competition has been costly for both irrespective of who got the contract. There are enough grounds for them to work together in the energy space and jointly manage their energy security goals. In fact, there are examples of energy cooperation between India and China.

Their bidding competition in Angola is a good example. India had proposed a US$600 million bid for a 50 per cent stake in Shell's Angola Block for 18 fields and had offered to assist Angola's railway project with an additional US$200 million. However, China outbid India's offer with a US$2 billion bid. Obviously, this was not a profit deal for any of the competitors (Singh 2009). Prior to the signing of the cooperation agreement in 2006, both were competing in the same country and same assets in Angola, Kazakhstan, Ecuador and Myanmar.

After their expensive bidding against each other on several occasions, both decided to cooperate rather than compete for their energy exploration abroad. In 2006, a 'Memorandum for Enhancing Cooperation in the field of Oil and Natural Gas' was signed in which both agreed to cooperate in the areas of E&P, refining and marketing of petroleum products and petrochemicals, research and development, conservation, and promotion of environment-friendly fuels. The deal also has the provision for both nations to trade in oil and jointly bid in third countries. Five major oil companies from India and China also concluded company-specific MoU. Since then both nations have collaborated in a number of projects and benefitted (Siddiqi 2011; Yang 2011, 153).

In February 2006, both signed a fifty-fifty joint venture (Himalaya Energy Syria) covering 36 production fields in Syria. The company was set up to buy the entire production shares of Canadian oil company Petro-Canada. This was the first instance of both nations coming together to acquire an oil asset (Siddiqi 2011; Yang 2011, 153).

The agreements were signed between OVL, OIL, and Indian Oil Corporation from the Indian side and China National Petroleum Corporation, Sinopec Corporation and China National Offshore Oil Corporation from the Chinese side. Both jointly bought 37 per cent stakes in Petro-Canada. This was aimed to avoid any competition between India and China for foreign energy assets and minimize the risk (Larkin 2005). The Indian and Chinese state oil companies have put their bid together in selected energy assets abroad, cutting the cost of feeding their oil-guzzling economies while making Asia's two fastest growing economies even stronger competitors in global energy markets. Both also agreed political and technological cooperation in the hydrocarbon sector.

Some of the examples are China's Sinopec with 51 per cent and India's ONGC (OVL) 29 per cent in Yadavaran oilfield in Iran, Sinopec and ONGC (OVL) with 50–50 per cent in Omimex de Colombia Limited in Colombia; China's CNPC 40 per cent and

ONGC (OVL) 25 per cent in Greater Nile Oil Project, Sudan; and exploration right of gas Block 155 (Singh 2009). Another example of India–China energy cooperation is the Chinese-built refinery in Khartoum, Sudan, and Indian-built pipeline to deliver refined products to a nearby port for export. This reduces the cost and helps both in the importation of oil. Likewise, there is an opportunity for both to cooperate in other gas pipelines. Given the security risk in the Strait of Malacca and increased cost and time in shipping the oil through the straits from the Middle East, both nations can benefit from considering each other's energy security need.

Gas pipelines are significant for both India and China as they move towards less carbon-emitting fuel. The construction of gas pipelines has been on the energy security agenda for both nations. Insufficient domestic gas reserves and high dependency on imported gas make it imperative for India and China to look for gas from the Caspian Basin, the Middle East and Russia. Both have focused on the construction of gas pipelines form these countries. China is building a gas pipeline which links it to its northwestern nations. India is constructing a number of gas pipelines both in its northwest and northeast front. These pipelines are also facing geopolitical challenges. India's gas pipelines from the northwestern region pass through Pakistan and its northeast gas pipelines cannot ignore China and Pakistan. Similarly China's gas pipelines in it south-west have to deal with India's geographical location and geopolitics. China's oil and gas pipeline in Myanmar, which passes through the Bay of Bengal, cannot undermine India's strategic location. These gas pipelines, on the one hand, open avenues for cooperation but, on the other hand, can aggravate strategic rivalry. These synergies necessitate both nations to collaborate and jointly invest in the energy sector. Collaborating on these pipelines can make these projects cost-effective and reduce the geopolitics tension.

In a recent development, the Chinese media *Global Times* proposed that China and India should jointly bid contracts to acquire and develop overseas oil and gas resources. This would lessen the competition and strengthen cooperation in investments in energy technologies and products. According to the state-run media, the Chinese and Indian oil companies are still behind in terms of management, technology and capital strength in comparison to developed countries' energy companies. India is behind China's energy technologies including in coal-fired power generation technology, nuclear energy, wind and solar energy technology as well as equipment manufacturing. On top of that, China has more financial power than India. China can benefit from India's superior expertise in energy information management, wind energy, globalized Indian firms and its large energy market. Moreover, China has advanced and low solar energy technologies which can be helpful for India which has abundant solar energy resources. Both can jointly invest in the solar PV industry through joint ventures and technology transfer to expand their solar energy market. Chinese expertise in the design and construction of large power stations, manufacturing of complete sets of hydropower equipment and coal-fired generation equipment as well as construction and management of power grids can be advantageous for India. The power distribution and grid connectivity have been hurdles in India's power sector (*Economic Times* 16 March 2017). Both nations have the potential to work in a range of areas in the energy sector ranging from joint bidding, gas pipelines, transmission and distribution, to scientific research and technological collaboration in renewable energy.

India–China collaboration could be seen on the climate change issue. Since the emergence of climate change concerns in the 1990s, both have jointly advocated for the onus for reducing carbon emissions is more on developed than developing countries. Since the beginning of the climate change negotiations in the 1990s, especially when the UNFCCC was opened for signature in Rio de Janeiro in 1992, India and China have jointly advocated for developed countries to bear the burden of the man-made emissions of GHGs. Both pitched for more time to be given to the developing countries to attain a comparable level of development and the developed nations' needs to step first and do more. Both coming together on the issue of sharing the burden of carbon emission was very much pronounced at the Kyoto Protocol.

Clearly, the Chinese official newspapers' pitch for China and India to collaborate in the importation of oil and gas from overseas to ensure stable supplies is a good idea. Nevertheless, energy bids abroad have not lasted and have not been frequent. The pitch from the Chinese media ignores the strategic issues between the two nations. In addition, China's pitiful engagement of India, encircling strategy and ignorance of India's security concerns, especially by blocking the resolution on terrorism emanating from Pakistan, have not been helpful. Despite the possible benefits by collaborating in the energy sector, the strategic issues are driving them in opposite directions. This is reflected in their energy security geopolitics in Africa and Latin America, the SCS and the Indian Ocean. India–China energy geopolitics is not confined to the exploration of oil and gas, and is also visible in the renewable and nuclear energy sectors.

The Geopolitics of Renewable Energy: India–China Competition

The energy future is renewable. If coal was the dominant source of energy for Britain to maintain its empire, and oil powered the US economy and its primacy in the post-war period to the second half of the 20th century, then renewable energy is going to be significant in the 21st century. Depleting sources of conventional energy sources, carbon emission concerns, local environmental degradation and climate change, and population growth are some of the important drivers that are pushing the world to shift to renewable energy. According to the IEA report, renewable energy is the fastest growing energy sector with an average ratio of 2.6per cent, followed by nuclear 2.3 per cent and fossil fuels less than 2 per cent, yearly.

Global climate change concerns have further triggered this shift towards renewable energy. The international community responded when 190 countries at COP21 in Paris on 12 December 2015 signed the landmark agreement to combat climate change and to accelerate and intensify the actions and investments needed for a sustainable low-carbon future. The 2015 Paris Climate Agreement was overwhelmingly supported globally with a common goal to undertake ambitious efforts to combat climate change and adapt to its effects, with enhanced support to help developing countries to reduce carbon emissions.[3] The Paris Agreement aims to increase the share of renewable energy such as wind and solar energy to attain a decarbonized world. Consequently, all the major countries including India and China, the two fastest growing major economies and the fastest growing CO_2 emitters, are taking concerted efforts towards this goal. New norms, values and rules are being created to support the movement toward renewable energy.

The growing importance and use of renewable energy gradually but certainly corrodes the dominance of hydrocarbons. The quest for renewable energy is not free from geopolitics. As renewable energy will grow and gain a higher percentage of the energy mixes in countries, it will also alter their geopolitical positions. Irrespective of how the renewable energy geopolitics turn out in the future, whether it is going to be similar to the geopolitics of oil and gas, and given the scant availability of literature on the future geopolitical outcome of the transition towards renewables, geopolitics are already visible in the unfolding competition between big energy-consuming nations. The United States and China, and also some individual EU countries such as Germany, have focused on the renewable energy sector by investing enormously in technology and have the ability to play a leading role.[4]

The renewable energy geopolitics is visible between India and China too as both nations move towards non-fossil fuels, especially solar energy. The renewable energy is more decentralized in nature compared to conventional energy and could potentially be more trusting and it could also empower people and regional authorities in relation to central governments and interests. This is more relevant in the developing countries, especially in countries such as India, given about one-fourth of its population are still waiting for adequate electricity supply. The move towards renewable energy has been one of the top agendas of the current Indian government. PM Modi committed to the reduction of carbon emissions at the Paris Climate

[3] The United Nations Climate Change, 'What is the Paris Agreement?': https://unfccc.int/process-and-meetings/the-paris-agreement/what-is-the-paris-agreement.

[4] For trends of renewable energy geopolitics, see Criekemans (2018); Criekemans (2011).

Accord and the Modi government has reaffirmed on many occasions India's unwavering commitment to shift to renewable energy. It has taken unflinching actions on climate change. India is steadily working on enhancing clean energy with a target of 100 GW of installed solar energy by 2022. India's quest for renewable energy is in line with the conceptual framework of four 'A's—availability, accessibility, affordability and, more prominently, acceptability, amidst the need to combat climate change.

China has made significant progress in the renewable energy sector. But, India over the past 5–7 years has made significant progress too. From the green airport, the largest solar park in India's Southern state of Karnataka to the ISA, India has made progress in this direction. ISA move could be seen as the first such move in India–China competition where China has been snubbed by India in its move towards solar energy. The ISA is headquartered in New Delhi and India is acting as one of the leading normative players in this newly formed treaty-based organization. Abundant sunlight is available in many parts of the world. The ISA is one such move to tap the potential of sunlight in 121 developing countries in Asia and Africa. The latent India–China renewable competition is visible in India's renewable energy effort at the ISA.

Besides, the geopolitics of renewable energy also creates geotechnical opportunities and limitations. One of the major problems in the coming years will be the availability of the rare earth materials that are needed in the technological advances of renewable energy technology. Renewable energy geopolitics might reflect a similar pattern to those of conventional energy. There might be green protectionism in the Western world, but also the condition of oil-producing countries might be problematic in a world where renewable energy is growing fast (Rothkopf 2009). Today, renewable energy consumption is growing at a pace three times faster than the overall demand for energy. It is estimated that by 2040–2050, renewable energy will constitute one-third of the world's total energy requirements. This will accelerate the importance of research and innovation, finance and technology in the green energy sector. This will have no less an impact on geopolitics than coal or oil had in the last century.

The geopolitics of renewable energy is likely to be reflected in the nations aiming to control the finance, supply chain, technology, production methods and cyberspace in the sector. The countries, which have dominated the conventional sources of energy, especially the Middle East, Russia and Venezuela, might continue to be the dominant player in the energy field. However, countries such as the United States and Germany stand to benefit, China is most likely to be dominant and India a significant player with influence in the developing countries. The renewable energy sources of sun, wind and water are abundant over the world except for a few places. The bigger renewable energy projects are likely to involve the geopolitics in the future similar to the risk to hydrocarbon sources, especially oil and gas. Nevertheless, renewable energy geopolitics is going to be for the rare earth materials used in the manufacturing of renewable energy products.

China has huge reserves of rare earth material which is used in many modern appliances, more importantly solar panels and wind turbines. Beijing almost has a monopoly over the rare earth materials cobalt and lithium since China overtook the United States in the early 1990s as the world's biggest producer and exporter of rare earth materials, with some rare earth products, coming only from China (Hongqiao 2016). This gives China an advantage over other countries in the trade of these materials. It will not be surprising if China also controls administration and technology advancement in these fields. China's renewable energy strategy already reflects this trend. China has focused on the development of green energy in the 13th Five-Year Plan. Over the past decade or so, China has emerged as the leading producer and exporter of renewable energy materials and products such as solar panels, wind turbines and batteries. China accounts for almost 60 per cent of the trade of solar energy products, is the biggest miner and supplier of rare earth materials, has the largest renewable energy instalment capacity and is the number 1 market for electric vehicles. In addition to its own reserves of rare earth materials, China significantly controls the mining of rare earth materials abroad including in Africa, Europe and South America. China is today the leading renewable energy player as the investor and supplier of products, services and technology.

Chinese actions indicate the unfolding geopolitics of renewable energy and its strategy to dominate this sector in the future. Between 2008 and 2013, China managed to bring a decline in the global price of the solar panels sector by 80 per cent which clearly indicates the Chinese financial hold on the solar panel industry. In 2010, China stopped the export of rare earth materials ostensibly citing environmental concerns, but this was a strategy to compel the renewable energy companies around the world to finance manufacturing in the Chinese renewable energy sector. China is employing cyber-warfare tactics that include corporate espionage to dominate the renewable energy market. A Chinese firm was charged with theft of proprietary software in the United States in 2013, and in 2014, US law enforcement agencies indicted five Chinese military officials for hacking five American companies, including SolarWorld, to manipulate and dominate the solar energy international trade (Deo 2018; Harris 2014).

China has the capability to make the renewable energy sector cost-effective. In the future, it is most likely to be visible in commercial competition and to an extent in the cyber warfare strategy. Commercial competition is likely to be of similar strategy by generous trade concessions, low rate loans, subsidies, tax relief, high tax on imports, incentives to invest in

Chinese manufacturing and firms, and cheap credit to encourage domestic manufacturing of renewable energy products. Certainly, this is going to be similar to the conventional energy geopolitics but more complex and likely to be reflected in trade and cyberspace, and diplomatic influence in the nations to control this fast-emerging new energy source. China's move to dominate and monopolize this energy sector global market and thwart commercial competition is going to be a challenge for India's green energy leadership initiatives through the ISA.

Renewable energy is also one of the main agendas of China's OBOR initiatives. China's plan under the green investment strategy is to fund and supply overcapacity to countries with energy needs of the nations along new Silk Route. Its proposed plan to integrate electricity markets throughout Asia with up to 3,000 km of high voltage transmission falls in its OBOR strategy. This route clashes with many of the nations in Asia that have joined the ISA.

The ISA is aimed at tapping solar energy for the growing energy need of developing nations, especially in Africa and Asia amidst the climate change concerns. It puts India in the driving seat for normative and economic leadership in the solar power market. In this context, it is worth mentioning the renewable energy efforts from India and China where both nations are trying to keep the other out. Recently held ISA in which more than 121 nations participated and joined the ISA does not include China as its members. Only countries with abundant sunlight have been included. France and India launched the ISA as the joint leaders of this consortium of nations aimed at promoting solar energy across the countries falling within the Tropics of Cancer and Capricorn.

Through the ISA India's place is recognized in the renewable energy international regime. Its joint leadership with France and growing collaboration with the European countries in the renewable energy sector gives India access to latest technology and much-needed investment in its renewable energy sector. India's plan to give aid to the Asian and Africa members of the ISA provides it much needed diplomatic influence in these countries. This could help India to balance growing Chinese economic influence in African and Asian nations. In addition, it gives India a strong foothold in administration with the power to shape procurement, production and investment planning.

India's domestic renewable energy sector is facing competition from China as it faces the dumping of cheap Chinese exports, which has forced the weak companies to shut down their ventures. The Indian government has responded by increasing the import duty on foreign solar modules. In response to the petition filed by the Indian Solar Manufacturers which sought protection from the growing import of cheaper solar cells and modules, on 30 July 2018, the Indian government responded by imposing a 25 per cent duty on imported solar products (Indian Gazette 2018). The anti-dumping duties are aimed to check the foreign companies' predatory pricing, market distortions and import of cheap solar products, of which 90 per cent came from China and Malaysia (Sushma 2017).

There have been talks of the United States and China joining the ISA which might help the ISA to achieve its objectives. This could lead to cooperation. Though the ISA permits the countries even if they don't fall fully or partially between the Tropics of Cancer and Capricorn, but whether they will have voting rights has yet to be decided. Given the Chinese strategy of monopolizing this sector, this might become an issue in the future.

The ISA is a much-needed platform for India which will not only enhance the quality of its manufacturing but also give a diplomatic reach to counter Chinese moves in the emerging geopolitics of the renewable energy sector. The challenge remains how India manages finance to its solar energy market, both domestic funding and foreign investment. The significance of the ISA lies in the fact that most of the countries which have abundant sunlight are in Africa, and India by leading the alliance can have meaningful relations with African nations and increase its sphere of influence in the region. As discussed in Chapter 4 in more detail, India has ramped up its foreign policy to build strong energy ties with the African nations and by providing economic aid. The ISA gives India a platform to enhance its presence and goodwill in the region and offset Chinese influence. However, the Chinese intention of monopolizing the sector poses a challenge to India's emerging normative and economic leadership in the energy sector, energy security pursuits, technology development, and manufacturing and diplomatic reach in the developing countries.

The Geopolitics of Nuclear Energy: India–China in the Great Game

Nuclear energy is becoming significant in the energy mix. It is growing faster than fossil fuels at the rate of 2.3 per cent annually. The move towards less carbon-polluting energy sources is making many countries shift towards nuclear energy, which is affordable and clean. Nuclear energy is deeply mired in geopolitics because the uranium and civilian nuclear technology used to generate nuclear power can be diverted for making atom bombs. The nuclear trade is controlled and monitored by the international nuclear non-proliferation regime in which the IAEA and the NSG play a significant role. India's nuclear power programme suffered for a long time because of the embargo on its nuclear programme (This has been discussed in Chapter 5 in detail).

The US–India civilian nuclear pact not only addresses India's goal of addressing energy security by reducing the fossil fuel share from its energy mix, but it opens India to acquire technology in its defence industry. Before the US–India nuclear deal, India was constrained by the nuclear embargo. The US–India defence agreement and the US–India nuclear deal need to be seen in the context of the geopolitics in Asia. The US design of engaging and aiding India in its quest for great power status would not have been possible without the US–India civilian nuclear deal. The first time it was reflected in Bush' presidential election campaign in 1999 when he expressed the desire to engage India by praising India's credentials as a functioning democracy. When the Bush administration surveyed the strategic landscape, it saw China looming large. Consequently, in May 2001, US Deputy Secretary of State Richard Armitage came to India to explain President Bush's strategic framework that included a missile defence programme and hinted at a new beginning with India (Trikha 2001). In a speech delivered in New Delhi in September, Robert Blackwill, the US Ambassador to India, repeated the US position, 'President Bush has a global approach to US–India relations, consistent with the rise of India as a world power' adding that this was 'because no nation can promote its values and advance its interests without the help of allies and friends' (*Washington File* 2001). There was now a noticeable sense of a transformed perception about India in the US.

The roadmap for US–India engagement gained a new sense of urgency after terrorists attacked the Twin Towers and Pentagon in America on 11 September 2001, and the Indian Parliament on 13 December 2001. Since these attacks, both nations have been intensely working jointly to counter Islamic terrorism. India also responded to the positive gesture shown by the Bush administration. It responded positively to the Bush administration missile defence initiatives. Even before the president announced his missile defence initiative on 1 May 2001, Foreign Minister Jaswant Singh, in his meeting at the White House on 6 April 2001, expressed his understanding of the imperatives of the coming transformation of the deterrence calculus in favour of defensive technologies. India supported the Bush administration's missile defence plan, which was attacked ferociously in China, Russia and Europe. India's support for missile defence systems and full cooperation with the US after 9/11 was viewed positively by the Pentagon (A. Sharma 2017b).

PM Atal Bihari Vajpayee of India's centre-right party the BJP-led NDA government envisioned India and the United States as natural allies. This was further consolidated the way the centre-right Republican Party President George W. Bush and PM Vajpayee saw the security threat environment amidst the growing Islamic terrorism and the rising communist China. Both nations' security interests converged and calculated well in both leaders' perception of a strong partnership between the two democracies.

Finally, the process towards nuclear deal began by the launching of the NSSP in January 2004 by President Bush and PM Vajpayee in January 2004. Both nations agreed to expand cooperation in three specific areas—civilian nuclear activities, civilian space programmes and high-technology trade. Both nations agreed to expand the dialogue on missile defence. These areas of cooperation were designed to progress through a series of reciprocal steps that build on each other. On 17 September 2004, the United States and India announced significant progress towards the implementation of the NSSP plan with an aim to elevate the economic ties and set the pace for collaboration in civilian nuclear activities, civilian space programmes and high-technology industries. The US–India joint statement said that the implementation of the NSSP would lead to significant economic benefits for both nations and pave the way for enhanced regional and global security (A. Sharma 2008; The US Embassy 2004; US Embassy in India 2004). The NSSP was hailed as a step towards the beginning of a new era in US–India relations. The nuclear non-proliferation was a constant irritant in the India–US relations for almost three decades. India's growing economy projection further necessitated India to address its lingering energy crisis and that too amidst global climate change concerns.

To achieve its nuclear energy potential, India's nuclear energy programme required more advanced nuclear reactors and enriched uranium. The Nuclear Non-Proliferation Act of 1978 was a hindrance as it put a US ban on sale or transfer of sensitive and dual-use technology to India and constrained nuclear energy and defence collaboration.[5] President George W. Bush was convinced that without solving the nuclear issue the US–India relationship could not be transformed. President Bush vision of India as befitting great power in the Asian century needed a militarily strong India.

As a result, on 28 June 2005, the US Defence Secretary Donald Rumsfeld and the Indian Defence Minister Pranab Mukherjee inked a 10-year defence partnership agreement known as the 'New Framework for the US–India Defence Relationship' outlining planned collaboration in multilateral operations, expanded two-way defence trade, increasing opportunities for technology transfers and co-production, expanded collaboration related to missile defence, and establishment of a bilateral defence procurement and production group. The defence pact unveiled mechanisms to promote a long-term bilateral defence industrial relationship and the possible outsourcing of research and production to India. The significance of the deal could be judged from the fact that co-production of the F-16 currently occurs in several key countries allied with the United States: Turkey, Belgium, the Netherlands and South Korea. But the US offer of the F-18 fighter aircraft for overseas co-production has not been offered to any country except India. This defence cooperation showed the Bush administration's seriousness about developing long-overdue defence ties with India (A. Sharma 2008). It was heavily criticized in China as the US move to balance China in Asia (*People's Daily Online* 2005).[6] But the

[5] United States Nuclear Regulatory Commission, *Nuclear Nonproliferation Act of 1978*: http://www.nrc.gov/reading-rm/doc-collections/nuregs/staff/sr0980/v3/sr0980v3.pdf.

[6] This article was first published on the first page of the *Global Times*, 1 July 2005, and was translated by *People's Daily Online*. The *Global Times* is the state-run newspaper of China.

defence pact paved the way for enhanced opportunities for the Indian defence industry to place itself better in the era of the developed countries' push towards globalization. This gave India a bargaining power in technology transfers and co-production from its three main defence acquisition sources—the United States, Russia and Europe. The defence agreement was extended for another 10 years on 3 June 2015 as the US Defense Secretary Ashton Carter and the Indian Defence Minister Manohar Parrikar signed a strategic 'Ten Year Defense Framework Agreement' in New Delhi (*Economic Times* 11 July 2018). India and the United States have made significant progress in the defence sector. Today India conducts more military exercises with the United States than any other country. Though Russian armoury still dominates India's arsenal and will continue to have a significant share in India's defence industry, the US defence giants such as Lockheed Martin and Boeing have made significant inroads into the Indian defence market. The United States has emerged as the top arms import source for India since 2012. The defence agreement has elevated and given depth to the India–US defence cooperation in the context of the balance of power in the Indo-Pacific region amidst the rise of militarily assertive China.

India has not been able to develop its indigenous defence industry. Internal factors such as obstinate bureaucracy, lack of a clear-cut vision and policy-level fallacy are cited as the reasons for this, but in the external context, it is lack of international collaboration that has slowed its defence industry development and modernization. Though India has made excellent progress in its missile technology and space programme, it has struggled to keep up with international ambitions and strategic requirements.

It is not surprising that India became the largest importer of arms every year since 2010 (*New Indian Express* 2018). India is still struggling to upgrade its arms manufacturing sector, despite this being a priority for over a decade and a half. India's defence preparedness and industry are finding it hard to keep pace with external security challenges and expanding strategic interests. For years, India has been relying on foreign suppliers, mainly the Soviet Union. After a humiliating defeat in the Indo-China War in 1962 and the US refusal to help with weapons technology, India looked towards Soviet Union for its defence preparedness. Eventually, the Indian armoury was full of Russian defence products and missiles. India's indigenous defence industry remains underdeveloped. Defence modernization continues to take place in a slapdash manner. Indian defence firms still struggle to move from a buyer–seller relationship with foreign firms to one based on equal partnership and joint production (Cohen and Gupta 2015; A. Sharma 2013).

Over the past two decades, India has diversified its defence source acquisition. In the process, there has been an increase in its arms imports from the United States, Israel and Western Europe, especially France. But it is the United States which has made significant inroads in India's defence industry over the past decade. This works both ways: The US defence industry is in need of export destinations. American defence giants such as Lockheed Martin and Boeing have been exploring potential business partners in India, attracted by the low-cost, well-educated, English-speaking and technically sound workforce. The United States has emerged as the biggest import destination for India's defence needs in recent years. Though India's defence imports constrain indigenous innovation, overall it has helped India's defence industry as these arms purchases are done under offset arrangements and in some cases joint-production and transfer of technology. Overall, India has been able to augment its defence capabilities. Today, India has more suppliers for its defence needs. Though it is still dependent on its foreign sources, it has the option to choose the best and these arms are being transferred to India in the offset manner with a major share of manufacturing done in India under the transfer of technology clause. These developments have allowed India to beef up its defence capabilities and enter into close security and strategic partnerships with arms-importing nations.

This would not have been possible without the US–India Civilian Nuclear Agreement signed on 18 July 2005. It was rightly observed by Air Commodore Jasjit Singh in 2006 that it is difficult to say whether the defence cooperation agreement helped the NCA to materialize or the other way round. But it is evident that without the nuclear road-block being bypassed, the defence industry relationship would have remained still-born since US laws and non-proliferation policy do not permit cooperation with a non-NPT nuclear weapon state that has been under sanctions for three decades (Singh 2006). Primarily aimed at addressing India's energy security crisis, the nuclear deal has put an end to the nuclear row between India and the United States. Prior to the deal, the United States would not trade with India in any sensitive technology that could be used for both civilian and military purpose.

Consequently, the amendment to the Nuclear Non-Proliferation Act 1978 allowed the 'two-way defence trade' and has increased the US defence industry's stake in strengthening US–India relations across the board. Today India's relationship with the United States has transformed drastically. It is comprehensive and reflected in all aspects of bilateral relations, including collaborations in economics, science and high-technology, joint military exercises, defence commerce, defence industry, defence agreement, counter-terrorism cooperation, civilian nuclear deal, energy security, civil society interaction and global commons.

Not surprisingly, the United States and India nuclear deal alarmed China and Pakistan, and the non-proliferation community. They criticized the Bush administration for the nuclear exception to India by offering full civilian nuclear energy cooperation despite India's non-signatory status to NPT. China perceived the defence agreement and nuclear deal as part of a US-designed containment policy to balance China's rise.[7] China and Pakistan opposed the deal from the beginning. To demonstrate its desire to maintain parity with nuclear rival India, Pakistan test-fired a nuclear-capable, surface-to-surface ballistic missile with a range of 2,000 km (*Times of India* 2006). The anti-nuclear deal groups tried their best to block the deal in Congress on the pretext of the arms race. In addition, there were consistent lobbying efforts by Pakistan together with other opponents of the US–India nuclear deal to block it.

The Republicans having a majority in the US Congress and the nuclear deal being backed by the Bush administration was not enough

---

[7] For the detail insight into the US–India nuclear deal, see Pant (2011).

to guarantee its passage. There were still deeply entrenched mindsets within the US Department of State and the non-proliferation lobby that were against the US decision of nuclear exception to India (Haniffa 2005). The anti-US–India nuclear lobby groups and non-proliferation activists like David Albright resisted the nuclear pact. The deal faced resistance in the 49-member House International Relations Committee (HIRC) headed by Illinois Republican Henry Hyde. Out of 26 Republicans on the committee, prominent Republicans like Indiana's Dan Burton and California's Dana Rohrabacher were not ready to support the deal, unless changes were made. China strongly lamented the deal and demanded a similar kind of deal for Pakistan.

However, the opponents of the US–India Civilian Nuclear Agreement and the non-proliferation lobbyists in both the US and in India could not compete with the well-funded efforts by Indian Americans, the business associations, the Indian embassy and the Political Action Committees formed by affluent Indian Americans. Given the opposition from various interests, the Indian government and the Indian American community knew that easy passage of the nuclear agreement would not be possible. As a result, various combined pro-nuclear deal lobbying activities were taken during both legislative steps of the nuclear deal resolution in Congress. First, in 2006, the bill was introduced to make a nuclear exception to a non-NPT member. This step also introduced some of the amendments to kill the nuclear deal bill by attaching the conditions under which Congress was to grant this exemption. The second step in September 2008 put the bill for final voting in Congress, where the executive branch had to negotiate the final, technical aspects of the nuclear agreement with India and get endorsement from the IAEA and NSG. During both stages, the Indian lobbying was well-coordinated, unrelenting, well-funded, intense and massive. Finally, the US–India civilian nuclear deal was signed by President Bush in October 2008 as the United States–India Nuclear Cooperation Approval and Non-proliferation Enhancement Act.

The Bush Administration altered the nature of the US–India relationship by nuclear exception to India. The difficulty in doing so was the conflict between two competing goals for the United States Foreign Policy—great power politics versus nuclear non-proliferation.

Great powers' status to a great extent depends upon their membership and influence to powerful international institutions. Great powers use their influence in the institutions either to maintain their status quo or enhance their power in the international system. India is not a permanent member of the UNSC, not a legitimate nuclear power, not in 'nuclear have' club of the NPT and has been consistently denied membership to many of the formal economic and political institutions. The nuclear deal in addition to India's rapid economic growth and growing military capability has helped India's elevation to great power status and is now paving the way for its inclusion in many important formal international institutions. India's rapid economic growth, growing military prowess and its deepening strategic partnership with the United States have raised India's international standing as an emerging great power. The nuclear issue was important for close India–US ties. As observed by Harsh Pant (2011, 273), 'the two democracies [have to go] through nuclear energy cooperation.' After the nuclear deal, China's perception of India as a great power began to change. Before the nuclear deal, India's great power aspiration and the idea of India as a great power were even ridiculed by China. Though India demonstrated its nuclear weapon capability as nuclear weapons state itself in 1998, the nuclear agreement has been seen by China and Pakistan as helping India to advance its nuclear weapon capability. The nuclear energy deal intensified India–China strategic competition.

After the unsuccessful attempt by China to block India's nuclear deal with the United States and waiver in NSG in 2008, China turned towards India's bid for the NSG membership. The NSG membership is significant for India. The NSG has the membership of 48 countries which trade in nuclear material, equipment and technology. Given India's aim of enhancing nuclear energy share in its energy mix to 25 per cent by 2050, India needs to import enriched uranium. India's nuclear power is very low in its energy mix. It lacks enriched uranium and needs more advanced technology and reactors as well.

The NSG was formed after India conducted a nuclear test in 1974. The NSG was designed to monitor specifically India's nuclear programme. India restricted itself to a peaceful nuclear programme till 1998 when it declared itself as nuclear weapons state. Though India abided by the non-proliferation norms, it has refused to sign the NPT and ratify the CTBT. But all the 48 members of the NSG have signed the NPT and CTBT. Out of 48 members, five are nuclear weapons states and also permanent members of the UNSC, and 43 members are NNWS, committed to not possess atomic weapons. The five nuclear weapon states have legally committed themselves to nuclear disarmament and work towards preventing the proliferation of nuclear weapon states. NSG members have ratified the CTBT, and are committed to achieving the goals of non-proliferation and nuclear weapon-free world by adhering to the treaties and agreements that work towards nuclear weapons-free zones, and by allowing the IAEA to validate their nuclear programmes are not working for making nuclear bombs (Hibbs 2016).

As its commitment to implement the 2008 NSG waiver guidelines, India went ahead with the separation plan clause and placed two-thirds of its civilian nuclear power reactors under the IAEA safeguards. Out of 22 nuclear reactors, 14 were placed under the IAEA safeguard and 8 nuclear reactors meant for the nuclear weapons programme were outside the IAEA safeguards. India has maintained its self-imposed ban on atomic test explosions. In June 2014, India approved a protocol that gave the IAEA right of entry to its atomic sites.

India's membership to the NSG was questioned mainly on the ground that its separation plan essentially created an additional category—reactors that are connected to the electrical grid but are declared part of the military programme—that is at odds with the practices of other states with nuclear weapons. Further, India's Additional Protocol lacks standard provisions regarding information sharing, even as compared to the protocols adopted by other nuclear-armed states. Similarly, in its negotiations with the US and other suppliers of nuclear fuel and technology, India refused to accept standard international procedures for tracking imported uranium throughout the fuel cycle.

India's case for NSG membership has been lingering for years. Before the NSG voting in Seoul in 2016, the United States and India lobbied to end its semi-nuclear power status. The United States supported India's bid to the NSG membership unconditionally, urged the members to support India's entry into the NSG and argued that India's entry into the club would strengthen nuclear non-proliferation goals. The US unconditional support to India's NSG, however, could not convince some of the sceptics' objections that India's nuclear exception at NSG would undermine the international commitments to non-proliferation regime. Some of the NSG members' negative perceptions about India's actions on the NPT since the NSG waiver in 2008, and their bad reminiscences of Bush administration demands for India's waiver at the NSG in 2008 also did not go in India's favour.

While the United States and India were able to convince the majority of the members for India's entry into the NSG, China, unlike in 2008, decided not to support India's further advancement. China opposed the Indian bid to the NSG membership on the grounds of India not being a signatory to the NPT, and that special status would undermine international non-proliferation efforts. China also contended that India's entry would encourage others to be defiant of the NPT regime and efforts. Despite India's non-proliferation commitments as a responsible nuclear power, the June 2016 NSG plenary failed to bring entry to the NSG. China was successful in blocking India's entry into the NSG in 2016.

China's support to the NSG waiver to India in 2008 could be seen from the Chinese perception about India's rise 10 years ago. China did not consider India a serious strategic rival. China also considered comparisons with India as demeaning. But China's perception has changed with India maintaining an average growth rate of 8 per cent, surpassing China as the fastest growing major economy, expanding strategic moves in the Indo-Pacific region, expanding foreign policy in Africa and Latin America, emerging as one of the main players in the Afghanistan reconstruction efforts, further strengthening and consolidating strategic partnership with the United States and becoming closer to the Political West. China sees India as an emerging great power which must be contained. Though India's bid to the

NSG membership is for addressing its energy security, the NSG is also one such international platform which can further elevate India's international standing.

Contrary to the claims of China that its decision at the NSG was based on the principles of non-proliferation, it was purely based on geopolitical consideration. India's NSG membership would give India the same privileges as China, give greater access to the international nuclear market and give the same economic benefits. It can further enhance India as a legitimate nuclear power outside of the NPT and can be used for power projection in the region and beyond. Equating Indian membership with Pakistan could also allow China to balance the scale by having another powerful voice oppose India's commercial moves in the nuclear sphere, and bring India's rising great power status down by clubbing with Pakistan (Neog 2016; also see Hibbs 2016; William 2016).

Beijing led the opposition to India's NSG bid and argued that it would further undermine efforts to prevent proliferation. However, China's stand was exposed when Pakistan, another non-signatory country to the NPT, applied for its bid to the NSG with the backing of Beijing. China supported Pakistan's bid despite Pakistan's involvement in nuclear proliferation activities. A. Q. Khan, the father of Pakistan's nuclear programme, was involved in black marketing of nuclear secrets to North Korea and Iran.

Today 'China–Pakistan All Weather Friendship' matters more than ever for China. By supporting India's NSG bid China did not want to antagonize Pakistan. Islamabad today has become an important ally for China in its India-encircling strategy, an important nation in its reach to the Middle East and a crucial nation in its ambitious OBOR project. Among Chinese strategic thinkers Pakistan is described as 'one real ally' and a relationship 'model to follow'. China was determined to not undermine its ties with Pakistan, and the NSG was also the test for China's credential as an ally. The China–Pakistan relationship has depended with significant geopolitics implications.[8] They are

---

[8] For a detailed and deep insight on the China–Pakistan ties, see Small (2015, 2016).

collaborating in the nuclear and defence sectors. China is the biggest importer of Pakistan's arsenal: More than 50 per cent of Chinese arms go to Pakistan. This alliance is both ways. After the end of the Cold War, Pakistan lost its strategic importance in the US strategic landscape. India has emerged as strategically important for the United States to maintain its primacy and balance China's rise in the Indo-Pacific region. Consequently, the growing US–India relationship has further pushed Pakistan towards China. Above all, China is all set to become the world's biggest economy in the coming decade and it has become more assertive in great power ambition. China's strategic competition with the United States and its allies is becoming more intense. In its growing tension with the United States and India, China sees Pakistan as a valuable ally.

Conclusion

The energy geopolitics of India and China do not reflect the traditional militaristic approach of the balance of power. India–China energy geopolitics is so far limited at commercial level, reflected in outbidding each other for oil and gas contracts, and diplomatic tussles visible during India's civilian nuclear energy deal with the United States and India's bid to the NSG.

India–China energy geopolitics is more visible in the Indo-Pacific region, exclusively the SCS and the Indian Ocean. India's Look East Policy started with economic interaction and is now being pursued under the Act East Policy. However, India's China-balancing strategy is not purely militaristic. India's balancing strategy is reflected in its military capability and expanding naval presence in its Eastern waters; it has been limited to occasional naval exercises with the United States, Japan, Vietnam, Australia or occasional forays by the Indian navy into the SCS and East Asian waters. India's balancing approach reflects more of a 'soft balancing' (Paul 2005) or a strategy similar to what Great Britain did in continental Europe during the 18th and 19th centuries, that is, an offshore balancer. Similarly, China's string of Pearls Strategy, or its occasional border violation on the Indo-China border, or its support to Pakistan by vetoing India-backed resolutions on Pakistani-based terrorist organization or terrorists reflect soft balancing instead of full-fledged military conflict. India–China competition in Latin America and Africa is confined to the commercial level.

This is further reflected the way India's eastern members see India's growing strategic engagement in the region. India's Act East Policy and balancing strategy in the Indo-Pacific region manifest imperatives from both India and neighbouring nations in eastern waters. This is more visible in the growing strategic engagement with ASEAN nations. Given no territorial disputes between India and the ASEAN nations, India's deepening strategic partnership with the United States amidst concerns of Chinese naval expansion in the region, and India's growing naval capability and presence in its Eastern waters make India a strategic asset to the China-wary countries in the Indo-Pacific region and ASEAN.

The benefits of cooperation in the conventional and renewable energy sectors, as well as overlapping sources of hydrocarbons give further ammunition for cooperation. Both know that collaboration on gas pipelines will be of great economic value and can lessen competition and strategic tension hampering energy security efforts. Both are the fastest growing polluters but need to continue economic growth and industrialization to raise the standards of living and alleviate poverty for millions of people. This has brought them together on the issue of climate change. Together both have advocated at the internal platforms for the developed countries to share the burden of carbon control more than new developing countries. These factors have lessened their strategic rivalry in their quest for energy security.

But, China's blocking of India's entry to the NSG in Seoul has activated India–China energy geopolitics. Despite summit-level meetings, ministerial and diplomatic-level talks, and growing bilateral trade, the India–China relationship continues to be a logjam. India sees China's move at the NSG driven by geopolitical interest, creating a hurdle for India's development. Though China hinted at a dialogue on India's NSG membership in recent meetings, the strategic rivalry has obstructed any such possibility.

It is unlikely that given the strategic tension heightened, China will budge from its stand. Though India has been denied the NSG membership, it has made entry into other three important groupings responsible for nuclear export control—MTCR in June 2016, Wassenaar Arrangement in December 2017 and Australia Group in January 2018. India's membership to the export control regimes has further helped it to prove its credential as a responsible nuclear power country (Sood 2018).[9] The membership to the NSG can enhance India's status strongly. India has been pursuing its effort to be an NSG member. China's position has not moved from the arguments made in India's bid to the NSG at the Seoul plenary in June 2016.

Recent developments have further intensified India–China strategic rivalry. China used its UNSC veto power to block a resolution aimed at declaring Masood Azhar, the Jaish-e-Mohammad chief, a terrorist. It continues to use its UNSC veto power to block the India-led resolution. From the Indian side, New Delhi has hosted the Dalai Lama in the Indian state of Arunachal Pradesh and remains strongly opposed to the CPEC. An almost three-month stand-off between Indian and Chinese military forces in the Doklam region of Bhutan has further revived their strategic tension.

China's continued defence and nuclear assistance to Pakistan, opposition to India's NSG membership conditional to Pakistan's entry and India-encircling 'String of Pearls' strategy are concerning for India. China is concerned about India's strategic posture towards China: India's military exercises with the United States, Japan and Vietnam, and its Act East Policy. India–China energy geopolitics has remained at the commercial competition, diplomatic manoeuvring level and naval exercises. Unlike the US–Soviet/Russia energy geopolitics, India–China energy geopolitics so far have not led to direct or indirect conflict, but both remain vulnerable.

---

[9] https://thediplomat.com/2018/06/can-india-make-headway-in-the-nuclearsuppliers-group-in-2018/

Conclusion

Energy security does not have a single definition. This is clear from the varying perceptions and interpretations discussed in Chapter 1. But there is clarity about the conceptual framework of four 'As'—availability, accessibility, affordability and acceptability. The framework is in line with the energy security policy of the majority of developed countries and developing countries with huge energy demand, such as India and China.

India's economic growth is unstoppable. Given its development goal to lift millions out of poverty, India has no option but to grow.[1] Growth has pushed energy security to the forefront of its domestic and foreign policy. In India's case, energy security is driven by the fact that its hydrocarbon sources are insufficient to fulfil a fast-growing economy and consumers' growing appetite for energy. India is heavily dependent on imports for oil and gas. India's thorium and coal reserves are not without issue. Coal, which is the main source of electricity generation in India, is not considered environment-friendly. Besides, India lacks high-end technology and best mining practices which could enhance quality and lessen environmental impacts. It is dependent on imports of clean coal from countries such as Australia. India's thorium reserves are the second largest in the world, but India has not been able to attain the processing technology to convert it to enriched uranium for nuclear reactors producing electricity. India's non-signatory status to the NNPT has been a constraining factor for it to acquire nuclear reactors and enriched uranium for electricity production.

As discussed in Chapter 2, India's energy demand has mounted over the past two decades after its economic liberalization in the

[1] For a detail insight in India's economic growth, see Jha (2018).

1990s. There is a big gap between demand and supply in India's energy sector. In the past, India failed to exploit its energy reserve potential due to various reasons including policy fallacy, obstinate bureaucracy and lack of advanced technology. However, over the past two decades, subsequent governments have taken steps to address the energy security crisis and reform the energy sector. India's energy security policy reflects an all-inclusive approach, exploring both fossil and non-fossil sources of energy at the domestic and international level.

At the domestic level, India's efforts have been mainly on the following:

- Overcome the organizational, financial and social challenges in the energy sector.
- Find a precise balance between fossil and non-fossil fuel in its energy mix.
- Shift from coal to natural gas in the hydrocarbon energy mix.
- Increase the share of nuclear and renewable energy.
- Build a smart system for drawing investment in energy infrastructure.
- Price energy to facilitate economic and environmental competence.
- Open access to energy for all at an affordable price.

India's renewable energy sector has made impressive progress, and this has been the hallmark of India's energy security efforts since PM Modi committed at the 2015 Paris Climate Summit to increase the share of renewables in India's overall energy mix.

The increased share of renewables will be facilitated by technological advancement at different stages of maturity, with wind and biomass energy already commercial, and solar PV well on the way to becoming so. Though India's energy basket will continue to be dominated by hydrocarbons in the near future, renewables will have an increased share in the coming decades. India's focus on renewables is not only based on environmental acceptability but because of abundance, accessibility and affordability.

India has taken steps to enhance its share of gas and is taking steps to tap the abundance in supply of natural gas by focusing on gas delivery infrastructure. This is further supported by the lower price of gas than oil. India has also been able to dictate and influence the global energy price market as being the third largest and fastest growing energy consumer. In addition to measures to streamline the energy sector at home, the Indian government has asserted internationally to determine the rational pricing of oil and gas. In recent years, PM Modi and India's Petroleum and Natural Gas Minister Dharmendra Pradhan have taken bold steps in regards to petroleum pricing reforms and frequently raised the issue of rational pricing of oil with oil-producing countries.

India's energy security efforts are focused on making energy accessible to all at an affordable price in a socially and politically acceptable carbon-controlled environment. This is reflected in some of the bold steps that have been taken by the Indian government in regulatory measures, including opening of the energy sector to private players, modernization and the technological evolution in the energy sector. They can be categorized as:

- Three major energy policy frameworks at the domestic level—IEP, Expert Group Reports on Low-Carbon Inclusive Growth and NEP.
- Liberalizing the energy sector by encouraging private participation by the Electricity Act 2003, improving the condition of power distribution through UDAY, and the opening of coal mines and the renewable energy sector to private companies.
- Reducing the share of hydrocarbons with emphasis on technology, conservation and energy efficiency through initiatives such as the NEMMP 2020 and FAME.
- For Oil and Gas E&P at the domestic level, taking major steps under the NELP and under the Modi government, under the HELP, marketing and pricing freedom for new gas production, and the policy for granting extension to PSCs.

However, energy security efforts at the domestic level will not make a major difference towards productivity. The continued import dependency which is about 80 per cent of its crude oil needs and nearly 40 per cent of its gas shows limits to India's domestic potential. This will continue to push India's energy security efforts to foreign nations' reserves. Despite all, hydrocarbons constitute the major share of India's energy basket.

For a long time, India's overseas destination for oil and gas has been the GCC region. This overdependence on the Gulf nations increases India's vulnerability to disruptions and price fluctuations which in turn affects the economic development and day-to-day life of consumers. In addition, the possibility of disruption from unseen political instability, religious extremism, terrorism and threats to supply lines have pushed India to look for new hydrocarbon destinations abroad. This has pushed India to compete globally, resulting in severe competition with major energy-consuming nations, especially China.

But in India's case, energy exploration abroad is multifaceted and not confined to and gas. It includes enriched uranium for its nuclear reactors and acquisition of innovative technology for domestic energy E&P, and its move towards renewable and alternative sources of energy. India has made a concerted effort to expand its option for gas sources and has taken a proactive foreign policy approach towards exploring for hydrocarbons, especially in the gas-rich nations in Africa, the Latin American and Caribbean (LAC) region, and the Caspian Basin. Consequently, to address the mounting energy demand, India's quest for these resources has pushed the energy security on its foreign policy agenda which is reflected in bilateral agreements, intense diplomacy and geopolitics.

Nuclear energy has emerged as an important source of energy. India has been taking steps to increase the share of nuclear energy in its energy basket. Over the past two decades, India has consistently pursued nuclear issues in its foreign policy and concluded nuclear agreements with several countries which required intense diplomacy, lobbying and negotiations. The share of nuclear energy in India's energy mix for a long time had been almost negligible at below 3 per cent. India's nuclear energy potential was held back because of the international nuclear embargo. India's atomic explosion in 1974 resulted in the 1978 Nuclear Non-Proliferation Act that barred trade in nuclear technology and enriched uranium. India needed to change this regime in its favour to exploit the full potential of its civilian nuclear programme. The beginning in this direction was India's atomic test in 1998. This was landmark from two aspects. The first aspect was the process of its engagement with the United States more seriously, which finally led to the conclusion of the US–India civilian nuclear deal, India's safeguard agreement with the IAEA and waiver at the NSG. Today India has concluded nuclear agreements with many countries. India's nuclear programme is getting advanced nuclear reactors, civilian nuclear technology and enriched uranium from the United States, France, Russia, Canada, Australia and UK.

The second aspect is that India's nuclear test triggered and revived India–China strategic rivalry. India's atomic explosion in 1998 was strongly condemned by China, which blamed India for triggering a nuclear race in South Asia. When the United States began the process of nuclear exception to India by signing of the US–India Civilian Nuclear Agreement on 18 July 2005, China opposed the deal strongly as it saw the deal in a geopolitical context—US strategy along with India to balance China's rise. China, along with Pakistan and Nuclear ayatollahs, tried its best to block the nuclear deal in the US Congress, but could not stand in front of active and powerful Indian lobbying. After the nuclear deal became an Act in 2008, China intensified its India containment strategy by signing a nuclear agreement with Pakistan in 2010. China decided to assist Pakistan with two more nuclear reactors. China has been supporting Pakistan's nuclear weapon and missile programme since the 1980s. When President Obama declared US support to India's bid to the NSG in 2010, China's strategy to block Indian expansion beyond South Asia intensified. The post-nuclear deal phase coincided with a steady increase in China's arms sales to Pakistan and an intensified China–Pakistan all-weather friendship. This was obvious when China vetoed India's bid to the NSG in the 2016 Seoul meeting on the principled stand of nuclear non-proliferation, but backed Pakistan's entry despite its illicit nuclear technology trade and a bad record of nuclear proliferation.

India's nuclear defiance by not signing the NPT was rooted in its argument that NPT created an unequal nuclear regime and undermined its security concerns as China is a nuclear weapon state. Since then, India's nuclear energy programme has been an issue of foreign policy and geopolitics. India's agreements with the United States, the waiver by the NSG, safeguard agreement with the IAEA and its agreements with France, Russia, Australia, Canada and UK are significant not only in the energy security but also the strategic context. India's growing acceptance in the international nuclear regime and as a de facto nuclear weapon state has alarmed China. Not surprisingly China has taken India's energy security exploration abroad seriously. Despite some examples of energy security cooperation, their energy security pursuit is intensifying their strategic rivalry. This is visible in their attempts to diversify energy sources abroad in Latin America, Africa and in the waters of the Indo-Pacific region.

As demonstrated in Chapters 4 and 6, to diversify the sources of hydrocarbons abroad, India has taken a concerted foreign policy to engage African, Latin American and Caspian Basin nations. China is also hunting energy in these regions. Overlapping destinations of oil and gas exploration have intensified traditional rivalry. India's steps to diversify the sources of its hydrocarbon imports and the exploration of new oil and gas reserves is also reflected in the waters of the Indo-Pacific region which is estimated to be abundant in energy sources. Its strategic location is critical for the security of sea routes through which energy imports flow.

China's naval expansion and military build-up in the SCS and its activities in the Indian Ocean have been worrying for India. India is concerned about China's growing military assertiveness in the Indo-Pacific region. India considers the Indian Ocean its special sphere of influence, and the area is significant for its economic prosperity and security; India is concerned about the security of energy sea routes and Chinese strategic moves in the Indian Ocean which India does not see entirely from the commercial angle claimed by China. India sees the Chinese strategy of 'String of Pearls' and OBOR as a grand design to dominate the energy sea route and possibly various choke-points in the Indian Ocean. As a result, India has moved to build-up strategic partnerships and ties with countries which share common views on the rule-based and free navigation principles. This is also rooted in democratic values and principles which bind India and the countries in the Indo-Pacific region together. One of the significant developments in this context has been a robust and deepening India–US strategic and defence partnership. Since the first Malabar Exercise in 1992 and then in the early 2000s, India and the United States have elevated their naval collaboration in the Indian Ocean to a level that today India conducts more military exercises with the United States than any country in the world.

Democratic values and the resolve that international law and rule-based order should be followed in SLOC, brings maritime democracies together. India shares common perspectives with the United States, Japan and Australia. These converging values have underpinned the idea of the Quad. The idea of the Quad emerged in the wake of Tsunami disaster rescue operations by the four maritime democracies and the first Quad naval exercise in 2007. After the Chinese demarche, the Quad went backstage (A. Sharma 2010). The Quad is again reviving amidst the Chinese militarily aggressive posture in the Indo-Pacific region. Another strategy is India, Japan and US trilateral exercises now named as JAI (Japan, America and India) with its literal meaning 'victory' in Hindi. JAI nations have been doing military exercises in the Indian Ocean regularly, and the first summit-level meeting of the heads of JAI nations was held in the sidelines of 2018 G20 meeting (Joshi 2018). These have enhanced interoperability and built trust and confidence between the navies of these three countries. In addition, India has taken concerted efforts to build strategic and security partnerships with China-wary countries in the Indo-Pacific region. India has taken steps to elevate security partnerships with three major countries in the region, namely Australia, Indonesia and Vietnam. Vietnam has offered India a foothold in the SCS and PM Modi's signing of strategic partnerships with Australia and Indonesia, in which defence and security are significant components, are crucial balancing strategic steps taken by India. India's other two important trilateral arrangements in the developing stage are India–Australia–Japan and India–Australia–Indonesia.

Though these efforts could be seen as India coming together with major democratic nations in the region to balance China, these efforts do not reflect purely military aspects of strategy. Strategic partnerships between India and major maritime democracies are driven by a common goal to find a balance of power in the Indo-Pacific. This is more of a soft balancing aimed at ensuring a rule-based order in the face of China's growing strategic assertiveness. How quickly such a new equilibrium can be established will depend in part on how forcefully the rules in the Indo-Pacific are challenged. These groupings can become more active and attain a greater military overtone depending on the intensity of the Chinese military assertiveness and threat to the equilibrium (Varghese 2018).

In recent years, the term Indo-Pacific has emerged as a new geopolitical construct for the term 'Asia-Pacific'. The term acknowledges the economic and strategic significance of India in the emerging strategic scenario in the region. The region has been witnessing a complete reconfiguration of the existing order: the relative decline of US power and influence, the steady rise of China as a comprehensive national power and the simultaneous rise of India, a dozen or so years behind China. China's dramatically expanding power has generated growing assertiveness on Beijing's part, on the one hand, and a pushback from the others, on the other (Thakur and Sharma 2018). As Peter Varghese, one of the leading proponents of the Indo-Pacific, says that is an act of imagination and recognition of an emerging structural shift in the strategic environment. At its heart, the Indo-Pacific reflects two propositions. First, that the maritime environment is likely to be the primary focus of strategic planning and competition over several decades. Second, that India's strategic focus will, over this period, shift well beyond the immediate neighbourhood and embed India in the strategic dynamics of the region.

In the United States, the term Indo-Pacific was frequently used by then Secretary of State Hillary Clinton. The name of the US-Pacific Command was changed to the US Indo-Pacific Command under the Donald Trump administration in 2018. The term has been endorsed and frequently used in strategic and political circles in Japan and Australia, as reflected in various Defence White Papers, and in China-wary countries including Vietnam and Indonesia. The term acknowledges India's economic and strategic significance in the emerging strategic realities of the great power game in the Indo-Pacific region.

India's strategic convergence with great powers, middle powers and also small island nations on the common ground of wrestling the Chinese domination of the Indo-Pacific will continue to shape the strategic geometry of the region. Amidst the declining US hegemonic power and its ambiguity on historical security commitments and rise of economically and militarily assertive China, India and concerned nations have been developing various units of Indo-Pacific strategy. These unfolding balancing strategies, and India's centrality in these strategic groupings, have further elevated the India–China strategic rivalry.

India and China are not involved directly in an arms race, and the balance of power between them differs in the various dimensions of military capability. Though both are vulnerable to military conflict, the situation of multi-level soft deterrence leads to a stronger security interdependence, and hence a reduced probability of armed conflict (Holslag 2009). Parallel to diplomatic efforts, confidence-building measures and increased interaction of armed forces, economic interactions between India and China have eased tensions and infused confidence.

In a globalized world, knowledge-based power and economic interdependence are becoming significant factors. Nuclear deterrence will likely prevent a war or military solution for resolving and settling border disputes or the potential scramble for resources. Today, both share the concerns such as climate change, security of the SLOC, terrorism, energy security, regional security and ensuring prosperity for their more than one billion population and these would make India–China overall relations in general and energy quest in particular more accommodative. This makes their competition more at the commercial and diplomatic level as demonstrated in Chapter 4–6.

From border disputes and great power aspirations to energy geopolitics, India–China strategic competition is inevitable. Neither is ready to play second fiddle. The continuing strategic rivalry now reflected in their quest for hydrocarbons in the waters of the Indo-Pacific keeps them vulnerable to military conflicts. Their strategic rivalry is also reflected in the shift towards renewable sources of energy.

Renewable energy is going to be a game changer for international relations. Its geographic and technical characteristics are fundamentally different from those of coal, oil and natural gas. Renewable energy sources are abundant and intermittent; production lends itself more to decentralized generation and involves rare earth materials in clean tech-equipment; their distribution, finally, is most electric in nature and involves stringent managerial conditions and long-distance losses.[2] The transition towards renewables is shaping the strategic dimension of international relations.

As the international norms are being built towards renewable energy, India and China energy geopolitics will also be reflected in this new energy space. Both are trying to have their sphere of influence in the renewable energy space. Their energy competition is moving from soft balancing of hard power to the soft power influence. This is about playing a normative leadership and/or influence in the renewable energy space.

Today, energy security involves geopolitics, bilateral relations with other nations, diversification of sources of supply and diversified types of energy, both renewable and non-renewable. The inevitability of renewable energy is undeniable and countries are gearing up for an increased share in their energy mix. Those that dominate the markets in these new technologies will likely have the most influence over the development patterns of the future. As other major powers find themselves in climate denial or atrophy, China and India both are moving to use and exploit the global renewable energy space, though China is ahead.

India is emerging as a new normative player in the renewable energy regime. PM Modi has pursued unwavering commitment to energy security by constructing 175 GW of renewable capacity by 2022. India's two important ministers responsible for the energy sector, the union Petroleum and Natural Gas Minister Dharmendra Pradhan, and

---

[2] For trends of renewable energy geopolitics, see Scholten, (2018).

the Minister of State for Power, Coal, New and Renewable Energy and Mines, Piyush Goyal, have time after time reaffirmed India's unflinching commitment to combat global warming irrespective of what happens in the rest of the world.

PM Modi being awarded the 'Champion of Earth' by the UN Secretary-General in 2018 is additional power to India's growing normative leadership in this new energy source regime. The award to Modi and praise for India's shift to renewable efforts give it advantage in soft power and that translates into geopolitics as well as in building relations with many influential Western nations. India's pursuit of energy security has pushed it into competition with China and from a strategic front to scramble for energy sources. The two Asian giants remain vulnerable to any deterioration; and this is likely to continue in the coming decades.

Bibliography

Acharya, A. 'India's "Look East" Policy.' In *The Oxford Handbook of Indian Foreign Policy*, edited by David M. Malone, C. Raja Mohan and Srinath Raghavan. Oxford: Oxford University Press, 2015, 452–465.

Ahmad, T. 'Advantages of Transnational Gas Pipelines.' *The Hindu*, 24 April 2006.

An African Energy Industry Report: 2018. Bolton: ISPY Publishing Limited, May 2018. Available at: https://www.futureenergyafrica.com/media/1751/1-mir-africa-mir-18-2-es_685804715-05-2018.pdf (accessed on 25 August 2018).

Asia Pacific Energy Research Centre. *A Quest for Energy Security in the 21st Century: Resources and Constraints*. Tokyo: Asia Pacific Energy Research Centre, 2007. Available at: https://aperc.ieej.or.jp/file/2010/9/26/APERC_2007_A_Quest_for_Energy_Security.pdf (accessed on 7 January 2017).

Australian Government. 'Framework for Security Cooperation between Australia and India 2014.' Department of Foreign Affairs & Trade, 18 March 2014. Available at: https://dfat.gov.au/geo/india/Pages/framework-for-security-cooperation-between-australia-and-india-2014.aspx (accessed on 25 November 2014).

Australian Trade Commission. *Report Highlights the Business Opportunities in India's Automotive Sector*. 2015. Available at: http://www.austrade.gov.au/News/Latest-from-Austrade/2015/report-highlights-the-business-opportunities-in-india-s-automotive-sector (accessed on 30 July 2018).

Badrinarayana, D. 'India's Integrated Energy Policy: A Source of Economic Nirvana or Environmental Disaster?' *Environmental Law Reporter*, 40 ELR 10709, July 2010. Available at: http://elr.info/sites/default/files/articles/40.10706.pdf (accessed on 20 May 2018).

Bakshi, J. 'Russia and India: From Ideology to Geopolitics, 1947–1998.' *Journal of Cold War Studies* 12, no. 3 (2010): 50–90.

Bansal, S. 'On Paper, Electrified Villages—In Reality, Darkness.' *The Hindu*, 26 March 2016. Available at: http://www.thehindu.com/opinion/op-ed/On-paper-electrified-villages-%E2%80%94-in-reality-darkness/article14176223.ece (accessed on 12 November 2017).

Barnes, J., A. Jaffe, and E. L. Morse. 'The New Geopolitics of Oil.' *The National Interest*, 17 December 2003. Available at: https://nationalinterest.org/article/the-new-geopolitics-of-oil-2513 (accessed on 7 January 2017).

Basrur, R. M. *Minimum Deterrence and India's Nuclear Security*. California: Stanford University Press, 2006.

BBC. 'Japan and India Sign Civilian Nuclear Agreement.' 11 November 2016. Available at: https://www.bbc.com/news/world-asia-37948246 (accessed on 12 October 2018).

————. 'India Will Build 10 New Reactors in Huge Boost to Nuclear Power.' 18 May 2017. Available at: https://www.bbc.com/news/world-asia-india-39958299 (accessed on 30 September 2018).

Bennett, J. 'Adani Coal "Welcome" Part of India's Future Electricity Supply, Says Country's Power Minister.' *ABC News*, 13 June 2017a. Available at: http://www.abc.net.au/news/2017-06-13/adani-coal-imports-relunctantly-welcomed-in-india/8611970 (accessed on 12 November 2017).

————. 'Australia Quietly Makes First Uranium Shipment to India Three Years after Supply Agreement.' *ABC News*, 19 July 2017b. Available at: https://www.abc.net.au/news/2017-07-19/australia-quietly-makes-first-uranium-shipment-to-india/8722108 (accessed on 15 October 2018).

Blackwill, R. D. 'The Future of US–India Relations.' Address at the Confederation of Indian Industry, New Delhi, U.S. Department of State. 17 July 2003. Available at: http://2001-2009.state.gov/p/sca/rls/rm/22615.htm (accessed on 23 July 2003).

Bowles, C. 'America and Russia in India.' *Foreign Affairs* 49, no. 4 (1971). Available at: https://www.foreignaffairs.com/articles/asia/1971-07-01/america-and-russia-india (accessed on 10 October 2018).

BP. *BP Statistical Review of World Energy*. June 2016. Available at: https://www.bp.com/content/dam/bp/pdf/energy-economics/statistical-review-2016/bp-statistical-review-of-world-energy-2016-full-report.pdf (accessed on 20 April 2018).

————. BP *Statistical Review of World Energy*. June 2017. Available at: https://www.bp.com/content/dam/bp/en/corporate/pdf/energy-economics/statistical-review-2017/bp-statistical-review-of-world-energy-2017-full-report.pdf (accessed on 12 April 2018).

————. 'Country and Regional Insights–India.' In *Energy Outlook*, 2018. Available at: https://www.bp.com/content/dam/bp/en/corporate/pdf/energy-economics/energy-outlook/bp-energy-outlook-2018-country-insight-india.pdf (accessed on 12 April 2018).

————. 'Regional Insight—Africa.' In *Statistical Review of World Energy*. Available at: https://www.bp.com/en/global/corporate/energy-economics/statistical-review-of-world-energy/country-and-regional-insights/africa.html (accessed on 22 August 2018).

————. *Statistical Review of World Energy 2016*. 2016. Available at: http://www.bp.com/en/global/corporate/energy-economics/statistical-review-of-world-energy.html (accessed on 30 October 2018).

————. *BP Statistical Review of World Energy 2018*. 2018. Available at: https://www.bp.com/content/dam/bp/en/corporate/pdf/energy-economics/

statistical-review/bp-stats-review-2018-full-report.pdf (accessed on 30 October 2018).

Brands, H. W. *India and the United States: The Cold Peace*. Boston: Twayne Publishers, 1990.

Brown, M. A., and B. K. Sovacool. *Climate Change and Global Energy Security: Technology and Options*. Cambridge: MIT Press, 2011.

Bushuev, V. V., and A. A Troitskii. 'The Energy Strategy of Russia until 2020 and Real Life: What Is Next?' *Thermal Engineering* 54, no. 1 (2007): 1–7.

Business Standard. 'India–Britain Civil Nuclear Agreement Signed.' 12 November 2015. Available at: https://www.business-standard.com/article/news-ians/india-britain-civil-nuclear-agreement-signed-115111201293_1.html (accessed on 20 October 2018).

———. 'Russia, Islamabad Sign MoU to Build Gas Pipeline from Iran to Pak and India.' 28 September 2018. Available at: https://www.business-standard.com/article/international/russia-islamabad-sign-mou-to-build-gas-pipeline-from-iran-to-pak-and-india-118092800044_1.html (accessed on 15 September 2018).

Business Wire. 'Waste to Energy Opportunities in India 2017–2022—Research and Markets.' 26 April 2017. Available at: https://www.businesswire.com/news/home/20170426006844/en/Waste-Energy-Opportunities-India-2017-2022---Research (accessed on 15 May 2018).

Carpenter, J. W. 'The Biggest Oil Producers in the World.' *Investopedia*, 13 October 2015. Available at: https://www.investopedia.com/articles/investing/101315/biggest-oil-producers-latin-america.asp (accessed on 14 August 2018).

Casier, T. 'The Rise of Energy to the Top of the Eu–Russia Agenda: From Interdependence to Dependence.' *Geopolitics* 16, no. 3 (2011): 536–552.

Central Electricity Authority, Ministry of Power, Government of India. *Draft National Electricity Plan*. Volume 1. December 2016. Available at: http://www.cea.nic.in/reports/committee/nep/nep_dec.pdf (accessed on 10 May 2018).

———. *Executive Summary*. New Delhi: Power Sector, April 2017. Available at: http://www.cea.nic.in/reports/monthly/executivesummary/2017/exe_summary-04.pdf (accessed on 17 May 2018).

Chakraborty, D., A. Shiryaevskaya, and W. Harry. 'U.S. Taps India as Asia's Debut Buyer of American Shale Gas.' *Bloomberg*, 2 April 2016. Available at: https://www.bloomberg.com/news/articles/2016-04-01/india-s-gail-buys-cheniere-s-second-lng-cargo-from-sabine-pass (accessed on 30 September 2018).

Chaudhary, D. R. 'Why is India–Japan Nuclear Deal Important?' *The Economic Times*, 12 July 2018. Available at: https://economictimes.indiatimes.com/news/defence/why-is-japan-nuclear-deal-important/articleshow/55423473.cms (accessed on 12 October 2018).

Checchi, A., A. Behrens, and C. Egenhofer. (2009). 'Long-Term Energy Security Risks for Europe: A Sector-Specific Approach.' CEPS working paper no. 309, Centre for European Policy Studies. Available at: https://www.ceps.eu/publications/long-term-energy-security-risks-europe-sector-specific-approach (accessed on 15 December 2016).

Cherp, A., and J. Jewel. 'The Concept of Energy Security: Beyond the Four As.' *Energy Policy* 75, December (2014): 415–421.

Chester, L. 'Conceptualising Energy Security and Making Explicit its Polysemic Nature.' *Energy Policy* 38, no. 2 (2010): 887–889.

China Daily. 'Solar Power Plant to be Built in North China.' *China Daily*, 3 June 2006.

———. 'China, Pakistan Reaffirm All-Weather Friendship.' *China Daily*, 18 May 2011. Available at: http://www.chinadaily.com.cn/china/2011-05/18/content_12535978.htm. (accessed on 25 October 2018).

Chu, B. 'Partition at 70: Can India Become the World's Largest Economy?' *Independent*, 14 August 2017. Available at: http://www.independent.co.uk/news/world/asia/india-world-largest-economy-partition-70-years-anniversary-growth-a7892886.html (accessed on 20 October 2017).

Clarke Energy. 'Electricity from Sewage in India: Waste-to-Wire.' 5 August 2014. Available at: http://www.clarke-energy.com/2014/electricity-sewage-india/ (accessed on 15 May 2018).

Climate Policy Initiative. 'Global Climate Finance: An Updated View on 2013 & 2014 Flows.' October 2016. Available at: http://climatepolicyinitiative.org/wp-content/uploads/2016/10/Global-Climate-Finance-An-Updated-View-on-2013-and-2014-Flows.pdf (accessed on 30 July 2018).

Cohen, S. *India: Emerging Power*. Washington, DC: Brookings Institution Press, 2001.

Cohen, S., and S. D. Gupta. *Arming without Aiming: India's Military Modernization*. Washington, DC: Brookings Institution Press, 2015.

Colglazier, E. W., and D. A. Deese. 'Energy Security in the 1980s.' *Annual Review of Energy* 8, no. 1(1983): 415–449.

Commission of the European Communities, 'A European Strategy for Sustainable, Competitive and Secure Energy', *Green Paper*, Brussels, 8 March 2006. Available at: http://europa.eu/documents/comm/green_papers/pdf/com2006_105_en.pdf (accessed on 25 May 2017).

Congressional Budget Office. 'Energy Security in the United States.' 9 May 2012. Available at: https://www.cbo.gov/publication/43012 (accessed on 10 February 2017).

Council on Foreign Relations. 'The US–India Nuclear Deal.' 5 November 2010. Available at: https://www.cfr.org/backgrounder/us-india-nuclear-deal (accessed on 6 October 2018).

Created by LBNL, AEEE and FICCI with inputs from stakeholders. *Facilitating Energy Efficiency to Reduce Energy Costs, Enhance Energy Security, and Minimize Environmental Impact: Summary of Recommendations on Energy*

Efficiency for the National Energy Policy. Submitted to NITI Aayog, New Delhi, 19 November 2015.

Criekemans, D. 'Geopolitics of Renewable Energy Game and its Potential Impact upon Global power Relations.' In *The Geopolitics of Renewables,* edited by Daniel Scholten, 37–73. Cham: Springer, 2018.

———. *The Geopolitics of Renewable Energy: Different or Similar to Conventional Energy?* Exploring Geopolitics, April 2011. Available at: http://www. exploringgeopolitics.org/publication_criekemans_david_geopolitics_of_ renewable_energy_technology_desertec_north_seas_countries_offshore_grid_ initiative_co2_emissions_investments_germany/ (accessed on 25 October 2018).

Curtis, L., A. Cohen, and O. Graham, 'The Proposed Iran–Pakistan–India Gas Pipeline: An Unacceptable Risk to Regional Security.' The Heritage Foundation. 30 May 2008. Available at: https://www.heritage.org/sites/ default/files/~/media/images/reports/2008/bg2139_map1sm/bg2139_map1. jpg (accessed on 10 September 2018).

Dannreuther, R. *Energy Security.* Cambridge: Polity Press, 2017.

———. 'Energy Security.' In *Handbook of New Security Studies,* edited by J. P. Burgess, 44–54. Abingdon: Routledge, 2010.

Davies, R. W., M. Harrison, and S. G. Wheatcroft (eds.). *The Economic Transformation of the Soviet Union, 1913–1945.* Cambridge: Cambridge University Press, 1994.

Davis, L. W. 'Prospects for Nuclear Power.' *Journal of Economic Perspectives* 26, no. 1 (2012): 49–66.

D'Cunha, S. D. 'Modi Announces "100% Village Electrification," but 31 Million Indian Homes Are Still in the Dark.' *Forbes,* 7 May 2018. Available at: https://www.forbes.com/sites/suparnadutt/2018/05/07/modi-announces-100-village-electrification-but-31-million-homes-are-still-in-the-dark/#3adb866063ba (accessed on 6 August 2018).

Deo, A. *The International Solar Alliance's China's Conundrum.* Observer Research Foundation, 20 March 2018. Available at: https://www.orfonline.org/ research/international-solar-alliance-china-conundrum%E2%80%8B/ (accessed on 15 November 2018).

Department of Atomic Energy (DAE). 'Important Agreements.' Available at: http://www.dae.nic.in/?q=node/75 (accessed on 25 July 2018).

Department of Energy & Climate Change. 'Energy Markets Outlook.' 16 December 2009. Available at: https://assets.publishing.service.gov.uk/gov-ernment/uploads/system/uploads/attachment_data/file/247999/0176.pdf (accessed on 14 December 2016).

Dhir, R. 'Exploration and Production Opportunity in Africa.' In *India and Africa: Forging a Strategic Partnership,* edited by Subir Gokarn, W. P. S. Sidhu and Shruti Godbole. Washington, DC: Brookings, 2015.

Director General of Hydrocarbons. *National Exploration & Licensing Policy.* Ministry of Petroleum and Natural Gas, Government of India. 1 April 2015.

Available at: http://dghindia.gov.in/index.php/page?pageId=59 (accessed on 5 August 2018).

Diwanji, A. K. 'Geopolitical Issues Set to Dominate Proposed Gas Pipeline from Iran to India.' *The Rediff Business Special*, 13 April 2000.

Dixit, J. N. *India–Pakistan in War and Peace*. New York, NY: Routledge, 2006.

Dudau, R., and A. C. Nedelcu. 'Energy Security: Between Markets and Sovereign Politics.' *Management & Marketing: Challenges for the Knowledge Society* 11, no. 3 (2016): 544–552.

EAI. India Waste to Energy. Available at: http://www.eai.in/ref/ae/wte/wte.html (accessed on 15 May 2018).

Ebinger, C. K. *Energy and Security in South Asia: Cooperation or Conflict*. Washington, DC: Brookings Institution Press, 2011.

———. 'India's Energy and Climate Policy: Can India Meet the Challenge of Industrialization and Climate Challenge.' Policy Brief, 16-01, Brookings Institution, June 2016. Available at: https://www.brookings.edu/wp-content/uploads/2016/07/india_energy_climate_policy_ebinger.pdf (accessed on 25 March 2017).

Effimoff, E. 'The Oil and Gas Resource Base of the Caspian Region.' *Journal of Petroleum Science and Energy* 28, no. 4 (2000): 157–159.

Embassy of India. 'India, United States Sign Historic Civil Nuclear Agreement.' Available at: https://www.indianembassy.org/India_Review/2008/Nov%2008.pdf%20india.pdf (accessed on 5 October 2018).

Embassy of India, Riyadh, Kingdom of Saudi Arabia. 'India–GCC Relations.' 2 July 2017. Available at: http://www.indianembassy.org.sa/india-saudi-arabia/india-gcc-relations (accessed on 14 August 2018).

Emmanual, W. 'Energy Alternatives India.' 5 March 2012. Available at: http://www.eai.in/ref/ae/wte/wte.html (accessed on 15 May 2018).

Energy Charter Secretariat. 'International Energy Security: Common Concept for Energy Producing, Consuming and Transit Countries.' March 2015. Available at: http://www.energycharter.org/fileadmin/DocumentsMedia/Thematic/International_Energy_Security_2015_en.pdf (accessed on 14 December 2016).

European Commission. 'Energy Security Strategy.' 2014. Available at: https://ec.europa.eu/energy/en/topics/energy-strategy-and-energy-union/energy-security-strategy (accessed on 20 February 2017).

Feller, G. 'China's Wind Power.' Available at: http://www.ecoworld.com/Home/Articles2.cfm? (accessed on 28 April 2018).

Financial Express. 'IEA Applauds India For Successful Auction Of Discovered Oil & Gas Fields.' 3 December 2016. Available at: https://www.financialexpress.com/market/commodities/iea-applauds-india-for-successful-auction-of-discovered-oil-gas-fields/461859/ (accessed on 7 August 2018).

———. 'Modi Government May Consider Ways to Lower Hydropower Tariffs to Help Compete against Cheaper Forms of Electricity.' 22 January 2018.

Available at: http://www.financialexpress.com/industry/modi-government-may-consider-ways-to-lower-hydropower-tariffs-to-help-compete-against-cheaper-forms-of-electricity/1025021/ (accessed on 17 May 2018).

Findlater, S., and P. Noel, (2010). 'Gas Supply Security in the Baltic States: A Qualitative Assessment.' Working paper, Electricity Policy Research Group [EPRG], Cambridge. Available at: https://www.emeraldinsight.com/doi/full/10.1108/17506221011058713 (accessed on 15 December 2016).

France in India: French Embassy in New Delhi. 'Bilateral Civilian Nuclear Cooperation.' 10 October 2018. Available at: https://in.ambafrance.org/Bilateral-Civilian-Nuclear, 7474 (accessed on 12 October 2018).

Fredholm, M. 'A New Energy Policy of Russia: Implementation Experience.' Conflict Studies Research Centre, Defence Academy of the United Kingdom. September, 2005.

Froome, C. 'India's Energy Future: Australian Coal or Renewable Revolution?' *The Conversation*, 5 June 2014. Available at: https://theconversation.com/indias-energy-future-australian-coal-or-renewable-revolution-26569 (accessed on 20 October 2017).

Froome, C., and R. Dargaville. 'Fact Check Q&A: Will India No Longer Buy Australian Coal?' *The Conversation*, 24 August 2015. Available at: https://theconversation.com/factcheck-qanda-will-india-no-longer-buy-australian-coal-46256 (accessed on 12 November 2017).

Galbraith, P. 'Nuclear Proliferation in South Asia: Whose Business?' In *Conflicting Images: India and the United States*, edited by Sulochana Raghavan Glazer and Nathan Glazer, 67–81. Riverdale, MD: The Riverdale Company Publishers, 1990.

Ganguly, G., A. Scobell, and B. Shoup, B. *US–Indian Strategic Cooperation into the 21st Century.* London: Rutledge, 2006.

Gelb, B. A. 'Caspian Oil and Gas: Production and Prospects.' CRS Report for Congress. 4 March 2015. Available at: https://www.globalsecurity.org/military/library/report/crs/45467.pdf (accessed on 25 August 2018).

Gokaran, K. 'Stepping on the Gas: Indo-Russian Energy Cooperation.' *Observer Research Foundation*, 26 February 2018. Available at: https://www.orfonline.org/expert-speak/steppig-on-the-gas-indo-russian-energy-cooperation/ (accessed on 10 October 2018).

Government of Canada. *Agreement between the Government of Canada and the Government of the Republic of India for Co-operation in Peaceful Uses of Nuclear Energy.* Treaty E105192, Global Affairs Canada, 3 March 2014. Available at: http://www.treaty-accord.gc.ca/text-texte.aspx?id=105192 (accessed on 20 October 2018).

———. 'Nuclear Cooperation Agreement between Canada and India.' Nuclear Safety Commission, 28 May 2015. Available at: https://nuclearsafety.gc.ca/eng/resources/educational-resources/feature-articles/nuclear-agreement-canada-india.cfm (accessed on 20 October 2018).

Government of India, Cabinet. 'Major Policy Initiatives to Give a Boost to Petroleum and Hydrocarbon Sector.' Press Information Bureau, 10 March 2016. Available at: http://pib.nic.in/newsite/printrelease.aspx?relid=137661 (accessed on 5 August 2018).

Government of India. 'Revision of Cumulative Targets under National Solar Mission from 20,000 MW by 2021–22 to 1, 00, 000 MW.' Press Information Bureau, 17 June 2015. Available at: http://pib.nic.in/newsite/PrintRelease. aspx?relid=122566 (accessed on 10 May 2018).

Granados, U. 'India's Approach to the South China Sea: Priorities and Balances.' *Asia & the Pacific Studies* 5, no. 1 (2008): 122–137.

Gulf Times. 'Qatar Reiterates Commitment to Meet India's Energy Demand.' 23 March 2018. Available at: https://www.gulf-times.com/story/585997/ Qatar-reiterates-commitment-to-meet-India-s-energy (accessed on 15 September 2018).

Gupta, T. 'Making Mining Great Again: Understanding Modi's Big Bet on Coal Privatisation.' *Swarajya*, 21 February 2018. Available at: https://swarajyamag. com/politics/making-mining-great-again-understanding-modis-big-bet-on-coal-privatisation (accessed on 25 July 2018).

Guruswamy, M. 'Pakistan–China Relations: Higher than the Mountains, Deeper than the Oceans.' *CLAWS Journal*, Summer (2010): 92–107.

Guruswamy, M., and Z. S. Daule. *India China Relations: The Border Issues and Beyond.* New Delhi: Viva Books, 2011.

Haniffa, A. 'Massive Campaign on against Indo–US N-Deal.' *Rediff.com.* 2005. Available at: http://in.rediff.com/news/2005/dec/02aziz.htm. (accessed on 20 November 2018).

Harris, S. 'Exclusive: Inside the FBI's Fight against Chinese Cyber-Espionage.' *Foreign Policy*, 27 May 2014. Available at: https://foreignpolicy. com/2014/05/27/exclusive-inside-the-fbis-fight-against-chinese-cyber-espionage/ (accessed on 15 November 2018).

Hathaway, R. M. 'The US–India Courtship: From Clinton to Bush.' In *India as an Emerging Power*, edited by Sumit Ganguly. London: Frank Cass Publisher, 2003.

Heine, J., and R. Viswanathan. 'The Other BRIC in Latin America: India.' *Americas Quarterly*, Spring (2011). Available at: http://www.americasquarterly. org/india-latin-america (accessed on 20 August 2018).

Hibbs, M. 'The Nuclear Suppliers Group's Critical India Decision.' *The Diplomat*, 18 June 2016. Available at: https://thediplomat.com/2016/06/the-nuclear-suppliers-groups-critical-india-decision/ (accessed on 25 November 2018).

———. *Pakistan Deal Signals China's Growing Nuclear Assertiveness.* Washington, DC: Carnegie Endowment for Peace, 27 April 2010. Available at: http:// www.carnegieendowment.org/publications/index.cfm?fa=view&id=40685. (accessed on 25 October 2018).

Hill, F., and R. Spector. 'The Caspian Basin and Asian Energy Markets.' *Brookings*, 1 September 2001. Available at: https://www.brookings.edu/ research/the-caspian-basin-and-asian-energy-markets/ (accessed on 5 September 2018).

Hindustan Times. 'Nuclear Power Corp Wants to Renegotiate Haripur Plant with Mamata Banerjee Government.' 3 April 2017. Available at: https://www.hindustantimes.com/kolkata/nuclear-power-corp-wants-to-renegotiate-haripur-plant-with-mamata-banerjee-government/story-mGRujjuFNXcO4YjqTx5NaK.html (accessed on 25 April 2018).

———. 'From March 6, India to Host Naval Exercise Amid China's Manoeuvring in High Seas.' 26 February 2018. Available at: https://www.hindustantimes. com/india-news/from-march-6-india-to-host-naval-exercise-amid-china-s-manoeuvring-in-high-seas/story-DGHYOofWu2M5IOnf9eNb7I.html (accessed on 5 November 2018).

———. 'Odisha and Bengal Will Be Two Pillars of the 2019 Govt, Says Dharmendra Pradhan.' 9 July 2018. Available at: https://www.hindustantimes. com/india-news/odisha-and-bengal-will-be-two-pillars-of-the-2019-govt-says-dharmendra-pradhan/story-cDDERGoP1s8D2KN5mDxXcI.html (accessed on 6 August 2018).

Ho, J. 'Indian Ocean Region: Critical Sea Lanes for Energy Security.' *RSIS Commentary*, 18 May 2011. Available at: https://www.rsis.edu.sg/rsis-publication/idss/1553-indian-ocean-region-critical/#.W_QRnzSUfTU (accessed on 5 November 2018).

Hoffman, A. R. 'The Connection: Water and Energy Security.' In *Energy Security*. Institute for the Analysis of Global Security, 13 August 2004. Available at: http://www.iags.org/n0813043.htm (accessed on 15 January 2017).

———. 'Energy Poverty and Security.' *Journal of Energy Security*, 23 April 2009. Available at: http://ensec.org/index.php?option=com_content&view=article &id=185:energy-poverty-and-security&catid=94:0409content&Itemid=342 (accessed on 15 January 2017).

Holslag, J. 'The Persistent Military Security Dilemma Between China and India.' *Journal of Strategic Studies* 32, no. 6 (2009): 811–840.

Hongqiao, L. 'The Dark Side of Renewable Energy.' *Earth Journalism Network*, 25 August 2016. Available at: https://earthjournalism.net/stories/the-dark-side-of-renewable-energy (accessed on 15 November 2018).

Horner, D. 'NSG Still Mulling Indian Membership: Arms Control Today.' Arms Control Association. July/August 2012. Available at: http://www.armscontrol. org/2012_07-08/NSG_Still_Mulling_Indian_Membership (accessed on 5 October 2018).

Hossain, J., V. Sinha, and V. V. N. Kishore. 'A GIS Based Assessment of Potential for Wind Farms in India.' *Renewable Energy* 36, no. 12 (2011): 3257–3267.

Hughes, L., and P. Y. Lipscy. 'The Politics of Energy.' *Annual Review of Political Science* 16, no. 1 (2013): 449–469.

IAGS. 'India's Energy Security Challenge.' In *Energy Security*. Available at: http://www.iags.org/n0121043.htm (accessed on 30 August 2018).

iGovernment. 'India to Assess Integrated Energy Policy Impact.' 1 September 2009. Available at: http://igovernment.in/site/India-to-assess-Integrated-Energy-Policy-impact (accessed on 22 May 2018).

India Today. 'Where is Doklam and Why It Is Important for India?' 27 March 2018. Available at: https://www.indiatoday.in/education-today/gk-current-affairs/story/where-doklam-why-important-india-china-bhutan-1198730-2018-03-27 (accessed on 15 September 2018).

India Water Review. 'India Remains Dominant Player in Global Hydropower.' 2 March 2016. Available at: http://www.indiawaterreview.in/Story/Features/india-remains-dominant-player-in-global-hydropower/1890/2#. Wq2tPDSUfTV (accessed on 17 May 2018).

India's Ministry of Commerce. 'Export and Import Data.' 13 April 2009. Available at: http://commerce.nic.in/eidb/irgn.asp (accessed on 20 August 2018).

Indian Embassy in the United States. 'India and Disarmament.' Washington, DC, 2008. Available at: www. indianembassy.org (accessed on 30 September 2008).

Indian Express. 'Civil Nuclear Cooperation: India–Japan Pact Comes into Force.' *Indian Express*, 21 July 2017. Available at: https://indianexpress.com/article/india/civil-nuclear-cooperation-india-japan-pact-comes-into-force-4760426/ (accessed on 25 July 2017).

Indian Gazette. 'Ministry of Finance Notification: In the Matter of Import of Solar Cells Whether or Not Assembled in Modules or Panels.' *EQ International*, 30 July 2018.

International Atomic Energy Agency. 'Agreement between the Government of India and the International Atomic Energy Agency for the Application of Safeguards to Civilian Nuclear Facilities.' INFCIRC/754, Information Circular, 29 May 2009. Available at: http://www.dae.nic.in/writereaddata/ncpw/infcirc754.pdf (accessed on 25 September 2018).

———. 'Power Reactor Information System, India.' Available at: https://www.iaea.org/PRIS/CountryStatistics/CountryDetails.aspx?current=IN (accessed on 25 April 2018).

International Energy Agency. 'Energy Security.' Available at: https://www.iea.org/topics/energysecurity/ (accessed on 25 January 2017).

———. 'Towards a Sustainable Energy Future.' 28 May 2001. Available at: http://www.iea.org/textbase/nppdf/free/2000/future2001.pdf. (accessed on 15 December 2016).

———. 'World Energy Outlook 2017.' 2017. Available at: https://www.iea.org/weo2017/ (accessed on 20 April 2018).

———. 'National Energy Policy.' NITI Aayog (National Institution for Transforming India), Government of India. February 2018. Available at:

https://www.iea.org/policiesandmeasures/pams/india/name-168042-en.php (accessed on 22 May 2018).

International Energy Forum. 'IEF Overview.' Available at: https://www.ief.org/about-ief/ief-overview.aspx (accessed on 25 January 2017).

International Energy Foundation. 'About Us.' Available at: http://www.energy-ief.org/about.html (accessed on 25 January 2017).

Interview with Lalit Mansingh, New Delhi, June 2004–June 2005.

IOL. 'Huge Scope for Indian Vendors in Africa's Energy Sector.' 15 May 2018. Available at: https://www.iol.co.za/business-report/energy/huge-scope-for-indian-vendors-in-africas-energy-sector-14984074 (accessed on 25 August 2018).

Jai, S. 'Govt Plans Push for Hydro Power.' *Business Standard*, 11 April 2016. Available at: http://www.business-standard.com/article/economy-policy/govt-plans-push-for-hydro-power-116041100037_1.html (accessed on 17 May 2018).

Jain, P. 'India and Japan Scale New Heights.' East Asia Forum, 2012. Available at: http://www.eastasiaforum.org/2016/11/25/india-and-japan-scale-new-heights/ (accessed on 26 July 2017).

Jha, R. *Facets of India's Economy and Her Society.* Volume II. London: Palgrave Macmillan, 2018.

Jiali, M. *Notice India: A Big Rising Country on the Rise.* Tianjin: Tianjin People's Press, 2002.

Johnston, I. 'India Cancels Plans for Huge Coal Power Stations as Solar Energy Prices Hit Record Low.' *Independent*, 23 May 2017. Available at: http://www.independent.co.uk/environment/india-solar-power-electricity-cancels-coal-fired-power-stations-record-low-a7751916.html (accessed on 20 October 2017).

JooAhn, S., and D. Graczyk. *Understanding Energy Challenges in India: Policy, Players and Issues.* International Energy Agency, 2012. Available at: https://www.iea.org/publications/freepublications/publication/India_study_FINAL_WEB.pdf (accessed on 12 December 2016).

Joshi, M. 'Japan–America–India: No Reason to Say "JAI" Unless Beijing Listens.' *The Quint*, 11 December 2018. Available at: https://www.thequint.com/voices/opinion/japan-america-india-summit-beijing-modi-shinzo-abe-trump (accessed on 12 December 2018).

Joshi, Y. 'Between Principles and Pragmatism: India and the Nuclear Non-Proliferation Regime in the Post-PNE Era, 1974–1980.' *The International History Review* 40, no. 5 (2018): 1073–1093.

Joskow, P. L. 'Problems and Prospects for Nuclear Energy in the United States.' In *Energy, Economics and the Environment*, edited by Gregory A. Daneke. Lexington, DC: Heath and Company, 1982.

———. 'The Future of Nuclear Power in the United States: Economic and Regulatory Challenges.' MIT CEEPR Working Paper 06-019, Center for

Energy and Environmental Policy Research, Massachusetts Institute of Technology, 2006.

———. 'The US Energy Sector, Progress and Challenges, 1972–2009.' *Dialogue: Journal of the US Association of Energy Economics* 17, no. 2 (2009): 7–11.

Judge, A., T. Maltby, and J. D. Sharples, 'Challenging Reductionism in Analyses of EU–Russia Energy Relations.' *Geopolitics* 21, no. 4 (2016): 751–762.

Kanti, P. K., and N. Sanathkumar, 'A Review Paper on Nuclear Power Plant and Its Importance in Indian Economy.' National Conference on Advances in Mechanical Engineering Science, NCAMES-2016. Available at: http://ijettjournal.org/Special%20issue/NCAMES-2016/NCAMES-152.pdf (accessed on 30 September 2018).

Kanwal, G. *Scant Progress: India–China Boundary Dispute and Sustained Dialogue.* New Delhi: CLAWS, 7 May 2008. Available at: http://www.claws.in/index.php?action=master&task=87&u_id=7. (accessed on 25 October 2018).

Katoch, P. C. 'China: A Threat or Challenge.' *CLAWS Journal*, Summer (2010): 79–91.

van Kemenade, W. *Détente between China and India: The Delicate Balance of Geopolitics in Asia.* Clingendael: Netherland Institute of International Relations. July 2008. Available at: http://www.clingendael.nl/publications/2008/20080700_cdsp_diplomacy_paper.pdf. (accessed on 25 October 2018).

Kenning, T. 'India Surpasses 1GW Rooftop Solar with Grid Parity for Most C&I Consumers.' *PV-Tech*, 19 October 2016. Available at: https://www.pv-tech.org/news/india-surpasses-1gw-rooftop-solar-with-grid-parity-for-most-ci-consumers (accessed on 10 May 2018).

Kerr, P. K. 'U.S. Nuclear Cooperation with India: Issues for Congress.' *Congressional Research Service*, 4 February 2010.

Kim, S. 'The World's First Fully Solar-Powered Airport.' *The Telegraph*, 21 August 2015. Available at: https://www.telegraph.co.uk/travel/destinations/asia/india/articles/The-worlds-first-fully-solar-powered-airport/ (accessed on 10 May 2018).

Kim, S. P. 'An Anatomy of China's String of Pearl's Strategy.' *The Hikone Ronso* 387, Spring (2011). Available at: http://www.biwako.shiga-u.ac.jp/eml/Ronso/387/Kim.pdf (accessed on 27 October 2018).

KPMG. 'Energy and Natural Resources: Union Budget 2017–18—A Post-budget Sectoral Point of View.' 2017. Available at: https://home.kpmg.com/content/dam/kpmg/in/pdf/2017/02/Energy-and-Natural-Resources.pdf (accessed on 21 July 2018).

Krishnan, A. 'Behind China's India Policy, a Growing Debate.' *Hindu*, 5 April 2010. Available at: http:////www.thehindu.com/opinion/lead/article388895.ece (accessed on 30 October 2018).

Kruyt, B., D. P. van Vuuren, H. J. M de Vries, and H. Groenenberg. 'Indicators for Energy Security.' *Energy Policy* 37, no. 6 (2009): 2166–2181.

Kumar, D. 'Securing India's Energy Future.' Department of Defence, Government of Australia, October 2012. Available at: http://www.defence.gov.au/ADC/Publications/Commanders/2012/05_Col%20Devindar%20Kumar%20SPP.pdf (accessed on 20 May 2018).

Kumar, J.S., K.V. Subbaiah, and P. Rao, 'Municipal Solid Waste Management Scenario in India" *Australian Journal of Engineering and Technology*, 2 (2014): 1–8.

Kumar, S. 'India's Energy Supply Security: Prospects and Challenges.' Modern Diplomacy, 7 January 2017. Available at: https://moderndiplomacy.eu/2017/01/17/india-s-energy-supply-security-prospects-and-challenges/ (accessed on 5 September 2018).

Kurian, A. L., and C. Vinodan. 'Energy Security: A Multivariate Analysis of Emerging Trends and Implications for South Asia.' *India Quarterly: A Journal of International Affairs* 69, no. 4 (2013): 383–400.

Kux, D. *India and the United States: Estranged Democracies.* Washington, DC: National Defense University Press, 1992.

———. *United States and Pakistan 1947–2000: Disenchanted Allies.* Oxford: Oxford University Press, 2001.

Ladislaw, S., and S. Bellur. *India's Hydrocarbon Exploration and Licensing Policy (HELP): Will it Help India's Upstream Oil and Gas.* Centre for Strategic and International Studies, 21 March 2017. Available at: https://www.csis.org/analysis/indias-hydrocarbon-exploration-and-licensing-policy-help-will-it-help-indias-upstream-oil. (accessed on 5 August 2018).

Lama, M. P. (2005). Integrating Stakeholders in Energy Cooperation. *South Asia Journal* 3, no. 1 (2005): 6–20.

Larkin, J. 'Indian, Chinese Oil Firms to Issue Joint Bids.' *The Washington Post*, 18 August 2005. Available at: https://www.washingtonpost.com/archive/business/2005/08/18/indian-chinese-oil-firms-to-issue-joint-bids/242b83d6-8594-42b1-bb8b-19e9eea56ccf/?noredirect=on&utm_term=.72907de7493f (accessed on 5 November 2018).

Laughland O, Weaver M (2014) Indian election result: 2014 is Modi's year as BJP secures victory. The Guardian. Available at https://www.theguardian.com/world/2014/may/16/india-election-2014-results-live. Accessed on 21 August 2016.

Levy, A. 'India Builds a Nuclear Plant amid Protests and Scandals.' *Newsweek*, 20 February 2016. Available at: http://www.newsweek.com/nagercoil-tamil-nadu-india-428740 (accessed on 25 April 2018).

Lin, C. Y. *Militarization of China's Energy Security Policy—Defence Cooperation and WMD Proliferation along Its String of Pearls in the Indian Ocean*, 3–4. Berlin: Institut für Strategie-Politik-Sicherheits-und Wirtschaftsberatung (ISPSW), 18 June 2008. Available at: http://www.isn.ethz.ch/isn/Digital-Library/Publications/Detail/?id=56390. (accessed on 27 October 2018).

Livemint. 'India, France Pledge to Push Forward on Jaitapur Nuclear Power Project.' 10 March 2018. Available at: https://www.livemint.com/Industry/

X05B03Vct56Y72VqAzr4fM/India-France-pledge-to-push-forward-on-Jaitapur-nuclear-pow.html (accessed on 12 October 2018).

Lock, A. 'World Coal Association Reacts to TERI Report.' World Coal, 13 February 2017. Available at: https://www.worldcoal.com/power/13022017/world-coal-association-reacts-to-teri-report/ (accessed on 20 October 2017).

Lockie, A. 'Iran Threatened to Cut Off a Key Oil Shipping Waterway—but the US Would Blow It Out of the Water.' *Business Insider*, 25 July 2018. Available at: https://www.businessinsider.com.au/iran-threatens-close-of-strait-of-hormuz-us-navy-response-oil-price-2018-7?r=UK&IR=T (accessed on 5 November 2018).

Löschel, A., U. Moslener, and D. T. G. Rübbelke. 'Energy Security—Concepts and Indicators.' *Energy Policy* 38, no. 4 (2010): 1607–1608.

Luft, G., A. Korin, and M. Martel. *Petropoly: The Collapse of America's Energy Security Paradigm.* Wood Dale, IL: Create Space Independent Publishing Platform, 2012.

Madan, T. 'Energy Security Series: India.' *The Brookings Foreign Policy Studies*, November 2006. Available at: http://www.brookings.edu/~/media/Files/rc/reports/2006/11india_fixauthorname/2006 (accessed on 20 May 2017).

Mahapatra, C. *Indo-US Relations in the 21st Century.* New Delhi: Knowledge World Publishers, 1998.

Malhotra, T. C. 'India Intends to Open Shale Reserves to Private Companies.' *E&P*, 21 December 2017. Available at: https://www.epmag.com/india-intends-open-shale-reserves-private-companies-1674946#p=full (accessed on 25 May 2018).

Malik, M. *China and India: Great Power Rivals.* Boulder, CO: First Forum Press, 2011.

Manish Vaid, 'TAPI pipeline progresses, but future uncertain', *Observer Research Foundation'* 5 February 2016. Available at: https://www.ogj.com/articles/print/volume-114/issue-5/transportation/tapi-pipeline-progresses-but-future-uncertain.html

Mastny, V. *The Soviet Union's Partnership with India.* Cambridge, MA: MIT Press, 2010.

Mathur, V., A. Pandey, and A. Ray. 'Mobilising Private Capital for Green Energy in India.' ORF Special Report, 4 December 2017. Available at: https://www.orfonline.org/research/mobilising-private-capital-green-energy-india/ (accessed on 25 July 2018).

McCallion, K. *Shoreham and the Rise and Fall of the Nuclear Power Industry.* Westport, CT: Praeger, 1995.

McHugh, B. 'Uranium Price Increase Around Corner as China and India Look to Nuclear to Reduce Carbon Emissions.' *ABC News*, 10 March 2016. Available at: https://www.abc.net.au/news/rural/2016-03-09/uranium-future-price-set-to-improve-as-new-plants-built/7232944 (accessed on 30 September 2018).

Medcalf, R. 'Australia's Uranium Puzzle: Why China and Russia but not India.' *The Fearless Nadia Occasional Papers.* Volume 1. 2011a. Available at https://

www.aii.unimelb.edu.au/documents/32/Australias-uranium-puzzle_Why-China-and-Russia-but-not-India.pdf (accessed on 28 September 2015).

———. 'Uranium Sales to India Would Spread Trust, Not Nuclear Arms.' *Sydney Morning Herald*, 2 December 2011b. Available at: https://www.smh.com.au/politics/federal/uranium-sales-to-india-would-spread-trust-not-nuclear-arms-20111201-1o94z.html (accessed on 10 July 2015).

Miglani, S., and G. De Clercq. 'Russia Signs Pact for Six Nuclear Reactors on New Site in India.' *Reuters*, 6 October 2018. Available at: https://www.reuters.com/article/us-india-russia-nuclear/russia-signs-pact-for-six-nuclear-reactors-on-new-site-in-india-idUSKCN1MF217 (accessed on 10 October 2018).

Millhone, J. H., S. Greene, and A. Vatansever. 'Russia's Neglected Energy Reserves.' Carnegie Endowment for International Peace, 18 May 2010. Available at: https://carnegieendowment.org/2010/05/18/russia-s-neglected-energy-reserves-event-2937 (accessed on 26 February 2017).

Ministry of Coal, Government of India. 'Vision.' Available at: https://www.coal.nic.in/content/vision (accessed on 11 January 2018).

———. *The Year 2016–17 at a Glance—Ministry of Coal: Annual Report 2016–17*. 2017. Available at: https://coal.nic.in/sites/upload_files/coal/files/coalupload/chap1AnnualReport1617en.pdf (accessed on 11 January 2018).

Ministry of Energy of the Russian Federation. 'Energy Strategy of Russia for the Period up to 2030.' Moscow, 2010. Available at: http://www.energystrategy.ru/projects/docs/ES-2030_(Eng).pdf (accessed on 26 February 2017).

Ministry of External Affairs (MEA), Government of India. 'India–Russia Joint Statement on the Outcome of the Official Visit of Prime Minister Dr Manmohan Singh to the Russian Federation.' 7 December 2005. Available at: https://mea.gov.in/bilateral-documents.htm?dtl/7105/IndiaRussia+Joint+Statement+on+the+outcome+of+the+Official+Visit+of+Prime+Minister+Dr+Manmohan+Singh+to+the+Russian+Federation (accessed on 30 August 2018).

———. 'Inaugural Address by External Affairs Minister at the 3rd Roundtable of ASEAN India Network of Think Tanks.' 25 August 2014a. Available at: https://www.mea.gov.in/Speeches-Statements.htm?dtl/23948/Inaugural+Address+by+External+Affairs+Minister+at+the+3rd+Roundtable+of+ASEAN+India+Network+of+Think+Tanks (accessed on 10 August 2018).

———. 'Strategic Vision for Strengthening Cooperation in Peaceful Uses of Atomic Energy between the Republic of India and the Russian Federation.' Press Information Bureau, 11 December 2014b. Available at: http://pib.nic.in/newsite/PrintRelease.aspx?relid=113165 (accessed on 10 October 2018).

———. 'India–Australia Relations.' February 2016. Available at: https://www.mea.gov.in/Portal/ForeignRelation/Australia_Jan_ENG_2016.pdf (accessed on 22 July 2018).

———. 'India–Japan Consultations on Non-proliferation & Disarmament.' Press release, 26 August 2017a. Available at: http://www.mea.gov.in/press-releases.

htm?dtl/27339/india++japan+consultations+on+nonproliferation+amp+disarmament (accessed on 12 October 2018).

———. *India France Relations*. October 2017b. Available at: https://www.mea.gov.in/Portal/ForeignRelation/2_France_November_2017.pdf (accessed on 6 October 2018).

———. 'Official Visit of the Prime Minister of Bhutan to India.' 6 July 2018a. Available at: https://www.mea.gov.in/press-releases.htm?dtl/30036/Official+Visit+of+the+Prime+Minister+of+Bhutan+to+India (accessed on 15 September 2018).

———. 'Remarks by External Affairs Minister at the 3rd Indian Ocean Conference, Vietnam.' Media Centre, 27 August 2018b. Available at: https://www.mea.gov.in/Speeches-Statements.htm?dtl/30327/Remarks+by+External+Affairs+Minister+at+the+3rd+Indian+Ocean+Conference+Vietnam+August+27+2018 (accessed on 30 October 2018).

Ministry of Foreign Affairs, Government of Japan. 'Agreement between the Government of Japan and the Government of the Republic of India for Cooperation in the Peaceful Uses of Nuclear Energy.' 11 November 2016. Available at: http://www.mofa.go.jp/files/000202920.pdf (accessed on 12 October 2018).

Ministry of Heavy Industries & Public Enterprises, Government of India. 'National Electric Mobility Mission Plan.' Press Information Bureau, 10 March 2015. Available at: http://pib.nic.in/newsite/PrintRelease.aspx?relid=116719 (accessed on 2 August 2018).

Ministry of New and Renewable Energy, Government of India. 'Year End Review 2017–MNRE.' Press Information Bureau, 27 December 2017. Available at: http://pib.nic.in/newsite/PrintRelease.aspx?relid=174832 (accessed on 28 April 2018).

Ministry of Non-conventional Energy Sources, Government of India. 'Solar Energy Programmes at a Glance.' Available at: http://mnes.nic.in/frame.htm?majorprog.htm (accessed on 30 April 2018).

Ministry of Petroleum & Natural Gas, Government of India. '4th India–Africa Hydrocarbons Conference Ends; Paves the Way for Strengthening India Africa Relations.' Press Information Bureau, 22 January 2016. Available at: http://www.pib.nic.in/newsite/PrintRelease.aspx?relid=135729 (accessed on 10 August 2018).

———. 'First 2G (Second Generation) Ethanol Bio-refinery in India to Be Set Up at Bathinda (Punjab): Foundation Stone Laying Ceremony to be Held on 25th December, 2016.' Press Release, 23 December 2016. Available at: http://pib.nic.in/newsite/PrintRelease.aspx?relid=155782 (accessed on 6 August 2018).

———. 'Cabinet Approves Survey of Un-appraised Areas of Sedimentary Basins of India.' Policy brief, August 2017. Available at: http://petroleum.nic.in/cabinet-approves-survey-un-appraised-areas-sedimentary-basins-india (accessed on 12 September 2017).

Ministry of Power, Government of India. 'Domestic Efficiency Lightening Programme (DELP).' 25 November 2015. Available at: https://powermin. nic.in/en/content/domestic-efficiency-lighting-programme-delp (accessed on 30 July 2018).

———. 'Ujjwal Discom Assurance Yojana.' 25 June 2018. Available at: https:// www.uday.gov.in/home.php (accessed on 25 June 2018).

Ministry of Statistics and Programme Implementation, Government of India. *Energy Statistics 2017*. 2017. Available at: http://www.mospi.gov.in/sites/ default/files/publication_reports/Energy_Statistics_2017.pdf (accessed on 11 January 2018).

Mishra, A. 'Achieving Energy Security in Country: Insights Based on Consumption of Petroleum Products.' NITI Aayog, Government of India, 2017a. Available at: http://niti.gov.in/content/achieving-energy-security-country-insights-based-consumption-petroleum-products (accessed on 2 August 2018).

Mishra, T. 'Jaitapur Nuclear Power Project Back on Negotiation Table.' *The Hindu Business Line*, 23 March 2017b. Available at: https://www.thehin-dubusinessline.com/economy/jaitapur-nuclear-power-project-back-on-negotiation-table/article9598582.ece (accessed on 20 April 2018).

Mistry, D. *The US–India Nuclear Agreement: Diplomacy and Domestic Policy*. New Delhi: Cambridge University Press, 2014.

Mitchell, J. V. 'Renewing Energy Security.' Royal Institute of International Affairs, 2002. Available at: http://www.chathamhouse.org.uk/files/3038_ renewing_energy_s ecurity_mitchell_july_2002.pdf (accessed on 7 January 2017).

Modi, Narendra. 'India–Saudi Arabia Joint Statement during the visit of Prime Minister to Saudi Arabia.' 3 April 2016. Available at: https://www.naren-dramodi.in/india-saudi-arabia-joint-statement-during-the-visit-of-prime-minister-to-saudi-arabia-439963 (accessed on 20 August 2018).

Mohan, C. R. *Crossing the Rubicon: The Making of India's New Foreign Policy*. New Delhi: Penguin Books, 1998.

Moreira, S. 'India's Expanding Role in Latin America: Promise and Challenges.' Available at: http://paperroom.ipsa.org/papers/paper_3267.pdf (accessed on 20 August 2018).

Mukherjee, M. *Private Participation in the Indian Power Sector: Lesson from Two Decades of Experience*. Washington, DC: World Bank Group, 2014.

Muni, S. D., and Girijesh Pant. *India's Energy Security Prospects for Cooperation with Extended Neighbourhood*. New Delhi: Rupa, 2005.

Nagao, S. 'The Significance of the Japan–India Nuclear Deal.' East Asia Forum, 25 December 2015. Available at: http://www.eastasiaforum.org/2015/12/25/ the-significance-of-the-japan-india-nuclear-deal/ (accessed on 26 July 2017).

Nakhle, C. 'Caspian Oil and Gas in a World of Plenty.' Crystol Energy, 24 June 2017. Available at: https://www.crystolenergy.com/caspian-oil-gas-world-plenty/ (accessed on 30 August 2018).

National Interest Analysis. 'Agreement between the Government of Australia and the Government of India on Cooperation in the Peaceful Uses of Nuclear Energy.' New Delhi. 5 September 2014. Available at: http://www.aph.gov. au/~/media/02%20Parliamentary%20Business/24%20Committees/244%20 Joint%20Committees/JSCT/2014/28%20October%202014/NIA%20-%20 India%20nuclear%20cooperation.pdf (accessed on 25 September 2018).

National Security Archive. 'Telephone Conversation Transcript.' 19 January 1972. Available at: https://nsarchive2.gwu.edu/NSAEBB/NSAEBB79/BEBB28. pdf (accessed on 8 October 2018).

Nayar, B. R., and T. V. Paul. *India in the World Order*. Cambridge: Cambridge University Press, 2003.

Neog, R. 'Why India Will Be Kept Out of the Nuclear Suppliers Group.' The National Interests, 2 June 2016. Available at: https://nationalinterest. org/blog/the-buzz/why-india-will-be-kept-out-the-nuclear-suppliers-group-16444 (accessed on 25 November 2018).

NITI Aayog, Government of India. 'Overseas Engagements.' Chapter 13. In *Draft National Energy Policy*. NITI Aayog, 27 June 2017a. Available at: http://niti.gov.in/writereaddata/files/new_initiatives/NEP-ID_27.06.2017. pdf (accessed on 5 July 2017).

———. *Draft National Energy Policy*. NITI Aayog, 27 June 2017b. Available at: http://niti.gov.in/writereaddata/files/new_initiatives/NEP-ID_27.06.2017. pdf (accessed on 5 July 2017).

———. 'Energy Demand: Efficiency and Conservation.' Chapter 3, p. 10. In Draft National Energy Policy. NITI Aayog, 27 June 2017c. Available at: http://niti.gov.in/writereaddata/files/new_initiatives/NEP-ID_27.06.2017. pdf (accessed on 5 July 2017).

Nuclear Power Corporation of India Limited (NPCIL). 'Kaiga Unit-1 Sets World Record in Continuous Operation.' Press release, 25 October 2018. Available at: http://www.dae.nic.in/writereaddata/KGS-1%20895%20days%20bilingual_0.pdf (accessed on 30 September 2018).

———. 'About Us.' Available at: http://www.npcil.nic.in/content/328_1_ AboutNPCIL.aspx (accessed on 30 September 2018).

Nuclear Power Industry in Japan. 'Nuclear Business and Reactor Production in Japan.' 2012. Available at: http://factsanddetails.com/japan/cat23/sub152/ item2307.html (accessed on 26 July 2017).

NucNet News No. 333/03, http://www.industrie.Gouv.fr as quoted in WNA Newsletter. January/February 2004.

Obama, Barack. *Remarks by the President on America's Energy Security*. Washington, DC: The White House, Office of the Press Secretary, Georgetown University, 30 March 2011. Available at: https://obamawhitehouse.archives.gov/ the-press-office/2011/03/30/remarks-president-americas-energy-security (accessed on 10 February 2017).

Office of Energy Policy and Systems Analysis. 'Valuation of Energy Security for the United States.' 19 January 2017. Available at: https://energy.gov/epsa/articles/valuation-energy-security-united-states (accessed on 10 February 2017).

Ogden, C. *China and India: Asia's Emergent Great Powers.* Cambridge: Polity Press, 2017.

Oilman Magazine. 'India to Diversify Crude Oil Import Policy.' Available at: https://oilmanmagazine.com/article/india-diversify-crude-oil-import-policy/ (accessed on 20 August 2018).

Öğütçü, M. 'Changing Dynamics and Risks in World Energy: The Way Forward.' In *Perspectives on Energy Risk*, edited by A. Dorsman, T. Gök, and M. Karan. Heidelberg: Springer, 2014.

OilVoice Press. 'West Africa's Gas Sector Becoming a Hive of Activity in 2018.' 16 November 2017. Available at: https://oilvoice.com/Opinion/10637/West-Africas-Gas-Sector-Becoming-a-Hive-of-Activity-in-2018- (accessed on 25 August 2018).

ONGC Videsh. 'ONGC Videsh Operating 41 Projects in 20 Countries.' Available at: http://www.ongcvidesh.com/our-assets-worldwide/ (accessed on 14 August 2018).

———. 'Operations: Sudan'. Available at: http://www.ongcvidesh.com/op_sudan.asp (accessed on 22 August 2018).

Lydia Powell. ORF Energy News Monitor II, 40 (1 April 2006).

———. ORF Energy News Monitor, II, 39 (24 March 2006).

Organisation for Economic Co-operation and Development. 'OECD Contribution to the United Nations Commission on Sustainable Development 15: Energy for Sustainable Development.' 2007. Available at: https://www.oecd.org/greengrowth/38509686.pdf (accessed on 15 January 2017).

Ortiz, G. 'Latin America Has One-Fifth of Global Oil Reserves.' *Inter Press Service*, 11 July 2015. Available at: http://www.ipsnews.net/2011/07/latin-america-has-one-fifth-of-global-oil-reserves/ (accessed on 20 August 2018).

Outlook (12 March 2018), 'India Is World's Largest Importer of Weapons with Insatiable Hunger, While Pakistan Slashes Import.' Available at: https://www.outlookindia.com/website/story/india-is-worlds-largest-importer-of-weapons-with-insatiable-hunger-while-pakista/309385 (accessed on 18 June 2018).

Overdrive. 'New Fuel Efficiency Norms Mandated by Indian Government to be Enforced by 2016–17.' 15 February 2015. Available at: http://overdrive.in/news/new-fuel-efficiency-norms-mandated-by-indian-government-to-be-enforced-by-2016-17/ (accessed on 15 July 2017).

Oxenstierna, S., and V. P. Tynkkynen. *Russian Energy and Security up to 2030.* London: Routledge, 2014.

Palmer, B. 'As India Goes, So Goes Civilization'. *on Earth*, 22 January 2016. Available at: http://www.onearth.org/earthwire/india-climate-changerenewable- (accessed on 20 March 2017).

Panagariya, A. 'India's Generous INDICs.' NITI Aayog, 2016. Available at: http://niti.gov.in/content/indias-generous-indics (accessed on 20 March 2017).

Pandey, H. 'Here Is Why the Nuclear Deal between Japan & India Matters.' CAPS in Focus, Centre for Air Power Studies, 7 June 2017. Available at: http://capsindia.org/files/documents/CAPS_Infocus_HP_15.pdf. (accessed on 25 July 2017).

Pant, G. 'India's Energy Security and the Gulf.' Paper presented in a seminar at the School of International Studies, Jawaharlal Nehru University, New Delhi, 19–20 March 2004.

Pant, H. V. *The US–India Nuclear Pact: Policy, Process, and Great Power Politics.* New Delhi: Oxford University Press, 2011.

———. *India's Afghanistan Muddle: A Lost Opportunity.* New Delhi: Harper Collins, 2014.

———. *Turbulence in Sino-India Relations.* Macdonald Laurier Institute, 2 November 2017. Available at: https://www.macdonaldlaurier.ca/turbu-lence-sino-indian-relations-harsh-pant-inside-policy/ (accessed on 30 October 2018).

Parliament of Australia. 'Visit to Australia by Mr Narendra Modi, Prime Minister of the Republic of India—Address to Members & Senators in HOR Chamber.' 18 November 2014. Available at: http://parlview.aph.gov.au/mediaPlayer.php?videoID=243743&operation_mode=parlview (accessed on 20 January 2015).

Paul, T. V. 'Soft Balancing in the Age of US Primacy.' *International Security* 30, no. 1 (2005): 46–71.

Pehrson, C. J. 'String of Pearls: Meeting the Challenge of China's Rising Power across the Asian Littoral.' July 2006. Available at: http//www.strategicstud-iesinstitute.army.mil/pdffiles/PUB721.pdf (accessed on 27 October 2018).

People's Daily Online. 'Washington Draws India in against China.' 7 July 2005. Available at: http://en.people.cn/200507/07/eng20050707_194676.html (accessed on 20 November 2018).

Perkovich, G. *India's Nuclear Bomb: The Impact on Global Proliferation.* Berkeley, CA: University of California Press, 1999.

Planning Commission, Government of India. 'Mid Term Appraisal.' Eleventh Five-Year Plan, 2007–2012. Planning Commission, 2011. Available at: http://planningcommission.nic.in/plans/mta/11th_mta/chapterwise/Comp_mta11th.pdf (accessed on 20 May 2018).

———. *Draft Report of the Expert Committee on Integrated Energy Policy.* Planning Commission, December 2005. Available at: http://planningcommission.nic.in/reports/genrep/intengpol.pdf (accessed on 12 September 2017).

———. *Integrated Energy Policy: Report of the Expert Committee.* New Delhi. August 2006. Available at: http://planningcommission.gov.in/reports/genrep/rep_intengy.pdf (accessed on 10 March 2017).

————. *12th Five Year Plan.* Volume 2, Table 14.5. New Delhi: Government of India, 2012.

————. 'The Final Report of the Expert Group on Low Carbon Strategies for Inclusive Growth.' *India Environment Portal,* 1 April 2014. Available at: http://www.indiaenvironmentportal.org.in/content/392618/the-final-report-of-the-expert-group-on-low-carbon-strategies-for-inclusive-growth/ (accessed on 22 May 2017).

PM India. 'PM's Address at the Inaugural Session of Petrotech—2016 Exhibition.' 5 December 2016. Available at: http://www.pmindia.gov.in/en/news_updates/pms-address-at-the-inaugural-session-of-petrotech-2016-exhibition/ (accessed on 2 August 2018).

Prasad, D. 'Reaping the Demographic Dividend: The Challenges in Creating Jobs for Young India.' *Next Billion,* 3 April 2012. Available at: https://nextbillion.net/reaping-the-demographic-dividend-the-challenges-in-creating-jobs-for-young-india/ (accessed on 20 October 2017).

Prasad, G. C. 'India Woos African Countries for Hydrocarbons.' *Livemint,* 31 December 2015. Available at: https://www.livemint.com/Industry/MEsuFnawO2fGzAvOFmqdhL/India-woos-African-countries-for-hydrocarbon-collaboration.html (accessed on 22 August 2018).

————. 'NITI Aayog Calls for Targeted Power Subsidy, Privatization of All Discoms.' *Livemint.com,* 24 November 2016. Available at: https://www.livemint.com/Politics/nULcNSObBnM1nMz82SfmBO/NITI-Aayog-calls-for-targeted-power-subsidy-privatization-o.html (accessed on 30 July 2018).

President of Russia. 'Speech at Meeting with the G8 Energy Ministers.' 16 March 2006, Moscow. Available at: http://en.kremlin.ru/events/president/transcripts/23488 (accessed on 26 February 2017).

Pressler, L. *Neighbors in Arms: An American Senator's Quest for Disarmament in a Nuclear Subcontinent.* India: Penguin/Viking Publishers, 2017.

Proctor, D. 'France, India Moving Forward with Massive Nuclear Project.' *POWER,* 5 January 2018. Available at: https://www.powermag.com/france-india-moving-forward-with-massive-nuclear-project/ (accessed on 22 September 2018).

Puntaru, C. 'Energy Security in Europe: How Is the EU Dealing with It?' *E-International Relations Students,* 3 December 2015. Available at: http://www.e-ir.info/2015/12/03/energy-security-in-europe-how-is-the-eu-dealing-with-it/ (accessed on 20 February 2017).

Quan, Y., and J. H. Liu. 'The Energy Security of China and India in the Rapidly Changing Global Energy Pattern.' *Peace and Development* 5 (2012): 61–69.

Rahim, S. A. 'Why Pakistan is TAPI'S Biggest Hurdle.' *The Diplomat,* 7 April 2018. Available at: https://thediplomat.com/2018/04/why-pakistan-is-tapis-biggest-hurdle/ (accessed on 15 September 2018).

Rajya Sabha. 'Q No. 180 Nuclear Pact with Japan.' Unstarred Question No. 180, 25 February 2017. Available at: http://www.mea.gov.in/rajya-sabha.

htm?dtl/26402/q+no180+nuclear+pact+with+japan (accessed on 25 July 2017).

Ramesh, M. 'Post-UDAY, Private Sector May Get a Greater Role in Power Distribution.' *The Hindu*, 11 August 2017. Available at: https://www.the-hindubusinessline.com/economy/macro-economy/postuday-private-sector-may-get-a-greater-role-in-power-distribution/article9814375.ece (accessed on 21 July 2018).

Rapier, R. 'Venezuela's Oil Reserves Are Probably Overstated.' *Forbes*, 1 July 2016. Available at: https://www.forbes.com/sites/rrapier/2016/07/01/venezuelas-oil-reserves-are-probably-vastly-overstated/ (accessed on 20 August 2018).

Remarks of Senator Barack Obama. 'Energy Security Is National Security.' Governor's Ethanol Coalition, Washington, DC, 28 February 2006. Available at: http://obamaspeeches.com/054-Energy-Security-is-National-Security-Governors-Ethanol-Coalition-Obama-Speech.htm (accessed on 10 February 2017).

Reuters. 'India's Changing Coal Imports Show Quality over Quantity: Russell.' 22 June 2016. Available at: http://www.reuters.com/article/us-column-russell-coal-india-idUSKCN0Z70GI (accessed on 11 January 2018).

———. 'Coal to Be India's Energy Mainstay for Next 30 Years: Policy Paper.' 16 March 2017. Available at: https://www.reuters.com/article/us-india-energy/coal-to-be-indias-energy-mainstay-for-next-30-years-policy-paper-idUSKCN18B1YE (accessed on 12 September 2017).

———. 'India Unveils $2.5 Bn Plan to Electrify All Households By End 2018.' 26 September 2017. Available at: https://af.reuters.com/article/energyOilNews/idAFL4N1M64KG (accessed on 6 August 2018).

Roland Dannreuther, *Energy Security*. (Cambridge: Polity Press, 2017)

Society of Indian Automobile Manufacturers (SIAM). 'Review of Automobile Mission Plan 2006–16.' Available at: http://www.siamindia.com/cpage.aspx?mpgid=16&pgid1=17&pgidtrail=76 (accessed on 30 July 2018).

Romm, J. 'Why the Renewables Revolution Is Now Unstoppable.' *Think Progress*, 1 February 2016. Available at: https://thinkprogress.org/why-the-renewables-revolution-is-now-unstoppable-698f8d08cf4c/ (accessed on 17 May 2018).

Rothkopf, D. 'Is a Green World a Safer World?' *Foreign Policy*, 22 August 2009. Available at: https://foreignpolicy.com/2009/08/22/is-a-green-world-a-safer-world/ (accessed on 15 November 2018).

Rotter, A. J. *Comrades at Odds: The United States and India, 1947–64.* Ithaca, NY: Cornell University Press, 2000.

Rovere, C. 'Text of Australia–India Nuclear Cooperation Agreement: Short-sighted, Unnecessary, Dangerous.' Submission to the Joint Standing Committee on Treaties, 7 November 2014. Available at: https://www.aph.gov.au/DocumentStore.ashx?id=515f34fe-2bbf-4dbd-af30-092969773fff&subId=301553 (accessed on 15 October 2018).

Rubinoff, A. G. 'Congressional Attitudes towards India.' In *The Hope and the Reality: Indo-American Relations from Roosevelt to Reagan*, edited by Harold Gould and Sumit Ganguly, 169–73. Boulder, CO: Westview Press, 1992.

———. 'Changing Perceptions of India in the US Congress.' *Asian Affairs: An American Review* 28, no. 1 (2001): 37–60.

———. 'Canada's Re-engagement with India.' *Asian Survey* 42, no. 6, (2012): 838–855.

Rublee, M. R. *Nuclear Non-Proliferation Norms: Why States Choose Nuclear Restraint.* Athens/London: University of Georgia Press, 2009.

Rumley, D., and S. Chaturvedi. *Energy Security and the Indian Ocean Region.* London: Routledge, 2015.

Ryall, J. 'Opposition to Nuclear Energy Grows in Japan.' *DW*, 21 October 2016. Available at: http://www.dw.com/en/opposition-to-nuclear-energy-grows-in-japan/a-36110302 (accessed on 10 March 2017).

Sarah Ladislaw and Sharmila Bellur, 'India's Hydrocarbon Exploration and Licensing Policy (HELP): Will it Help India's Upstream Oil and Gas', Centre for Strategic and International Studies, 21 March 2017. Available at: https://www.csis.org/analysis/indias-hydrocarbon-exploration-and-licensing-policy-help-will-it-help-indias-upstream-oil

Saran, S. 'Address at the FEC Forum on India-Japan Cooperation in Peaceful Uses of Nuclear Energy.' Q. 3346 Civil Nuclear Agreement with Japan, 16 March 2011. Available at: http://www.mea.gov.in/lok-sabha.htm?dtl/16892/q3346+civil+nuclear+agreement+with+japan (accessed on 27 July 2017).

Sarma, E. 'Don't Bank on Coal Exports to India: Future of Electricity Generation Focused on Renewable.' *Asia and the Pacific Policy Society* (27 February 2017). Available at: https://www.policyforum.net/dont-bank-coal-exports-india/ (accessed on 12 November 2017).

Scheffe, J. 'Solar Fuels: How Planes and Cars Could Be Powered by the Sun.' *The Conversation*, 23 June 2015. Available at: https://theconversation.com/solar-fuels-how-planes-and-cars-could-be-powered-by-the-sun-41938 (accessed on 30 April 2018).

Scholten, D., ed. *The Geopolitics of Renewables.* Cham: Springer, 2018.

Science Alert. 'India Establishes World's First 100 Percent Solar-Powered Airport.' *Science Alert*, 20 August 2015.

Sen, A. 'Gas Pricing Reform in India: Implications for the Indian Gas Landscape.' Oxford Institute for Energy Studies Paper NG96, Oxford Institute for Energy Studies, Oxford, 2015.

———. 'India's Upstream Revival—Help or Hurdle?' Oxford Energy Comment, Oxford Institute for Energy Studies, November 2016.

Sen, A., and T. Chakravarty. 'Auctions for Oil and Gas Exploration Leases in India: An Empirical Analysis.' Oxford Institute of Energy Studies Paper SP30, Oxford Institute for Energy Studies, Oxford, 2013.

Seshasayee, H. 'India's Rising Presence in Latin America.' *Americas Quarterly*. Available at: http://www.americasquarterly.org/content/indias-rising-presence-latin-america (accessed on 20 August 2018).

Sethi, M. 'Indo-Russian Nuclear Cooperation: Opportunities and Challenges.' *Strategic Analysis* 24, no. 9 (2000): 1757–1761.

———. 'Inputs for a Nuclear Energy Policy for India.' In *Nuclear Power and Non-Proliferation: Conflict or Convergence*, edited by Jasjit Singh, 80–101. New Delhi: Knowledge World, 2004.

Sethi, N. 'India targets 1,000mw solar power in 2013.' *The Times of India*, 18 November 2009. Available at: https://timesofindia.indiatimes.com/india/India-targets-1000mw-solar-power-in-2013/articleshow/5240907.cms (accessed on 10 May 2018).

Shaffer, B. *Energy Politics*. Philadelphia, PA: University of Pennsylvania Press, 2009.

Shafi, M. 'India Plans Nearly 60% of Electricity Capacity from Non-fossil Fuels by 2027.' *The Guardian*, 22 December 2016. Available at: https://www.theguardian.com/world/2016/dec/21/india-renewable-energy-paris-climate-summit-target (accessed on 10 May 2018).

———. 'Indian Solar Power Prices Hit Record Low, Undercutting Fossil Fuels.' *The Guardian*, 10 May 2017. Available at: https://www.theguardian.com/environment/2017/may/10/indian-solar-power-prices-hit-record-low-undercutting-fossil-fuels (accessed on 10 May 2018).

———. 'India Has Enough Coal without Adani Mine, Yet Must Keep Importing, Minister Says.' *The Guardian*, 13 June 2017. Available at: https://www.theguardian.com/environment/2017/jun/13/india-enough-coal-without-adani-mine-must-keep-importing-piyush-goyal (accessed on 20 October 2017).

Sharma, A. 'India and Energy Security.' *Asian Affairs* 38, no. 2 (2007): 158–172.

———. 'Indo-US Strategic Convergence: An Overview of Defence and Military Cooperation.' CLAWS Kartikeya paper no. 2, 1–32, 2008. Available at: http://www.claws.in/images/publication_pdf/CLAWS%20Papers%20No[1].2,%202008.pdf. (accessed on 28 November 2008).

———. 'Growing Indo-French Nuclear Bonhomie: Ensuring India's Energy Security.' Centre for Land Warfare Studies, 25 March 2009. Available at: http://www.claws.in/164/growing-indo-french-nuclear-bonhomie-ensuring-indias-energy-security-dr-ashok-sharma.html (accessed on 25 March 2009).

———. 'Quadrilateral Initiative: An Evaluation.' *South Asian Survey* 17, no. 2 (2010): 237–253.

———. 'The Enduring Conflict and the Hidden Risk of India–Pakistan War.' *SAIS Review of International Affairs* 32, no. 2 (2012): 129–142.

———. 'The US–India Strategic Partnership: An Overview of Defense and Nuclear Courtship.' *Georgetown Journal of International Affairs* (4 July 2013). Available at: https://www.georgetownjournalofinternationalaffairs.org/online-edition/the-u-s-india-strategic-partnership-an-overview-of-defense-and-nuclear-courtship-by-ashok-sharma (accessed on 4 October 2018).

————. 'US–India Defence Industry Collaborations: Trends, Challenges and Prospects.' *Maritime Affairs: Journal of the National Maritime Foundation of India* 9, no. 1 (2013): 129–147.

————. 'India's Expanding Foreign Policy in the Asia-Pacific Region: Implications and Prospects for the India–New Zealand Relationship.' *Maritime Affairs: Journal of the National Maritime Foundation of India* 10, no. 1 (2014a): 54–74.

————. 'Australia and India's Nuclear Deal: A New Beginning in the India–Australia Relationship.' *SAIS Review of International Affairs* (22 October 2014b). Available at: http://www.saisreview.org/2014/10/22/australia-and-indias-nuclear-deal-a-new-beginning-in-the-india-australia-relationship/#_edn2 (accessed on 10 October 2018).

————. 'Modi's Vision of the India–Australia Relationship: From Periphery to Centre.' *Australian Outlook*, 1 December 2014c. Available at: https://www.internationalaffairs.org.au/australianoutlook/modis-vision-of-the-india-australia-relationship-from-periphery-to-centre/ (accessed on 1 December 2014).

————. 'Why Is India Still the Largest Arms Importer?' *The Conversation*, 15 April 2014d. Available at: https://theconversation.com/why-is-india-still-the-worlds-largest-arms-importer-25462 (accessed on 15 April 2014).

————. 'What's behind the New US–India Defense Pact.' *The Conversation*, 17 June 2015. Available at: https://theconversation.com/whats-behind-the-new-us-india-defense-pact-42944 (accessed on 17 June 2016).

————. 'Australia–US Security Alliance and the Strategic Geometry in Indo-Pacific Region: An Evaluation.' *Artha—Journal of Social Sciences* 16, no. 4 (2017a): 39–60.

————. *Indian Lobbying and its Influence in US Decision Making: Post-Cold War.* New Delhi: SAGE Publications, 2017b.

Sharma, G. 'What Does Caspian Sea's New "Shared Usage" Convention Hold for Oil and Gas Exploration.' *Forbes*, 13 August 2018. Available at: https://www.forbes.com/sites/gauravsharma/2018/08/13/what-does-caspian-seas-new-shared-usage-convention-hold-for-oil-and-gas-exploration/#48288c1d2043 (accessed on 30 August 2018).

Sharma, R. '20 deals in 24 hours: Russia–India Relations Given $100 Billion-Worth Boost.' *RT*, 12 December 2014. Available at: https://www.rt.com/op-ed/213835-russia-india-contracts-nuclear/ (accessed on 10 October 2018).

————. 'India–Bangladesh Negotiate Gas Pipeline.' *The New Indian Express*, 23 January 2017. Available at: http://www.newindianexpress.com/nation/2017/jan/23/india-bangladesh-negotiate-gas-pipeline-1562721.html (accessed on 17 March 2019).

Shidoreand, A., and J. Busby. 'What Will It Take to Turn Natural Gas Around in India?' Council on Foreign Relations, 2 August 2016. Available at: https://www.cfr.org/blog/what-will-it-take-turn-natural-gas-around-india (accessed on 12 September 2018).

Siddiqi, T. 'China and India: More Cooperation than Competition in Energy and Climate Change.' *Journal of International Affairs* 64, no. 2 (2011): 73–90.

Siddiqui, H. 'Latin America Holds Huge Promise for Indian Trade & Investment.' Ministry of External Affairs, Government of India, Public Diplomacy, 20 July 2015. Available at: http://www.mea.gov.in/in-focus-article.htm?25492/ Latin+America+holds+huge+promise+for+Indian+trade+amp+investment (accessed on 22 August 2018).

Singh, A. 'The India–Canada Civilian Nuclear Deal: Implications for Canadian Foreign Policy.' *International Journal* 65, no. 1 (Winter 2009–10): 233–253.

Singh, B. K. 'Energy Security and India–China Cooperation.' *International Association for Energy Economics*, first quarter, 2009, 17–19. Available at: http://www.iaee.org/en/publications/newsletterdl. aspx?id=92 (accessed on 5 November 2018).

Singh, J. 'A Partnership for the Future.' *Daily News Analysis*, 10 July 2006. Available at: http://www.dnaindia.com/report.asp?NewsID=1040808&CatID=19 (accessed on 20 November 2018).

———. ed. *India–Russia Relations.* New Delhi: Knowledge World Publishers, 2012.

Sirhindi, M. 'Power Sector Experts Raise Concerns over Privatization of Distribution System.' *The Times of India*, 19 November 2017. Available at: https://timesofindia.indiatimes.com/city/chandigarh/power-sector-experts-raise-concerns-over-privatization-of-distribution-system/article-show/61711942.cms (accessed on 21 July 2018).

Small, A. *China–Pakistan Axis: Asia's New Geopolitics.* London: C Hurst & Co Publishers Ltd, 2015.

———. 'Why China Is Playing a Tougher Game on the NSG This Time Around.' *Herald*, 23 June 2016. Available at: https://thewire.in/diplomacy/ why-china-is-playing-a-tougher-game-on-the-nsg-this-time-around (accessed on 25 November 2018).

Smillie, D. 'UN Chief Guterres Highlights Importance of Sustainable Energy in Message to EXPO 2017.' UN News Centre, 10 June 2017. Available at: http://www.un.org/apps/news/story.asp?NewsID=56948#.WdVtHrIjGUl (accessed on 10 January 2017).

Sood, R. 'India and Non-proliferation Export Control Regimes.' ORF Occasional Papers, 9 April 2018. Available at: https://www.orfonline.org/research/india-and-non-proliferation-export-control-regimes/ (accessed on 25 November 2018).

Sourie, A. 'Iran–Pakistan–India Gas Pipeline Project and its Economic and Political Impact.' *ORF Energy News Monitor* II, no. 16 (14 October 2005).

Sovacool, B. K., and M. A. Brown. 'Competing Dimensions of Energy Security: An International Perspective.' *Annual Review* 35, no. 1 (2010): 77–108.

Squassoni, S. 'The U.S.–Indian Deal and Its Impact.' *Arms Control Today* (July/ August 2010). Available at: https://www.armscontrol.org/act/2010_07-08/ squassoni (accessed on 5 October 2018).

Stelletti, L. 'The Caspian Basin, Great Opportunities But High Costs.' About Energy, 25 August 2017. Available at: https://www.aboutenergy.com/en_IT/topics/caspian-basin-eng.shtml (accessed on 30 August 2018).

Stockholm International Peace Research Institute (SIPRI). 'South Asia and the Gulf Lead Rising Trend in Arms Imports, Russian Exports Grow, Says SIPRI.' Press release, 17 March 2014. Available at: https://www.sipri.org/media/press-release/2014/south-asia-and-gulf-lead-rising-trend-arms-imports-russian-exports-grow-says-sipri (accessed on 20 October 2017).

Sundria, S., and D. Chakraborty. 'India's Oil Consumption Grows at Fastest Pace in 14 Months.' *The Economic Times*, 13 February 2018. Available at: https://economictimes.indiatimes.com/industry/energy/oil-gas/indias-oil-consumption-grows-at-fastest-pace-in-14-months/articleshow/62896084.cms (accessed on 12 April 2018).

Sushma, U. N. 'India's Wind Energy Sector Is a Complete Mess Right Now—Thanks to the Modi Government.' *Quartz*, 2 August 2017. Available at: https://qz.com/1036577/indias-wind-energy-sector-is-a-complete-mess-right-now-thanks-to-the-narendra-modi-government/- (accessed on 28 April 2018).

———. 'India Is Ruining Its Grand Plans to Become a Solar Power House.' *Quartz*, 15 January 2018. Available at: https://qz.com/india/1179391/solar-safeguard-duty-the-modi-government-is-undermining-its-own-grand-plans/ (accessed on 15 November 2018).

Tellis, A. J. *India as a New Global Power: An Action Agenda for the United States.* Washington, DC: Carnegie Endowment for Peace, July 2005.

Thakur Institute of Management Studies and Research. *Analysis of GDP of India from 1990–2010.* Available at: https://www.scribd.com/doc/50598237/Analysis-of-GDP-of-India-from-1990-2010 (accessed on 10 March 2017).

Thakur, R. 'Follow the Yellowcake Road: Balancing Australia's National Interests against International Anti-nuclear Interests.' *International Affairs* 89, no. 4 (2013): 943–961.

Thakur, R., and A. Sharma. 'India and Australia's Strategic Framing in the Indo-Pacific.' *Strategic Analysis* 42, no. 2 (2018): 69–83.

The Economic Times. 'Automotive Mission Plan 2016–2026 Unveiled: Here Are the Key Highlights.' *The Economic Times*, 2 September 2015. Available at: http://auto.economictimes.indiatimes.com/news/industry/automotive-mission-plan-2016-26-unveiled-here-are-the-key-highlights/48772090 (accessed on 30 July 2018).

———. 'India Aims to Become 100% E-Vehicle Nation by 2030: Piyush Goyal.' *The Economic Times*, 26 March 2016. Available at: https://economictimes.indiatimes.com/industry/auto/news/industry/india-aims-to-become-100-e-vehicle-nation-by-2030-piyush-goyal/articleshow/51551706.cms (accessed on 2 August 2018).

———. 'India, Russia to Study Building $25 Billion Pipeline.' *The Economic Times*, 16 October 2016. Available at: https://economictimes.indiatimes.com/

industry/energy/oil-gas/india-russia-to-study-building-25-billion-pipeline/articleshow/54878729.cms (accessed on 10 September 2018).

————. 'Canada Back as Partner in Civilian Nuclear Programme after 42 Years.' *The Economic Times*, 9 November 2016. Available at: https://economictimes.indiatimes.com/news/politics-and-nation/india-steps-up-civil-nuclear-deal-with-canada-russia/articleshow/55307192.cms (accessed on 20 October 2018).

————. 'Full Text of Prime Minister Modi's Energetic Speech at Petrotech 2016.' *The Economic Times*, 5 December 2016. Available at: https://energy.economictimes.indiatimes.com/news/oil-and-gas/full-text-of-prime-minister-modis-energetic-speech-at-petrotech-2016/55806415 (accessed on 12 April 2018).

————. 'India Gas Demand Growth Stymied by Slow Infrastructure Development.' *The Economic Times*, 6 December 2016. Available at: http://economictimes.indiatimes.com/industry/energy/oil-gas/india-gas-demand-growth-stymied-by-slow-infrastructure-development/articleshow/55836029.cms (accessed on 20 April 2018).

————. 'India, China Should Jointly Bid for Oil, Gas Fields: Beijing Media.' *The Economic Times*, 16 March 2017. Available at: https://economictimes.indiatimes.com/industry/energy/oil-gas/india-china-should-jointly-bid-for-oil-gas-fields-beijing-media/articleshow/57672357.cms (accessed on 5 November 2018).

————. 'India's Solar Energy Capacity Expanded by Record 5,525 MW.' *The Economic Times*, 6 April 2017. Available at: https://economictimes.indiatimes.com/industry/energy/power/indias-solar-energy-capacity-expanded-by-record-5525-mw/articleshow/58037873.cms (accessed on 10 May 2018).

————. 'UDAY Scheme Improves Performance of Discoms in FY17: Report.' *The Economic Times*, 31 July 2017. Available at: https://economictimes.indiatimes.com/industry/energy/power/uday-scheme-improves-performance-of-discoms-in-fy17-report/articleshow/59850764.cms (accessed on 21 July 2018).

————. 'Coal Imports Decline 24 Per Cent in August, Auction Looms.' *The Economic Times*, 15 September 2017. Available at: https://economictimes.indiatimes.com/industry/indl-goods/svs/metals-mining/coal-imports-decline-24-per-cent-in-august-auction-looms/articleshow/60528186.cms. (accessed on 11 January 2018).

————. 'Civil Nuclear Cooperation "Important Pillar" of India–France Engagement: Sushma Swaraj.' *The Economic Times*, 17 November 2017. Available at: https://economictimes.indiatimes.com/news/politics-and-nation/civil-nuclear-cooperation-important-pillar-of-india-france-engagement-sushma-swaraj/articleshow/61693489.cms (accessed on 12 October 2018).

————. 'Energy Sector Key to India's Economic Growth, Says Sushma Swaraj.' *The Economic Times*, 12 April 2018. Available at: https://energy.economictimes.indiatimes.com/news/oil-and-gas/energy-sector-key-to-indias-economic-growth-says-sushma-swaraj/63730454 (accessed on 10 August 2018).

————. 'India, US Ink New Defence Framework Accord.' *The Economic Times*, 11 July 2018. Available at: https://economictimes.indiatimes.com/news/defence/india-us-ink-new-defence-framework-accord/articleshow/47532025.cms (accessed on 20 November 2018).

————. 'Narendra Modi Becomes First Indian PM to Set Foot in Rwanda.' *The Economic Times*, 24 July 2018. Available at: https://economictimes.indiatimes.com/news/politics-and-nation/narendra-modi-becomes-first-indian-pm-to-set-foot-in-rwanda/articleshow/65108475.cms (accessed on 5 November 2018).

The Energy and Resource Institute (TERI). *India's Energy Security: New Opportunities for a Sustainable Future*. New Delhi: TERI, 2009.

The Hindu. 'Sakhalin-I Starts Oil, Gas Production.' *The Hindu*, 2 October 2005.

————. 'Sulabh Know-How to Go International.' *The Hindu*, 11 November 2008. Available at: http://www.thehindu.com/todays-paper/tp-national/tp-newdelhi/Sulabh-know-how-to-go-international/article15339409.ece (accessed on 17 May 2018).

————. 'Media Note: Agreement between India and Canada for Cooperation in Peaceful Uses of Nuclear Energy.' *The Hindu*, 28 June 2010. Available at: https://www.thehindu.com/news/national/Media-Note-Agreement-between-India-and-Canada-for-Co-operation-in-Peaceful-Uses-of-Nuclear-Energy/article16271026.ece (accessed on 20 October 2018).

————. 'India Should Aim at Energy Independence by 2030: Kalam.' *The Hindu*, 3 April 2014. Available at: http://www.thehindu.com/news/cities/Mangalore/india-should-aim-at-energy-independence-by-2030-kalam/article5866031.ece (accessed on 20 October 2017).

————. 'Cabinet Nod for Natural Gas Cargo-Swap Deal with Japan.' *The Hindu*, 11 October 2017. Available at: https://www.thehindu.com/business/Economy/cabinet-nod-for-natural-gas-cargo-swap-deal-with-japan/article19840983.ece. (accessed on 5 September 2018).

————. 'Will Achieve 175 GW Renewable Energy Target Well Before 2022.' *The Hindu*, 5 June 2018. Available at: https://www.thehindu.com/business/Industry/will-achieve-175-gw-renewable-energy-target-well-before-2022/article24089941.ece (accessed on 6 August 2018).

The Indian Express 'India's Carbon Emission Grew over 5 Per Cent in 2015: Study.' *The Indian Express*, 14 November 2016. Available at: http://indianexpress.com/article/technology/tech-news-technology/indias-carbon-emission-grew-over-5-per-cent-in-2015-study-4374889/ (accessed on 20 October 2017).

————. 'Wind Power Tariff at Record Low at ₹2.6/kWh.' *The Indian Express*, 6 October 2017. Available at: http://www.newindianexpress.com/business/2017/oct/06/wind-power-tariff-at-record-low-at-rs-26kwh-1667430.html (accessed on 28 April 2018).

The Japan Times 'Japan–India Civilian Nuclear Pact.' *The Japan Times*, 19 May 2017. Available at: https://www.japantimes.co.jp/opinion/2017/05/19/

editorials/japan-india-civil-nuclear-pact/#.XAuC9zSUfTU (accessed on 12 October 2018).

The Ministry of Statistics and Programme Implementation. *Youth in India*. New Delhi: Government of India, 2017. Available at: http://mospi.nic.in/sites/default/files/publication_reports/Youth_in_India-2017.pdf (accessed on 20 October 2017).

The New Indian Express. 'Centre Okays 50,000-Strong Force to Tackle Chinese Threat.' *The New Indian Express*, 18 July 2013. Available at: http://www.newindianexpress.com/nation/Centre-okays-50000-strong-force-to-tackle-Chinese-threat/2013/07/18/article1688814.ece (accessed on 27 October 2018).

————. 'India is the World's Largest Arms Importer, Say SIPRI.' *The New Indian Express*, 13 March 2018. Available at: http://www.newindianexpress.com/nation/2018/mar/13/india-is-worlds-largest-weapons-importer-sipri-report-1786293.html (accessed on 20 November 2018).

The New York Times. 'India Passes Nuclear Deal.' *The New York Times*, 30 August 2010.

The Role of Energy in Achieving the Poverty Millennium Development Goals. 12 October 2007. Available at: https://ec.europa.eu/energy/intelligent/projects/sites/iee-projects/files/projects/documents/e-mindset_the_role_of_energy_in_achieving_mdg1.pdf (accessed on 20 January 2017).

The Times of India. 'Pak Testfires N-Capable Haft-VI Missile.' *The Times of India*, 30 April 2006.

————. 'Ola Readies Big Push for Electric Vehicles.' *The Times of India*, 7 April 2007. Available at: https://timesofindia.indiatimes.com/auto/miscellaneous/ola-readies-big-push-for-electric-vehicles/articleshow/58070227.cms (accessed on 2 August 2018).

————. 'Paris Summit: PM Modi Pulls up Rich Nations on Climate Change.' *The Times of India*, 1 December 2015. Available at: https://timesofindia.indiatimes.com/home/environment/global-warming/Paris-summit-PM-Modi-pulls-up-rich-nations-on-climate-change/articleshow/49980783.cms (accessed on 20 October 2017).

————. 'Government Working to Eliminate Coal Import: Piyush Goyal.' *The Times of India*, 22 September 2016. Available at: https://economictimes.indiatimes.com/industry/indl-goods/svs/metals-mining/government-working-to-eliminate-coal-import-piyush-goyal/articleshow/54459898.cms (accessed on 11 January 2018).

————. 'Indian Economy on Solid Growth Track, Says IMF Chief Lagarde.' *The Times of India*, 16 October 2017. Available at: https://timesofindia.indiatimes.com/business/india-business/indian-economy-on-solid-growth-track-says-imf-chief-lagarde/articleshow/61094948.cms?utm_source=facebook.com&utm_medium=social&utm_campaign=TOIMobile (accessed on 20 October 2017).

———. 'Indian Oil Cos to Drop Anchor in Israeli Waters as Iran Deal Hangs Fire.' *The Times of India*, 28 December 2017. Available at: https://timesofindia.indiatimes.com/business/indian-oil-cos-to-drop-anchor-in-israeli-waters-as-iran-deal-hangs-fire/articleshow/62272531.cms (accessed on 5 September 2018).

———. 'China Objects to Vietnam's Call for Indian Investment in South China Sea.' *The Times of India*, 11 January 2018. Available at: https://timesofindia.indiatimes.com/business/india-business/china-objects-to-vietnams-call-for-indian-investment-in-south-china-sea/articleshow/62457912.cms (accessed on 7 November 2018).

———. 'A Big Leap: Opening up Coal Mining for Private Sector Is Historic, Now a Regulator Is Needed.' *The Times of India*, 22 February 2018. Available at: https://blogs.timesofindia.indiatimes.com/toi-editorials/a-big-leap-opening-up-coal-mining-for-private-sector-is-historic-now-a-regulator-is-needed/ (accessed on 25 July 2018).

———. 'India, Vietnam Vow to Jointly Work for Open Indo-Pacific.' *The Times of India*, 3 March 2018. Available at: https://timesofindia.indiatimes.com/india/india-vietnam-vow-to-jointly-work-for-open-indo-pacific/articleshow/63146342.cms (accessed on 7 November 2018).

———. 'India, France Ink 14 Pacts: Major Boost to Defence, Nuclear Energy Cooperation.' *The Times of India*, 10 March 2018. Available at: https://timesofindia.indiatimes.com/business/india-business/india-france-ink-14-pacts-major-boost-to-defence-nuclear-energy-cooperation/articleshow/63246479.cms (accessed on 12 October 2018).

———. 'India Gets into Driver's Seat with Global Solar Alliance.' *The Times of India*, 12 March 2018. Available at: https://timesofindia.indiatimes.com/india/india-gets-into-drivers-seat-with-global-solar-alliance/articleshow/63260137.cms (accessed on 15 May 2018).

———. 'India Gets the Cheapest LNG as Russia's Gazprom Begins Supplies.' *The Times of India*, 4 June 2018. Available at: https://timesofindia.indiatimes.com/business/india-business/india-gets-cheapest-lng-as-russias-gazprom-begins-supplies/articleshow/64450442.cms (accessed on 5 September 2018).

———. 'India, Bhutan Reaffirm Commitment to Hydropower Cooperation.' *The Times of India*, 7 July 2018. Available at: https://energy.economictimes.indiatimes.com/news/power/india-bhutan-reaffirm-commitment-to-hydropower-cooperation/64892036 (accessed on 15 September 2018).

———. 'Reduce Crude Oil Prices or Expect Demand to Sink, India Warns OPEC.' *The Times of India*, 11 July 2018. Available at: https://timesofindia.indiatimes.com/business/india-business/reduce-prices-or-expect-demand-to-sink-india-warns-opec/articleshow/64942074.cms?from=mdr (accessed on 10 August 2018).

———. 'India, Russia Sign Eight Pacts after Modi–Putin Summit.' *The Times of India*, 5 October 2018a. Available at: https://timesofindia.indiatimes.

com/india/india-russia-sign-eight-pacts-after-modi-putin-summit/article-show/66085306.cms (accessed on 5 September 2018).

———. 'India, Russia Agree to Expand Nuclear Collaboration in Third Countries.' *The Times of India*, 5 October 2018b. Available at: https://timesofindia.indiatimes.com/india/india-russia-agree-to-expand-nuclear-collaboration-in-third-countries/articleshow/66091949.cms (accessed on 10 October 2018).

———. 'Foreign Assets and Ventures of Indian Oil Companies across the World.' *The Times of India*, 28 December 2018. Available at: https://timesofindia.indiatimes.com/business/india-business/foreign-assets-and-ventures-of-indian-oil-companies-across-the-world/articleshow/62280019.cms (accessed on 14 August 2018).

The Tribune. 'HPCL to Set up ₹600-Crore Bio-Ethanol Unit in Bathinda.' *The Tribune*, 18 May 2017. Available at: http://www.tribuneindia.com/news/business/hpcl-to-set-up-rs-600-crore-bio-ethanol-unit-in-bathinda/408603.html (accessed on 6 August 2018).

The United Nations, Climate Change. 'What is the Paris Agreement?' Available at: https://unfccc.int/process-and-meetings/the-paris-agreement/what-is-the-paris-agreement (accessed on 5 November 2018).

The United Nations, Sustainable Development. 'United Nations Forum on Energy Efficiency and Energy Security for Sustainable Development: Taking Collaborative Action on Climate Change.' Seoul, 17–18 December 2007. Available at: https://sustainabledevelopment.un.org/index.php?page=view &type=13&nr=359&menu=1634 (accessed on 10 January 2017).

The University of Texas at Austin. 'Energy Security: A Threat Multiplier.' Available at: https://www.strausscenter.org/energy-and-security/energy-poverty.html (accessed on 20 January 2017).

The US Embassy. 'The American Vision of US–India Economic Cooperation.' Press release, 16 September 2004. Available at: http://chennai.usconsulate.gov/wwwhprind.html (accessed on 15 November 2017).

The White House. 'US National Security Strategy.' 16 March 2016. Available at: http://www.whitehouse.gov/nsc/nss/2006/ (accessed on 20 April 2018).

Third India–Africa Forum Summit. 'Partners in Progress: Towards a Dynamic and Transformative Development Agenda.' 29 October 2015. Available at: http://www.mea.gov.in/Uploads/PublicationDocs/25981_framework.pdf (accessed on 22 August 2018).

Tongia, R., and S. Ali. 'Is the Draft National Energy Policy Actionable?' *Livemint*, 22 August 2017. Available at: https://www.livemint.com/Opinion/5h2buHAGm6ZrVEEAg6CBCM/Is-the-draft-national-energy-policy-actionable.html (accessed on 25 May 2018).

Touhey, R. 'Canada and India at 60.' *International Journal* 62, no. 4 (2007): 733–750.

Trikha, S. 'We Have Questions about Pakistan: Armitage.' *Indian Express*, 12 May 2001.

Tsao, J., N. Lewis, and G. Crabtree. 'Solar FAQs.' Working draft version, 20 April 2016. Available at: http://www.sandia.gov/~jytsao/Solar%20FAQs.pdf (accessed on 30 April 2018).

Tuteja, R. *China–India Territorial Dispute: Moving towards Resolution.* CLAWS seminar report, 22 May 2008. Available at: http://www.exploringgeopolitics. org/publication_criekemans_david_geopolitics_of_renewable_energy_technology_desertec_north_seas_countries_ http://www.claws.in/index. php?action=details&m_id=88&u_id=5 (accessed on 25 October 2018).

United Nations Development Programme. 'World Energy Assessment: Energy and the Challenge of Sustainability.' 1 December 2000. Available at: http:// www.undp.org/content/undp/en/home/librarypage/environment-energy/ sustainable_energy/world_energy_assessmentenergyandthechallengeofsustainability.html (accessed on 12 December 2016).

———. 'Energizing the Millennium Developmental Goals: A Guide to Energy's Role in Reducing Poverty.' August 2005. Available at: http://content-ext. undp.org/aplaws_publications/2679356/ENRG-MDG_Guide_all.pdf (accessed on 15 January 2017).

United Nations, Economic and Social Commission for the Asia-Pacific. 'Energy Security and Sustainable Development in Asia and the Pacific.' 31 August 2008. Available at: http://www.unescap.org/resources/energy-security-and-sustainable-development-asia-and-pacific (accessed on 10 February 2017).

United Nations Environment Programme. *Indian Solar Loan Programme.* Programme Overview and Performance Report. February 2005. Available at: https://siteresources.worldbank.org/EXTRENENERGYTK/ Resources/5138246-1238175210723/India0consumer0credit0program0for0 PV.pdf (accessed on 10 May 2018).

United States Nuclear Regulatory Commission. *Nuclear Nonproliferation Act of 1978.* Available at: http://www.nrc.gov/reading-rm/doc-collections/nuregs/ staff/sr0980/v3/sr0980v3.pdf (accessed on 5 October 2018).

US. Energy Administration. Table 1.3 and 10.1, *Monthly Energy Review.* April 2017.

US Embassy in India. 'US, India Announce Progress in Strategic Trade Talks.' Press release, 17 September 2004.

Valentine, S. V. 'The Fuzzy Nature of Energy Security.' In *The Routledge Handbook of Energy Security,* edited by Benjamin K. Sovacool, 56. Abingdon: Routledge, 2010.

van Zyl, W. 'All Roads Sustainable Energy Lead to the Sun.' *The Conversation,* 19 May 2017. Available at: https://theconversation.com/all-roads-to-sustainableenergy-lead-to-the-sun-77180 (accessed on 30 April 2018).Varadarajan, S. 'Myanmar-India Gas Pipeline on Anvil.' *The Hindu,* 13 January 2005.

Varghese, P. N. 'Chapter Nineteen: The Geopolitical Pillar.' In *An India Economy Strategy to 2035, Navigating from Potential to Delivery* (A Report to the Australian Government). 2018. Available at: https://dfat.gov.au/geo/

india/ies/pdf/dfat-an-india-economic-strategy-to-2035.pdf (accessed on 12 December 2018).

Verma, N., and S. Phartiyal. 'Oil and Gas: India Seeks Active Role in Africa.' *The Africa Report*, 26 January 2017. Available at: http://www.theafricareport. com/North-Africa/oil-a-gas-india-seeks-active-role-in-africa.html (accessed on 25 August 2018).

Waltz, K. N. 'The Emerging Structure of International Politics.' *International Security* 18, no. 2 (1993): 44–79.

Washington File. 'Text: Robert Blackwill on US–India Collaboration on International Issues.' *Washington File*, 4 September 2001.

Wesley, M. 'The Elephant in the Room.' *Monthly*, February 2012. Available at: https://www.themonthly.com.au/issue/2012/february/1328594251/michael-wesley/elephant-room (accessed on 10 October 2018).

William, L. 'A Path Forward on Indian NSG Membership.' *Diplomat*, 1 April 2016. Available at: https://thediplomat.com/2016/04/a-path-forward-on-indian-nsg-membership/ (accessed on 25 November 2018).

Winzer, C. 'Conceptualizing Energy Security.' EPRG working paper 1123, Cambridge Working paper in Economics, July 2011. Available at: https:// www.repository.cam.ac.uk/bitstream/handle/1810/242060/cwpe1151.pdf;jse ssionid=47DB591BDF8120627BD465710C2D9E54?sequence=1 (accessed on 15 December 2016).

Wong, C. M. L. 'Australian Uranium Deal Could Make Indian Nuclear Power Safer.' *The Conversation*, 5 September 2014. Available at: https://thecon-versation.com/australian-uranium-deal-could-make-indian-nuclear-power-safer-31290 (accessed on 6 September 2014).

World Atlas. 'Top Oil Producing Countries in Latin America and the Caribbean.' 21 September 2017. Available at: https://www.worldatlas.com/articles/top-oil-producing-countries-in-latin-america-and-the-caribbean.html (accessed on 20 August 2018).

World Nuclear Association. 'Nuclear Power in India: Country Profile.' 19 May 2017. Available at: http://www.world-nuclear.org/information-library/country-profiles/countries-g-n/india.aspx. (accessed on 26 July 2017).

———. 'Nuclear Power in India.' Available at: http://www.world-nuclear.org/ information-library/country-profiles/countries-g-n/india.aspx (accessed on 25 April 2018).

———. 'Plans for New Reactors Worldwide.' Available at: http://www.world-nuclear.org/information-library/current-and-future-generation/plans-for-new-reactors-worldwide.aspx (accessed on 20 April 2018).

World Nuclear News. 'India and UK Sign Civil Nuclear Agreement, Discuss Climate Change.' 13 November 2015. Available at: http://www.world-nuclear-news.org/Articles/India-and-UK-sign-civil-nuclear-agreement,-discuss. (accessed on 20 October 2018).

World Wind Energy Association. *World Wind Assessment Report.* WWEA Technical Paper Series TP-01-14, December 2014. Available at: http://www.

wwindea.org/download/technology/WWEA_WWRAR_Dec2014_2.pdf (accessed on 28 April 2018).

Y Charts. 'Ecuador Crude Oil Production: 531.32K bbl/day for 2017.' 27 March 2018. Available at: https://ycharts.com/indicators/ecuador_crude_oil_production_annual (accessed on 22 August 2018).

Yang, J. 'A Strategic Game: China's Energy Relations with India and Japan.' In *China's Energy Relations with the Developing World*, edited by Carrie Liu Currier and Manochehr Dorraj, 153. New York, NY: Bloomsbury, 2011.

Yergin, D. 'Ensuring Energy Security.' *Foreign Affairs* 85, no. 2 (2006): 69–82.

———. 'Oil Diplomacy.' Testimony before the House Committee on International Relations Hearings, 20 June 2012. Available at: https://www.brookings.edu/wp-content/uploads/2016/06/yergin20020620-1.pdf (accessed on 12 December 2016).

Yergin, Y. *The Prize: The Epic Quest for Oil, Money and Power*. New York, NY: Simon & Schuster, 1991.

Yueh, L. 'An International Approach to Energy Security.' *Global Policy* 1, no. 2 (2010): 216–217.

Zhijie Zhang and Wanli Xing, "Overseas Oil Cooperation between China and India Based on Crude Oil Trade Flow Analysis", 2nd International Workshop on Renewable Energy and Development (IWRED 2018), 2018, available at https://iopscience.iop.org/article/10.1088/1755-1315/153/3/032046/pdf

Zhang, Z., and W. Xing, W. 'Overseas Oil Cooperation between China and India Based on Crude Oil Trade Flow Analysis.' IOP Conference Series: Earth and Environmental Science, 153. 2018. Available at: http://iopscience.iop.org/article/10.1088/1755-1315/153/3/032046/pdf (accessed on 5 November 2018).

Zhou, R. *The Competitive Situation and Cooperative Tendency between China and India's New Source of Oil Supply*. Master dissertation, SiChuan University, China, 2006.

Websites

Caspian Studies. Available at: http://www.caspianstudies.com (accessed on 30 August 2018).

Director General of Hydrocarbons, http://www.dghindia.org/admin/Document/Topstory/13.pdf

Energy Information Administration, Official Energy Statistics from the US Government. Available at: http://www.eia.doe.gov/ (accessed on 13 April 2018).

Gas Authority of India Limited. Available at: http://gail.nic.in/homepage/home-new.htm (accessed on 10 August 2018).

Ministry of Coal, Government of India. Available at: http://www.coal.nic.in/ (accessed on 11 January 2018).

National Data Repository, Directorate General of Hydrocarbons. Available at: www.ndrdgh.gov.in/NDR/ (accessed on 5 August 2018).

New Exploration and Licensing Policy. Available at: http://www.arthapedia.in/ index.php?title=New_Exploration_and_Licensing_Policy_(NELP) (accessed on 5 August 2018).

Oil and Natural Gas Corporation (ONGC). Available at: http://www.ongcindia. com/ (accessed on 10 August 2018).

ONGC Videsh. Available at: http://www.ongcvidesh.com/ (accessed on 25 August 2018).

Index

About the Author

Ashok Sharma is currently a Visiting Fellow at the University of New South Wales (UNSW), Canberra, at the Australian Defence Force Academy (ADFA). Dr Sharma is also an Adjunct Associate Professor at the Institute of Governance and Policy Analysis at the University of Canberra. He is the Head of the South Asia Strategic, Security and State Fragile Program and Conjoint Head of the Indo-Pacific and Major Power Studies Program at the National Asian Studies Centre at the University of Canberra. He is also a Deputy Chair at New Zealand Institute of International Affairs, Auckland.

Prior to joining the UNSW at ADFA in 2016, Dr Sharma was a Fellow at Australia–India Institute, the University of Melbourne, Lecturer in Politics and International Relations at the University of Auckland, a Visiting Academic at the University of Waikato, Hamilton, and an Endeavour Post-Doctoral Fellow at the Australian National University in 2008.

Dr Sharma has a BA (Hons) in Political Science from Ramjas College, University of Delhi, an MA in Political Science and an MPhil and PhD in American Studies from Jawaharlal Nehru University, New Delhi. He was a Faculty at the University of Delhi and worked with think tanks based in New Delhi, namely Brookings Project at the Observer Research Foundation, Centre for Air Power Studies and Centre for Land Warfare Studies.

Dr Sharma specializes in great power politics with emphasis on US–India–China geopolitics. He is extensively published and is the author of well-received book *Indian Lobbying and its Influence in US Decision Making: Post-Cold War* (SAGE Publications, 2017).